NEUROCONTROL

Adaptive and Learning Systems for Signal Processing, Communications, and Control

Editor: Simon Haykin

Werbos / THE ROOTS OF BACKPROPAGATION: From Ordered Derivatives to Neural Networks and Political Forecasting

Krstić, Kanellakopoulos, and Kokotović / NONLINEAR AND ADAPTIVE CONTROL DESIGN

Nikias and Shao / SIGNAL PROCESSING WITH ALPHA-STABLE DISTRIBUTIONS AND APPLICATIONS

Diamantaras and Kung / PRINCIPAL COMPONENT NEURAL NETWORKS: Theory and Applications

Tao and Kokotović / ADAPTIVE CONTROL OF SYSTEMS WITH ACTUATOR AND SENSOR NONLINEARITIES

Tsoukalas and Uhrig / FUZZY AND NEURAL APPROACHES IN ENGINEERING

Hrycej / NEUROCONTROL: Towards an Industrial Control Methodology

Beckerman / ADAPTIVE COOPERATIVE SYSTEMS

NEUROCONTROL
Towards an Industrial Control Methodology

Tomas Hrycej

Daimler-Benz AG
Research Center Ulm

A Wiley-Interscience Publication

JOHN WILEY & SONS, INC.

New York / Chichester / Weinheim / Brisbane / Singapore / Toronto

This text is printed on acid-free paper

Copyright © 1997 by John Wiley & Sons, Inc.

All rights reserved. Published simultaneously in Canada.

Reproduction or translation of any part of this work beyond that permitted by Section 107 or 108 of the 1976 United States Copyright Act without the permission of the copyright owner is unlawful. Requests for permission or further information should be addressed to the Permissions Department, John Wiley & Sons, Inc., 605 Third Avenue, New York, NY 10158-0012

Library of Congress Cataloging in Publication Data:
Hrycej, Tomas, 1954–
  Neurocontrol : towards an industrial control methodology / Tomas Hrycej.
    p.   cm.
  Includes bibliographical references and index.
  ISBN 0 471 17628 1 (alk. paper)
  1. Neural networks (Computer science)   2. Automatic control.
  I. Title.
  QA76.87.H783   1997
  629.8'9–dc20   96–33490

Printed in the United States of America

10 9 8 7 6 5 4 3 2 1

To
Emily, Anita, Nathalie,
and my mother

CONTENTS

Preface xi

I Methods 1

1 **Important Concepts of Classical Control** 3
 1.1 Linear Control / 4
 1.2 Nonlinear Control / 6
 1.3 Optimal Control / 8
 1.4 Robust Control / 9
 1.5 Adaptive Control / 10

2 **Fundamental Approaches to Neurocontrol** 11
 2.1 Template Learning / 12
 2.2 Learning Plant Inversion / 13
 2.3 Closed-Loop Optimization / 18
 2.4 Critic Systems / 21
 2.5 Summary of Fundamental Approaches to Neurocontrol / 33

3 **Neural Networks for Control** 37
 3.1 Neural Networks Using Various Types of Nonlinearities / 38
 3.2 Symmetric or Asymmetric Networks / 47
 3.3 Feedforward and Feedback Networks / 54

4 **Optimization Methods for Neurocontrol** 63
 4.1 Local Optimization Methods / 64
 4.2 Global Optimization Methods / 75
 4.3 How to Make Optimization Easier / 83
 4.4 What Gets Lost with Incremental Computation? / 85

5 Plant Identification — 93

5.1 Classical View of Identification / 93
5.2 Plant Identification by Neural Networks / 99
5.3 Model Structure / 100
5.4 Model Quality Criterion / 120
5.5 Algorithms for Identification / 133
5.6 Sampled Data / 153
5.7 Using Explicit Models with Free Parameters / 156

6 Controller Training — 159

6.1 Classical View of Control / 160
6.2 Neurocontrol Formulation of the Control Problem / 170
6.3 Controller Structure / 171
6.4 Cost Function for Neurocontroller Training / 189
6.5 Training Methods / 194
6.6 Neurocontrol with Analytic Models / 203
6.7 Using Neural Network Models for Analytic Controller Design / 204

7 Robust Neurocontrol — 207

7.1 Classical View of Robust Control / 208
7.2 Robustness Aspects of Neurocontrol / 214
7.3 Neurocontroller Structure / 215
7.4 Training Examples / 216

8 Learning and Adaptiveness in Neurocontrol — 223

8.1 Adaptiveness, Learning, and Generalization / 223
8.2 Pragmatic Characteristics of Adaptive Control / 225
8.3 When Is Adaptiveness Really Needed? / 226
8.4 Incremental and Batch Schemes for Online Adaptiveness / 227
8.5 Data Accumulation / 230
8.6 Forgetting / 233
8.7 Global and Local Optimization Methods / 236
8.8 Computational Experiment / 237

9 Stability of Neurocontrollers — 243

9.1 Proving the Stability of a Neurocontroller / 244
9.2 Designing a Stable Neurocontroller / 251

10 A Neurocontrol Algorithm—The Easiest Way to the Goal — 253

10.1 Plant Identification Algorithm / 255
10.2 Neurocontroller Training Algorithm / 257
10.3 Optimization Algorithm / 261
10.4 Appropriate Choices for Incremental Adaptation / 265
10.5 Tools for Neurocontrol / 265

II Case Studies — 271

11 Introductory Remarks to Case Studies — 273

12 Elastomere Test Bench Control — 277

12.1 Test Benches / 277
12.2 Elastomere Test Bench / 278
12.3 Plant Model Identification / 279
12.4 Controller Training / 282

13 Drive Train Test Bench Control — 309

13.1 Plant Model Identification / 310
13.2 Controller Training / 320

14 Lateral Control of an Autonomous Vehicle — 325

14.1 An Autonomous Road Vehicle / 325
14.2 Modeling Lateral Control / 326
14.3 Controller Training / 326
14.4 Results / 328

15 Biological Wastewater Treatment Control — 331

15.1 Biological Wastewater Treatment Process / 332
15.2 Structure of a Wastewater Treatment Plant / 335
15.3 Plant Identification / 338
15.4 Identification of Hidden States / 346
15.5 Controller Training / 350
15.6 Results / 357

16 Conclusion: Application Potential of Neurocontrol — 363

16.1 Main Asset of Neurocontrol: Generality / 363
16.2 Neurocontrol Is a Numeric Method / 364
16.3 Neurocontrol Practice / 365
16.4 Method Selection for Industrial Applications / 366

16.5 Relationship to the Foundations of Neural Networks / 366
16.6 What the Future May Hold / 367

References 369

Index 379

PREFACE

Neurocontrol is a dynamic research field that has attracted considerable attention from the scientific and control engineering community in the last several years.

Among the possible applications of artificial neural networks, such as classification, filtering, and control, the most important is that of control for the following reasons:

- Applications such as filtering coincide well with linear solutions and approximations.
- Applications such as pattern recognition and classification relate to both nonlinear problems and provide well-established special nonlinear algorithms (e.g., polynomial or nearest-neighbor classification).
- Almost all difficult control problems are nonlinear, but few universally applicable nonlinear control design approaches are broadly used in practice.

Despite the many excellent results reported, industrial applications of neurocontrol are not yet routine. Some topical neurocontrol applications have been implemented in academic or industrial research, but the method has not found its way into the broader control engineering field.

An obvious contrast exists to another innovative method for nonlinear control—fuzzy control. Commercial tools for fuzzy control are available, and a considerable number of control engineers have at least attempted to work with them, many of them having succeeded in developing working controllers. Different development of both disciplines may be grounded in different requirements for starting work. Not much is required to start working with fuzzy control; it is sufficient to buy a fuzzy tool or to program a simple fuzzy inference system and start experimenting. However, fuzzy technology alone will not solve any of the control engineer's problems. It does not provide solutions for nonlinear state space design or for robust control, and it does not suggest stable adaptive algorithms. Its magic is in enabling the control engineer to implement quickly whatever he would like to try. Trying to do what you think may work is the fuzzy control philosophy.

Neurocontrol, in contrast at least at first glance, appears to be a monolithic task. It has the capacity to solve difficult control problems and to optimally solve arbitrary nonlinear control problems directly from sampled data. This automatic, algorithmic design method requires little analytical work and provides the possibility of automatic adaptation to plant or environment changes. Its character as an algorithmic method that does everything by itself does not allow implementation in small steps—it either works as a whole or not at all. However, such a method can be accepted only if it has reached the stage of complete maturity in which all steps necessary to implement a working algorithm are described in textbooks. This is the critical point: Neurocontrol as described in the literature is far from providing sufficient detail to be routinely applied to the difficult control problems for which it has potential. The published algorithms are only starting points; most industrial users starting from scratch may spend years filling in sufficient details to reach optimal, nonlinear, adaptive, and robust control design that requires nothing but sampling data. It is my opinion, and to a large extent also my personal experience, that this goal can be reached, and that the field is mature enough to provide the required details. All that remains is to collect them in a book! This book is intended to be a step in that direction.

One way to obtain the necessary details is by scanning related disciplines for how they can contribute to this goal. The most prominent two disciplines are:

1. Classical control and systems theory which provides extensive and detailed statements about what can be controlled and with which information.
2. Numerical optimization which is concerned with efficient algorithms to search for optima in multidimensional state spaces.

The problem in establishing a relationship between fundamental neurocontrol approaches and classical control is not that neurocontrol has no classical counterparts. Rather, the problem is that these counterparts are more or less exotic in control practice. Most control engineers have heard about optimal or adaptive control, but only a few have applied these approaches to real industrial control problems. This is why it is important to start from the basics of classical control, recall the important statements of these exotic disciplines, and integrate this knowledge with that of neurocontrol approaches.

This book is targeted toward the following groups:

- Researchers who seek the relationships between neurocontrol and classical control.
- Control theorists who want to identify the place of neurocontrol methods in the classical control framework.
- Control practitioners who intend to develop neurocontroller applications.

It is the last group on which the future of neurocontrol depends. It is up to control practitioners to find applications on a broad scale. Neurocontrol's most important practical asset is that it can dramatically reduce development time by providing a unified framework for a very broad (maybe even universal) class of nonlinear control problems. This book includes a recipe that demonstrates such a quick successful neurocontrol application with minimal developmental expense.

The book is structured in two parts. The first is concerned with theory and concepts. Chapter 1 very briefly scans through classical control disciplines for their potential to contribute to neurocontrol. It is largely descriptive in its overview, but formal results are presented throughout the book wherever they are needed.

What neurocontrol has to offer is presented in the two subsequent chapters. Chapter 2 is about fundamental approaches to neurocontrol. These approaches differ in the way the problem of learning optimal control from sampled data is transformed to concrete functional approximation problems that are then directly solved by neural network learning methods. The unified framework in which all the approaches are formulated is that of learning viewed as optimization of a defined cost function. The topic of Chapter 3 is the choice of neural network structures appropriate for these tasks.

The algorithmic means for solving a control problem represented in a neural network framework are called learning or training algorithms by the neural network community. They are, in fact, specialized optimization algorithms. In Chapter 4, some algorithms are surveyed and discussed with regard to those properties relevant to neurocontrol.

After the basic outline of how to solve neurocontrol problems with a numerical algorithm is understood, it is time to fill out the scheme with the details of control theory. Fundamental in designing any neurocontroller from sampled data are two complementary subtasks: identification of a neural-network-based plant model from sampled data (Chapter 5) and training of a neurocontroller optimal for the given identified plant (Chapter 6). Both chapters have a similar structure, starting with a brief overview of relevant classical concepts and continuing with separate treatment of three important constituents: structuring the plant model or neurocontroller mapping, the cost function for optimization, and the learning algorithm.

The following three chapters focus on important extensions of the basic scheme related to important fields of classical control. They are robust control (Chapter 7), adaptive schemes (Chapter 8), and stability proofs (Chapter 9).

The last chapter of this methodological part of the book presents a "recipe for quick success" (Chapter 10). In it, I suggest some alternative algorithms that are easy to implement to a broad class of industrial control problems. Although the commercial tools currently offered on the market are of limited utility for a neurocontrol application of realistic complexity, a brief section concerning these tools is included.

The second part of the book is concerned with case studies from my experience. A general framework for their description is given in Chapter 11. The particular applications presented are elastomere test bench control (Chapter 12), car drive train control (Chapter 13), lateral steering control of an autonomous road vehicle (Chapter 14), and wastewater treatment plant control (Chapter 15).

The case studies have been chosen with two objectives in mind. First, they illustrate the theoretical concepts developed in the methodological part of the book. Second, they are intended to show the scope and complexity of applications that have been solved with the help of a *single neurocontrol method*. They are important in supporting the statement that neurocontrol is really a universal method. The book is concluded by Chapter 16 which summarizes the benefits of neurocontrol from the point of view of current practice.

I would like to thank to the reviewers of Wiley-Interscience for many helpful comments on the manuscript. I am also indebted to the Czech company Art of Intelligence for providing a beta version of their tool NNcon (probably the only serious tool really dedicated to neurocontrol on the market) for tests and benchmarks (see also Section 10.5).

TOMAS HRYCEJ

I

METHODS

1

IMPORTANT CONCEPTS OF CLASSICAL CONTROL

The field of neurocontrol has rapidly expanded over the last several years by virtue of its incontestable success in solving practical control problems. However, every rapid expansion is connected with a certain amount of wild growth. The opportunities and possibilities in expanding research fields can leave systematic investigations to be postponed. This has been the case, to a certain extent, in neurocontrol.

Fortunately, neurocontrol is but a subfield of control. To put neurocontrol on a firm foundation, we can draw upon the huge body of a systematically evolved theory of classical control.

Classical control theory is strongly biased toward linear systems. The fact that neurocontrollers excel in nonlinear systems may produce a misleading impression that the relationship between classical and neurocontrol is limited. That leads to the next erroneous conclusion that most of the linear control theory completely loses its validity in the nonlinear domain.

This conclusion is problematic on two important counts: On the one hand, general nonlinear systems simply do not allow us, because of their analytical intractability, to formulate a theory that is as strong as that of linear systems. On the other hand, nonlinear systems can be qualitatively similar to linear ones under some circumstances. Let us illustrate this with an example. Examining the existence of solutions of a set of m linear equations of n variables, we find that there are three qualitatively different cases (assume that the coefficient matrix is full rank, which is often the case in nature).

1. For $m < n$, there are infinitely many solutions such that $n - m$ variables can be set to some arbitrary values determining the remaining m variable values in a unique way. This is the *underdetermined* case.
2. For $m = n$, there is a unique solution.
3. For $m > n$, there is no solution except in the case where the extended coefficient matrix has the same rank as the nonextended one. This is the *overdetermined* case.

These qualitative cases can be found also in nonlinear equation systems, except for the uniqueness of solution; nonlinear systems with $m = n$ may have one, more, or even no solution. The underdetermination and overdetermination principles apply, too.

Analogically, linear control theory provides valuable insight into the nature of control. It has developed fundamental frameworks and concepts for the investigation of stability, optimality, robustness, and adaptivity.

This chapter will briefly review the important fields of classical control theory for results that have the potential of being useful for neurocontrol. The choice of models has been motivated by the scope of the following chapters. The discussion of this chapter is mostly informal. More formal treatment is reserved to the remaining chapters of the book.

1.1 LINEAR CONTROL

Linear control is concerned with systems of the form

$$\dot{\mathbf{z}} = \mathbf{Az} + \mathbf{Bu} \tag{1.1}$$

with state **z**, input **u**, and measurable output

$$\mathbf{y} = \mathbf{Cz} \tag{1.2}$$

and controllers of the form

$$\mathbf{u} = -\mathbf{Kz} + \mathbf{Mw} \tag{1.3}$$

with **w** the reference state, that is, the state to which the plant is to be brought with the help of a controller. **A**, **B**, **C**, **K** and **M** are matrices of appropriate dimensions. The goals of linear design are as follows:

  Stability of the closed loop, that is, convergence back to an equilibrium point after being moved away from this point by a disturbance.
  Optimality of closed-loop behavior in some user-defined sense.

For linear systems, local stability is identical with global stability, which is the convergence to an equilibrium point from all points of the state space. A stability guided design is based on the analysis of closed-loop eigenvectors. The condition to be satisfied is that the real parts of all eigenvectors must be negative. Negative eigenvector real parts are synonymous with the trajectory of the closed loop experiencing either damped oscillations or exponential overshoot-free convergence.

However, with the stability condition satisfied, the controller is still underdetermined. There are two additional requirements:

1. A complete model for the closed-loop response (reference model)
2. Optimization

In using a reference model, the properties of a closed loop (consisting of the plant and a feedback controller) can be specified in terms of eigenvalues. Because eigenvalues completely describe the behavior of a linear system, a plant with arbitrary dynamic properties can be transformed, with the help of a feedback controller, into a closed loop with arbitrarily different dynamic properties. This is true under some conditions that, at least in principle and with limited precision, are relatively frequently satisfied in practice. In theory, with an appropriate sequence of control actions (deadbeat control), linear discrete systems with n state variables are guaranteed to be able to reach an arbitrary goal state within n steps. But the possibility of changing the dynamic properties of a closed loop can hardly be applied in practice.

Another, and perhaps more straightforward, way to close the underspecification gap is by defining a cost (or utility) function that is to be minimized (maximized). Linear systems can be analytically optimized by an arbitrary quadratic cost function (LQ control). However, it is important to note the scope of explicit solutions of optimal linear control. In addition to the linearity of the plant and the quadratic cost function, *control action cannot be bounded*. This is an important assumption. Whereas the first two assumptions are acceptable approximations, the last one is frequently violated. Whenever fast control is required, unbounded LQ-optimal control will consist in vigorous control actions. Control action limitations are inconsistent with the unbounded solution. This makes linearity the only acceptable assumption though it never applies exactly, so individual assertions of control theory should be interpreted with this in mind.

Most of the design methods can be applied under the following two conditions

Controllability. All state variables are affected by control action

Observability. All state variables affect the measurable variables and can thus be reconstructed with the help of consecutive observations

For both the plant model and the controller, the theory gives hints to the way in which the information about measurable output and input can substitute for the knowledge of a complete state. This defines the structure of such sufficient models and controllers—a valuable help for neurocontrol formulations.

Some basic theorems setting the limits of what can be done, under formal conditions, are as follows:

- A controllable and observable plant (i.e., even one that is unstable in an open loop) can be stabilized by complete state feedback.
- With p action variables, stationary goal states of p observable variables can be reached.
- Exact following of a reference trajectory is possible only if this trajectory has been produced by a system of order higher or equal to the order of the plant.
- The trajectory to be followed must be specified for p output variables by means of their derivatives of order corresponding to the relative degree of these variables.

1.2 NONLINEAR CONTROL

Nonlinear control theory is concerned with general systems of the form

$$\dot{\mathbf{z}} = F(\mathbf{z}, \mathbf{u}) \tag{1.4}$$

with measurable output

$$\mathbf{y} = H(\mathbf{z}) \tag{1.5}$$

and controllers of the form

$$\mathbf{u} = G(\mathbf{z}, \mathbf{w}) \tag{1.6}$$

An analytical solution is known only for a subclass of nonlinear systems described by

$$\dot{\mathbf{z}} = F_1(\mathbf{z}) + F_2(\mathbf{z})\mathbf{u} \tag{1.7}$$

The problem with the applicability of the analytical approach above has less to do with the special form of nonlinear systems; it rather describes a fairly broad class of real problems. The difficulty is in the computational expense for evaluating the controller (matrix inversion in every sampling period) and the dependence on the availability and precision of the plant model.

The possibilities of formulating control are, in principle, the same as for linear systems: They depend on a reference model or a cost function. The difference is in the feasibility of reference-model-based formulation. The behavior of simple linear reference models, in particular, first- and second-order models, can be described in terms of intuitively comprehensive concepts such as damping and time constants. By contrast, higher-order linear models have no such intuitive representation. For nonlinear systems, a lot of control engineering competence is necessary to find reference models that can be exactly followed by the plant and, in addition, can express controller design preferences.

This is why formulating control goals via cost function is frequently the better alternative. Then, some general optimal control approach such as dynamic programming is applied. I discuss this approach in Section 1.3.

The difficulties with genuine nonlinear controller designs suggest instead the *linearization approach*, also known as *gain scheduling*. It consists of the following steps:

1. The main working points of the plant are elaborated
2. The plant is linearized for each working point
3. Linear design is applied to each working point
4. An interpolation method is used for determining the controller action in states between the distinguished working points

1.2.1 Stability of Nonlinear Systems

One of the most important achievements of nonlinear control theory are the concepts and tools for investigating the stability of nonlinear systems.

There are several definitions of stability:

Lyapunov stability, in which the system is sure not to leave a certain region, or *Lyapunov asymptotic stability*, in which the system is sure to converge to an attractor point.

Input/output stability, with a closely related *bounded input/bounded output (BIBO)* stability, in which certain types of input behavior (e.g., bounded) of input behavior will lead to a certain type of output behavior.

Total stability, in which stability is guaranteed under certain types of disturbances.

The Lyapunov stability concepts seem to be the most flexible. A function of the closed-loop state, the Lyapunov function, is constructed with the following properties:

1. The function is positive definite and has its minimum in the point (or region) for which the stability is to be proved.
2. The time derivative of the function is negative. That is, the trajectory of closed loop follows a path along which the Lyapunov function value diminishes.

8　IMPORTANT CONCEPTS OF CLASSICAL CONTROL

Lyapunov stability is straightforward intuitive. It assigns to each state a value that can be interpreted as a distance from the goal state, or as a reversed sign value of the state with regard to reaching the goal state. The closed-loop state follows the trajectories along which the distance monotonically decreases.

The importance of the Lyapunov function can be made more obvious if we take into account that it defines, to a certain degree, the controller, or at least a class of stable controllers for a given problem. Suppose a Lyapunov function (for discrete time) is explicitly known. Then, the controller action is to be such that the next state of the plant has a lower Lyapunov function value than the present state.

1.3　OPTIMAL CONTROL

The topic of optimal control theory is to design controllers that are optimized to (or even can be proved to be the best with regard to) a certain performance criterion. For a linear plant and a quadratic performance criterion, the Ricatti controller represents an explicit and global solution of the problem.

Another setting in which a globally optimal control solution can be found is in a discrete state space with discrete actions. For each state, we select and assign to that state the action producing the best cost-function value resulting from immediate cost of the present state, the cost of the action, and the cumulative recursively computed cost of the next state and all subsequent states passed under the assumption of the optimal action sequence.

Dynamic optimization is a general scheme for state evaluation and selection of the optimal action. Alternatively, if each state at each sampling period is represented by a node in a directed graph and actions are represented by connecting edges of the subsequent states, then the task can also be transformed to the critical path problem of graph theory.

In a fixed time horizon, computational expense is proportional to the number of states per sampling period and to the the number of actions. Continuous problems can be approximated by discretization of state space and linear interpolation of continuous states. There is, of course, a trade-off between computational expense and precision. For state spaces of dimension higher than four or five, the disappointing outcome of this trade-off is usually that reasonable precision cannot be reached with reasonable computational expense. For example, with six state variables, taking a grid of only ten values for each state variable results in a million states. Then, even with algorithms of only quadratic complexity such as dynamic optimization, the task is intractable.

In addition to the global optimization approaches, there are theoretical results from the calculus of variations concerning local optima such as the Pontryagin maximum principle. Their locality makes them less interesting for nonlinear problems; even an accurate local optimum may still be a bad overall solution of the problem.

The optimal control problem can be formulated as an optimization problem. For example, a formulation developed by Tsypkin [138, see Section 7.3] is the following (notation changed). The plant is modeled as

$$\mathbf{z}_{t+1} = F(\mathbf{z}_t, \mathbf{u}_t) \qquad (1.8)$$

The goal is to find a controller

$$\mathbf{u}_{t+1} = G(\mathbf{z}_t) \qquad (1.9)$$

that minimizes

$$E[C(\mathbf{w} - \mathbf{z})] \qquad (1.10)$$

with $E[\cdot]$ being the mean value over time. If the functional G is represented by a function parametrized by parameter set ω, the task is to find ω_{min} for which function Eq. (1.10) has a minimum value.

1.4 ROBUST CONTROL

Robust control addresses the problem of controlling a plant whose behavior is slightly different from that of a plant model. The reasons for the difference may be the differing structure or parameters of exact and approximate models, a systematic disturbance, or a random disturbance.

A popular pragmatic classical approach to robust control is concerned with preserving stability. The closed-loop eigenvalues are chosen so that they remain in the stability region even if the plant model should change in a defined range.

Another approach is concerned with preserving the closed-loop equilibrium point under plant modifications or systematic disturbances. The means to reach this goal is error integration. Since the integrated error grows with time if the error itself does not converge to zero, a controller using an integral of error as an additional input can be designed for a closed loop to converge to the desired equilibrium point even under model imprecision or systematic disturbance.

Some classical results are available about systematic disturbances and necessary controller inputs:

1. Controller input including an error integral is necessary if stability under constant additive disturbance is to be reached.
2. Controller input including the error integral is necessary if stability under linearly changing additive disturbance (i.e., a ramp) is to be reached.

3. Controller input including the $(n-1)$th error integral is necessary if stability under nth-order polynomial additive disturbance is to be reached.

1.5 ADAPTIVE CONTROL

Adaptive control is another way to reach a goal similar to that of robust control. Instead of designing robust controllers that work under conditions different from those for which they have been designed, adaptive controllers recognize the difference between the assumption and reality and change to perform better in the new conditions.

Adaptation schemes can be based on both a reference model and a cost function. The approach called *reference model adaptive control* (MRAC) is, by its name, committed to the first alternative [e.g., see Parks, 113]. It is based on formulating the rules for computing the direction of change of controller parameters as a function of the difference between the behavior of the closed loop, and the reference model output. Controller parameters can be adapted either directly or via the estimation of plant model parameters. Problems with reference models for nonlinear systems have been addressed in Section 1.2.

A more general approach is that of *self-tuning regulators* (STR), [see Åström and Wittenmark, 8], which consists in adaptive estimation of a plant model and applying a formalized controller design method to the plant model. This design method can be based on cost function optimization. Various STR methods have been proposed for linear control, but they also have clear potential for general nonlinear problems.

An important aspect of adaptive control conerns guaranteeing the stability [Narendra and Annaswamy, 103]. The adaptation algorithm together with the controlled plant and the controller can be viewed as a single dynamic system. In this system, the closed loop itself must be stable, and the process of changing controller parameters must converge to the optimal state. The classical theory of stable adaptive control provides results that hint at which nonlinear controller structures and adaptation principles promise stability.

2

FUNDAMENTAL APPROACHES TO NEUROCONTROL

Neural networks have been shown to have diverse capabilities. Their applications range from functional approximation over associative memory to optimization. Functional approximation in a general sense searches for a function that is optimal with regard to a certain criterion or *cost function*. An important, and probably the most elaborated, special case is the search for a function that attains specified values for a given set of argument values (input/output pair approximation).

The neural network solution of a functional approximation task is based on a transformation to a numeric optimization task. The function searched is represented by a neural network with a set of free parameters such as weights, thresholds, or radial basis function centers. A numeric optimization method is used to find a set of parameter values for which the cost function has a minimum value. For many neural network applications, such as pattern recognition, prediction, and filtering, input/output pairs are given, and the optimization task consists in minimizing the difference between desired and computed outputs for given inputs.

Also neurocontrol is one of the application fields that benefit from the functional approximation capability of neural networks. The function to be approximated is the optimal controller. However, the function values to be approximated (i.e., control actions) are usually not explicitly known. What is known are general conditions or optimality criteria for the consequences of these actions.

For example, if the control task is driving a car, the optimality criterion might be following the road course safely but with minimum lateral acceleration that brings about discomfort for passengers. Individual control actions are steering wheel positions at discrete time periods of length (e.g., 0.1 s). No individual action can be directly evaluated by the optimality criterion: that is, driving off the road cannot be attributed to a single one of these actions.

The problem of finding the relationship between the functional values and their evaluated consequences is called *credit assignment problem*. This problem is central for neurocontrol methods. So it is appropriate to classify the fundamental approaches by the way they address the credit assignment problem. Such a classification is attempted in the following sections.

Much research in neurocontrol is devoted to the various ways to compute the error or cost function gradient. Although the error gradient provides useful information for proceeding in the direction of the optimum in *incremental learning schemes*, neurocontrol should not be reduced to a science of gradient computation. It is important that the diversity of methods for gradient computation does not prevent us from realizing that the term optimized is much less diverse. In other words, it is useful to abstract from the way the minimum is sought and to focus on what is minimized.

This is why for each of the following fundamental neurocontrol approaches, *a fundamental cost function* (or more alternatives thereof, if necessary) is formulated. Minimizing this function by searching for the optimal set of free parameters **w** may take place in an incremental or batch-oriented way, with the help of the gradient or without it. This leads to more transparence and better orientation in the numerous neurocontrol approaches.

Of course, defining the fundamental cost function is only the first step in solving the problem. The next basic steps are (1) choosing a concrete neural network that is to approximate the sought function and (2) choosing an optimization method (or training method, in neural network vocabulary) to find the optimum of the fundamental cost function. These aspects are discussed in Chapters 3 and 4, respectively.

2.1 TEMPLATE LEARNING

One way to circumvent the credit assignment problem is to take a template controller as a generator of control actions. Then, the functional values are known, and some of standard learning methods for neural network based functional approximation can be used.

> *The fundamental cost function for template learning measures the dissimilarity between template outputs and outputs computed by the neurocontroller for given controller inputs.*

More formally, it is the sum of squares (or other even functions) of deviation of action $\mathbf{u}_t$ of neurocontroller f from measured control actions of the template $\mathbf{v}_t$ for a given state $\mathbf{z}_t$

LEARNING PLANT INVERSION

$$C = \sum_t (\mathbf{v}_t - \mathbf{u}_t)^2 = \sum_t [\mathbf{v}_t - f(\mathbf{z}_t, \mathbf{w})]^2 \qquad (2.1)$$

with $\mathbf{w}$ being the vector of free network parameters.

The structural scheme of the method is given in Figure 2.1. Dashed lines indicate the sampling measurements for training, together with an implicit standard neural network training algorithm for mapping samples one to another.

This idea suggests the critical question of why one should train a neural controller if the template controller is already available—the trained neurocontroller can hardly be superior the template. However, there are some applications where this approach is reasonable:

1. *The template controller is not available in the target setting.* The most prominent subclass are problems consisting in mimicking a human controller. The interesting subcases are:

    The human controller has a good performance, for example, an expert with long experience in process control.

    Mimicking human controller is the goal of control, a frequent objective for applications such as autonomous road vehicle control, where the passengers should feel as comfortable as with a human driving in the vehicle.

2. *The template controller is too complex for the target setting.* Some controller types such as multidimensional look-up tables gained by dynamic optimization (see Section 1.3) are too large for implementation on microcontrollers.

2.2 LEARNING PLANT INVERSION

The idea of learning plant inversion with the help of a neural network has been developed by several researchers such as Widrow, McCool, and Medoff [154],

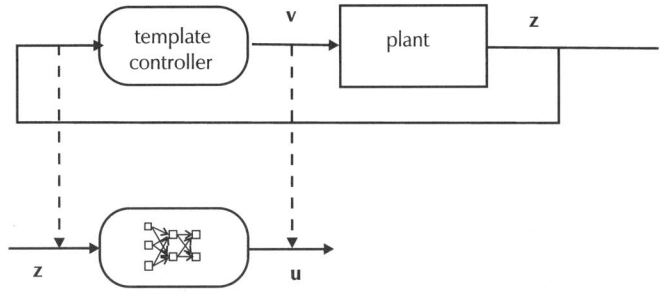

Figure 2.1 Template learning.

14 FUNDAMENTAL APPROACHES TO NEUROCONTROL

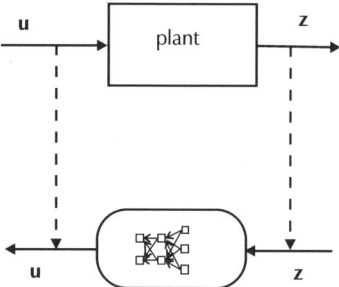

Figure 2.2 Plant inversion learning.

Psaltis, Sideris, and Yamamura [119], Grossberg and Kuperstein [48], and Kawato, Furukawa, and Suzuki [71]. The basic principle is the following: The plant output is viewed as a function of plant input; that is, the existence of mapping *output ← input* is assumed. To reach predefined reference values of plant output, inverse mapping *input ← output* is sought (Fig. 2.2). The hypothesis of existence of the mapping *output ← input* justifies learning the inverse mapping from the measurement pairs ⟨output,input⟩.

The fundamental cost function for plant inversion measures the similarity of measured and computed plant inputs for given plant outputs.

Observing the general discrete time model of the dynamic system

$$\mathbf{z}_{t+1} = f(\mathbf{z}_t, \mathbf{u}_t) \tag{2.2}$$

makes clear that such a straightforward type of inversion cannot be found in the general case. What may be possible in special cases is to find the mapping

$$\mathbf{u}_t = I(\mathbf{z}_t, \mathbf{z}_{t+1}) \tag{2.3}$$

The fundamental cost function for this simplest form of plant inversion amounts to the sum of squares (or other even functions) of the deviation of state forecast $\mathbf{z}_t$ received from the neural network model of inversion I from measured state $\mathbf{d}_t$ for a given control action $\mathbf{u}_t$

$$C = \sum_t (\mathbf{d}_t - \mathbf{z}_t)^2 = \sum_t [\mathbf{d}_t - I(\mathbf{u}_t)]^2 \tag{2.4}$$

Then, substituting the desired state for $\mathbf{z}_{t+1}$ would deliver the action variable vector $\mathbf{u}_t$ that moves the plant to this state. The problem is that the mapping is not necessarily defined for every pair of $\langle \mathbf{z}_t, \mathbf{z}_{t+1} \rangle$. The reason for this is simple: Not every state $\mathbf{z}_{t+1}$ is reachable. For linear plants of type

$$\mathbf{z}_{t+1} = \mathbf{A}\mathbf{z}_t + \mathbf{B}\mathbf{u}_t \tag{2.5}$$

this would be the case only for an invertible (i.e., square nonsingular) matrix **B**, a rather scarce occurrence.

A generalization of this principle that would be applicable to all observable plants (which is, by contrast, an abundant class) would consist in seeking the inverse mapping for a sequence of n actions (n being the order of the plant)

$$\langle \mathbf{u}_t, \mathbf{u}_{t+1}, \ldots, \mathbf{u}_{t+n-1} \rangle = I(\mathbf{z}_t, \mathbf{z}_{t+n}) \tag{2.6}$$

This mapping usually exists [it corresponds to the feedforward deadbeat controller, a solution of n subsequent state equations (2.2)] but still has three critical shortcomings:

1. Its arguments are the pairs (initial state, terminal state). To cover this huge space (the square of the state space, in topological terms) with acceptable density by measured data is a formidable task.
2. The action sequence received with help of mapping (2.6) does not contain state feedback. This may lead to insufficient robustness against disturbances and plant parameter variations. In particular, the stabilization of unstable systems requires state feedback. (This may be emulated by iterative application of the mapping. To which extent this scheme is equivalent to a true feedback with regard to stability would have to be investigated.)
3. The complete state is usually not measurable.

Note 2.1 A frequently cited problem of nonunique action appears only for redundant action variables. Then, for every input pair $(\mathbf{z}_t, \mathbf{z}_{t+n})$ some (although not unique) action sequence exists. For learning, it becomes a problem if the same, or very similar, input pair $(\mathbf{z}_t, \mathbf{z}_{t+n})$ occurs together with several different action sequences in the measured data.

A solution that uses only measurable output variables is based on the input/output plant model that explicitly accounts for relative degrees of individual output variables, as proposed by Narendra and Mukhopadhyay [104]. This general input/output model (of observable and controllable plants) is

$$\mathbf{y}^*_{t+1} = f(\mathbf{y}_t, \ldots, \mathbf{y}_{t-d_o}, \mathbf{u}_t, \ldots, \mathbf{u}_{t-d_o}) \tag{2.7}$$

where $\mathbf{y}^*_{t+1}$ is the vector of $y_{i,t+r_i}$ and r_i is the relative degree of the ith output variable. The observability index d_o is less than or equal to the plant order n. An inversion of this plant form would be

$$\mathbf{u}_t = I(\mathbf{y}^*_{t+1}, \mathbf{y}_t, \ldots, \mathbf{y}_{t-d_o}, \mathbf{u}_{t-1}, \ldots, \mathbf{u}_{t-d_o}) \tag{2.8}$$

This mapping uses the past output measurements $\mathbf{y}_t, \ldots, \mathbf{y}_{t-d_o}$, the past actions $\mathbf{u}_{t-1}, \ldots, \mathbf{u}_{t-d_o}$, and the future measurements $\mathbf{y}^*_{t+1}$ to compute the action $\mathbf{u}_t$ necessary to move the plant to the state characterized by $\mathbf{y}^*_{t+1}$.

A practical obstacle in applying this method on a broad scale is the necessity of knowing the relative degrees. A relative degree independent variant has been proposed by Åström and McAvoy [9] and by Malinowski et al. [95]. It is based on the traditional form of input/output mapping

$$\mathbf{y}_{t+1} = f(\mathbf{y}_t, \ldots, \mathbf{y}_{t-d_o}, \mathbf{u}_t, \ldots, \mathbf{u}_{t-d_o}) \qquad (2.9)$$

with inversion

$$\mathbf{u}_t = I(\mathbf{y}_{t+1}, \mathbf{y}_t, \ldots, \mathbf{y}_{t-d_o}, \mathbf{u}_{t-1}, \ldots, \mathbf{u}_{t-d_o}) \qquad (2.10)$$

However, this formal inversion works only for relative degrees equal to one. To see this, it has to be realized what the meaning of relative degree is. The relative degree of an output variable expresses the minimum time delay after which this variable can be influenced by a set of action variables. If the relative degrees are uniformly equal to one, the arbitrary desired state $\mathbf{y}_{t+1}$ can be (at least theoretically, if the action is unbounded) reached by action $\mathbf{u}_t$ of Eq. (2.10). By contrast, if the relative degree is greater than one, $\mathbf{u}_t$ cannot influence $\mathbf{y}_{t+1}$; it is too late to do something at time t if the result is expected at time $t+1$.

This can be illustrated on a simple example of a discrete double integrator plant

$$\begin{aligned} z_{1,t+1} &= z_{1,t} + u_t \\ z_{2,t+1} &= z_{2,t} + z_{1,t} \\ y_{t+1} &= z_{2,t} \end{aligned} \qquad (2.11)$$

The action u_t cannot obviously influence y_{t+1} but only y_{t+2}.

This is why the training of Eq. (2.10) is an ill-posed problem whenever the relative orders exceed one. This approach may still be useful if the relative order that is uniformly equal to one were a frequent special case or a good approximation of the general case. Unfortunately, this case is, in fact, a singularity—it is equivalent to the assumption that nonmeasurable variables are irrelevant to the measurable system dynamics. But under such assumption, the state description can be reduced to measurable outputs, so a simpler inversion of the form (2.3) can be used directly.

Besides, to complete inversion of dynamic systems, there are two more restricted, but also less problematic, scenarios. The first is the application to algebraic systems (or zeroth-order systems) of the type

$$\mathbf{z}_{t+1} = f(\mathbf{u}_t) \qquad (2.12)$$

In this setting, the input/output pairs are of reasonable dimension. The only problem connected with this scheme are possible action variable redundancies (see Note 2.1): actions that constitute equivalence classes not known a priori. Such problems are particularly frequent in robotics [see Kawato et al., 71].

Another, closely related scenario is based on representation of stationary states of dynamical plants. (It is, in fact, a generalization of the algebraic system scenario.) Writing the system equation (2.2) in continuous time form

$$\dot{\mathbf{z}} = f(\mathbf{z}, \mathbf{u}) \qquad (2.13)$$

makes obvious that the stationary states equilibrium points of the plant satisfy

$$\mathbf{0} = f(\mathbf{z}, \mathbf{u}) \qquad (2.14)$$

and can thus be, under some conditions, transformed to the form

$$\mathbf{u} = I(\mathbf{z}) \qquad (2.15)$$

The existence of mapping (2.15) is not guaranteed for all $\mathbf{z}$ of the state space, and the problem with redundant action variables is similar as for algebraic systems.

It has to be pointed out:

the plant inversion approach is closely connected with the concept of a control goal consisting in reaching a specified reference state.

Such reference states are available for usual stabilization or trajectory following tasks. Unfortunately, these are not the only types of industrial control problems. A good example is control of a wastewater treatment plant, as discussed in Chapter 15. In this case, some of the goals of control are only indirectly related to the state variables. They are formulated as a minimum expense for keeping pollution below a legally prescribed level. A straightforward implementation of more general control goals that either (1) specify the control goal only partially or (2) do not imply any explicit reference to a desired state is difficult with the inversion approach.

Although the plant inversion approach does not provide a solution for arbitrary control problems, it may be an efficient means for special tasks, particularly trajectory following with a carefully specified realistic reference trajectory. A special algorithm for this is explained in Section 6.5.1.

An important asset of the plant inversion approach is that it can use well-known, frequently implemented, and conceptually simple methods for training by input/output pairs such as backpropagation-based methods or their higher-order modifications.

2.3 CLOSED-LOOP OPTIMIZATION

The next group of control approaches is characterized by relating the control performance to an explicit cost function. The only condition this cost function has to satisfy is that it must be a function of plant states and inputs (i.e., of controller action) in a closed loop. Credit assignment is done by figuring out what effect free controller parameters have on the cost function in a mathematical way. Because the closed-loop behavior is determined by both the controller and the plant, this approach uses a mathematical description of the plant, referred to as the *plant model*.

The requirements on the precision of the plant model may vary with the particular role of the model in the individual method, but generally it can be said that the identification of the plant model is an essential constituent part of this approach. Although combining analytical models with neurocontrollers as well as of neural network models with classical nonlinear controllers would be a straightforward generalization, it is typical that both plant model and controller are represented by two neural networks.

The problem formulation usual in classical optimal control is minimizing the functional

$$\int_t c(\mathbf{z}, \mathbf{u}) dt \qquad (2.16)$$

under the constraint

$$\dot{\mathbf{z}} = f(\mathbf{z}, \mathbf{u}) \qquad (2.17)$$

that is, looking for trajectories of $\mathbf{z}$ and $\mathbf{u}$ (a) that are possible trajectories of the plant described by function f and (b) that minimize $\int_t c \, dt$.

In discrete time and with a finite time horizon T, this can be reformulated as a function-minimization problem. The parameter vector $\mathbf{w}_{\min}$ that minimizes the fundamental cost function is searched

$$C(\mathbf{z}_1, \mathbf{u}_0, \mathbf{z}_2, \mathbf{u}_1, \ldots, \mathbf{z}_T, \mathbf{u}_{T-1}, \mathbf{w}) \qquad (2.18)$$

where $\mathbf{z}_t$ is defined recursively as

$$\mathbf{z}_{t+1} = f(\mathbf{z}_t, \mathbf{u}_t) \qquad (2.19)$$

with initial state $\mathbf{z}_0$ being given. (Note that arguments $\mathbf{z}_t$ are, strictly speaking, obsolete; they can be computed from the initial state $\mathbf{z}_0$ and the sequence of $\mathbf{u}_t$.) In special, but widespread control tasks such as stabilization or trajectory following, the cost function will consist of square or other deviations of the state or state trajectory from the desired reference state or trajectory.

The fundamental cost function for closed-loop optimization is identical with the control quality criterion.

This direct and completely general implementation of arbitrary control goal is an obvious, and in many cases decisive, advantage of the closed-loop optimization approach.

The knowledge of mapping f, that is, of the mathematical model representing the plant, is an important prerequisite for closed-loop optimization. It constitutes a separate task. Only its construction goal is stated here: to represent the dynamic behavior of the plant *as closely as possible*. This defines a separate fundamental cost function for plant identification.

The fundamental cost function for plant model identification measures the dissimilarity between measured plant outputs and outputs computed by the neurocontroller for given plant inputs.

The discussion of various aspects of plant identification as well as concrete formulations of corresponding fundamental cost functions is postponed to Chapter 5. A general scheme for control loop optimization is given in Figure 2.3.

Several basic variants of this scheme have been proposed. The scheme using a *reference model of closed loop* [101] tunes the controller to minimize the deviation between the behavior of the closed loop made by a real plant and the controller, on the one hand, and the behavior of the predefined reference

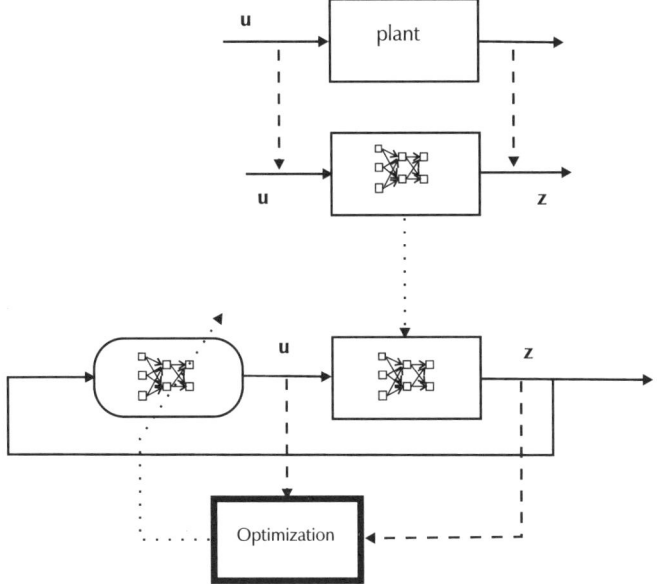

Figure 2.3 Closed-loop optimization.

20 FUNDAMENTAL APPROACHES TO NEUROCONTROL

model (Fig. 2.4), on the other. The reference model plays the role of generator of sample pairs for neurocontroller training. Here, the cost function for closed-loop optimization is specialized for measuring this deviation.

The procedure for approaching the cost function minimum can be organized as incremental or batch processing. Incremental procedures are almost exclusively based on gradient descent. If used together with reference models, they are analogous to the classical model reference adaptive control (MRAC) [103]. Some incremental procedures with explicit cost function can be found in the broad class of classical self-tuning regulators [8]. In incremental neural-network-based schemes, the weights of both the plant model network and the controller network are adapted simultaneously. Stability of the adaptation scheme is one of the crucial questions [101], as it is for classical incremental approaches. A unifying efficient algorithmic framework for analytical gradient computation is provided by backpropagation through time [109, 145] or the related dynamic backpropagation [105]. Publications concerning closed-loop gradient computation represent a large part of the neurocontrol literature.

Batch procedures [e.g., see Hrycej, 56] are started once or repeatedly after a sufficient number of samples is collected or after essential changes in plant behavior are detected. They require more storage to retain the history but provide considerable freedom in the choice of optimization method, using analytical, numerical, or no gradient. For those based on analytical gradient, the backpropagation-through-time scheme can be advantageously used.

Closed-loop optimization is the focus of the remaining parts of this book. A detailed discussion of incremental and batch schemes, plant model and

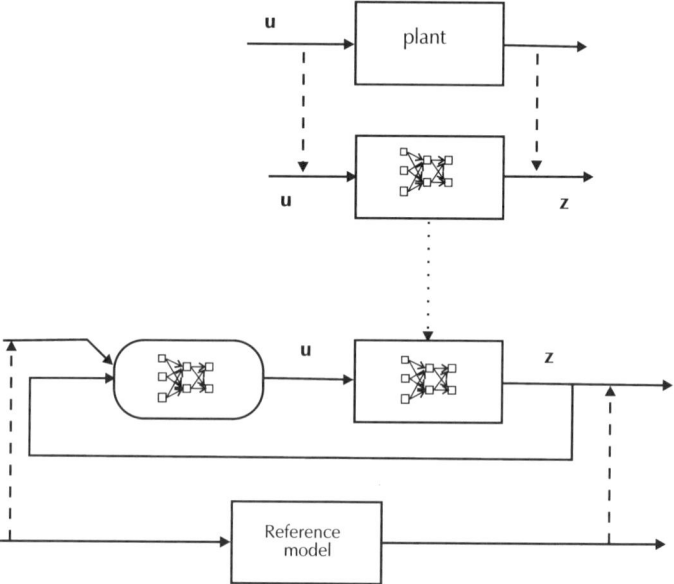

Figure 2.4 Closed-loop optimization with reference model.

controller networks, optimization methods and stability issues, is postponed to Chapters 4 to 6, 8, and 9.

2.4 CRITIC SYSTEMS

A large group of neurocontrol approaches, abundantly represented in the neurocontrol research community, are the *critic systems*.

The structure of critic systems is given in Figure 2.5. There are two neural networks with different functions. The first is the neurocontroller, usually called the *action network*. The neurocontroller is completely analogous to that of other neurocontrol approaches. The second network is the *critic*. Like the plant model network in closed-loop optimization, this network's existence is required for training the neurocontroller. In contrast to the closed-loop optimization, this network does not model the plant but models individual plant states from the viewpoint of the optimality criterion. In other words, credit assignment is defined as a subproblem that is solved simultaneously with the optimal control problem by learning.

In abstract terms, the critic of Figure 2.5 is the function

$$j = J(\mathbf{z}_t) \tag{2.20}$$

The meaning of function J corresponds to that of function C in Eq. (2.18) for the closed-loop optimization approach—it is the cost function as a criterion of control quality. The difference is that J is independent from the sequence of plant inputs. The function J thus represents the utility of state $\mathbf{z}_t$ (therefore the term *strategic utility function* was coined by Werbos [148]).

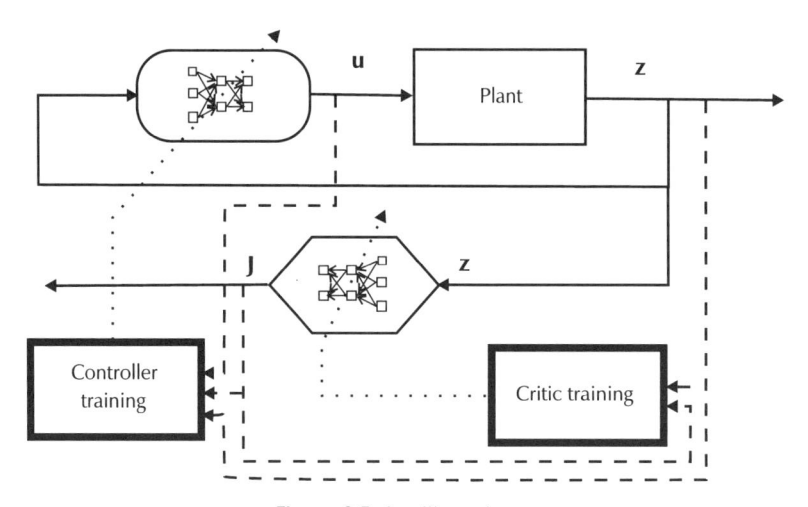

Figure 2.5 A critic system.

22 FUNDAMENTAL APPROACHES TO NEUROCONTROL

Note 2.2 It has been proposed that instead of utility, the critic represents the gradient of utility with regard to the state (DHP—dual heuristic programming). Its justification is that the gradient is needed for training the action network, not the value. This is true for training algorithms that use a gradient. The basic procedures remain analogous to those for critics representing utility, with less transparency of the underlying fundamental cost function. An overview of possible alternatives and combinations has been presented by Wunsch and Prokhorov [160].

The second part of a critic system is the action mapping

$$\mathbf{u}_t = g(\mathbf{z}_t, \mathbf{w}) \qquad (2.21)$$

The goal of action mapping is to improve the state, that is, to choose an action $\mathbf{u}_t$ that given system behavior

$$\mathbf{z}_{t+1} = f(\mathbf{z}_t, \mathbf{u}_t) \qquad (2.22)$$

leads to a state $\mathbf{z}_{t+1}$ with a higher utility value $J(\mathbf{z}_{t+1})$.

Let us now briefly discuss what alternatives there are to find the mappings J and g that lead to the best performance.

2.4.1 Training the Action Network

For fixed utility function, or critic J, several procedures have been suggested for finding the optimal mapping g. The all share the same basic utility function.

The fundamental cost function for action network learning is a negative strategic utility (as estimated by the critic network) of the state following after the action of the action network.

CASE 1 If the plant model (2.22) is known {a setting used for heuristic dynamic programming (HDP) [148]}, the action $\mathbf{u}$ leading to the $\mathbf{z}_{t+1}$ with the highest utility can be found by the search in the (frequently one-dimensional) space of actions. For training of the action network, these optimal actions $\mathbf{v}_t$, together with the state $\mathbf{z}_t$, provide explicit input/output pairs. The fundamental cost function is identical to that for template learning (2.1) and could be trained in incremental or batch mode.

Alternatively, the fundamental cost-minimization function can be directly formulated as

$$-J(\mathbf{z}_{t+1}) = -J[f(\mathbf{z}_t, \mathbf{u}_t)] = -J\{f[\mathbf{z}_t, g(\mathbf{z}_t, \mathbf{w})]\} \qquad (2.23)$$

For minimization of the fundamental cost function, its gradient for the parameter vector $\mathbf{w}$ is useful. It can be computed with the help of the chaining rule (or generalized backpropagation) and the network weights adapted correspondingly.

Note that, with the knowledge of the plant model, this approach is in direct competition with performing the closed-loop optimization of Section 2.3. The latter alternative is substantially more straightforward, and the chance for success is correspondingly higher.

CASE 2   If the critic is extended to be a function of both the present state and the next action (called action-dependent HDP, see Watkins [142] and Wunsch and Prokhorov [160]), then

$$j = J(\mathbf{z}_t, \mathbf{u}_t) \tag{2.24}$$

and the best action can be chosen directly. Therefore, input/output pairs for an action network can be generated, and the fundamental cost function of template learning (2.1) is minimized.

The direct approach would be to minimize the fundamental cost function

$$-J(\mathbf{z}_t, \mathbf{u}_t) = -J[\mathbf{z}_t, g(\mathbf{z}_t, \mathbf{w})] \tag{2.25}$$

This procedure extends the dimension of the critic input but, in most cases, not in a dramatic way. Typically, the number of action variables is substantially lower (frequently one) than that of state variables.

CASE 3   The last alternative is based on reinforcement learning. If separated from the underlying gradient search (or stochastic approximation in the stochastic case), it consists of the following:

Specifying a quality criterion, the *reinforcement signal*, for an action (or its consequence)
Collecting data about bad and good actions, that is, active exploration based mostly on random perturbations
Training the action network to perform good actions

Reinforcement learning has its roots in the field of learning automata [106]. A learning automaton is an object acting in an environment. The object performs actions upon the environment, and the environment responds by giving these actions a value, known as the reinforcement signal. The simplest model works with random binary actions and binary reinforcement signals. Automaton learning consists in increasing the probability of actions associated with positive reinforcement. This leads to a successive improvement of average reinforcement, or improvement of the performance of the automaton.

Of special interest for critic systems is a generalization of the learning automata technique to operation in a context space, that is, where actions and their value depend on a parameter vector, in our case on the state vector. Then, the automaton learns individual optimal actions for individual states. This concept has been studied by Williams [156].

The action network based on reinforcement learning is closely related to reinforcement learning with context-dependent automata. To collect data about its performance evaluation, the action network performs random actions. The data collected by the network are used to train the network so that it performs actions with high reinforcement-signal value. In the neurocontrol setting, the reinforcement signal is provided by the critic network; it is the assessed utility of the state z_{t+1} evoked by the action. Although the fundamental cost function definition from the beginning of this section also applies to reinforcement learning, its special form is worth formulating.

The fundamental cost function for reinforcement learning measures the dissimilarity between actions with highest utility and actual neurocontroller actions for given states.

This modified definition expresses the fact that the changes of actions are operationally motivated by the observation of (randomly found) more optimal functions, instead of optimizing the utility directly.

For the case where only zero and unity utility values are acquired, the fundamental cost function for learning is

$$C = \sum_{t \in S_1} [\mathbf{v}_t - g(\mathbf{z}_t, \mathbf{w})]^2 \qquad (2.26)$$

where $\mathbf{v}_t$ is the (random) action at time t, and S_1 is the set of time indices in which a utility of one has been signaled. A critic system optimized for Eq. (2.25) would, after a sufficient quantity of data has been collected, perform with a utility of one as frequently as possible.

To see the relationship between the fundamental cost function (2.26) and the adaptive procedures for minimizing it, the learning procedure for the action network of Barto et al. [12] can be investigated. It is a linear network of the form

$$u = \mathbf{wz} \qquad (2.27)$$

with additional random perturbation. If some features, such as discretization of the action, learning delays, and learning by penalty after actions with negative consequences, are omitted, the learning rule is

$$\Delta w = \alpha r u \mathbf{z} \qquad (2.28)$$

For binary reinforcement r, this is a well-known correlation learning rule that converges (for uncorrelated components of $\mathbf{z}$, see Kohonen [76]) to the state minimizing (2.26).

Note 2.3 Using adaptation-by-penalty for action with negative utility, as was done by Barto et al. [12], can speed up convergence because more data deliver information for adaptation, but it can lead potentially to unstable behavior [106].

Note that the random action of the action network is necessary only for covering the state and action space. The random action for exploration and training the action network for good performance might be separated completely (e.g., by distinguishing between an exploration action and a performing action). In many incremental reinforcement learning algorithms, the distribution of random actions is shifted toward actions so far evaluated as good. This measure probably improves the search efficiency in simple settings, such as linear action networks acting on linear systems, but may contribute to failure by an early restriction of the search space in the general case. The questions have to be answered separately of whether the state space and the action space for each state have been searched sufficiently.

Restricting to binary utility values is, of course, unrealistic for serious neurocontrol tasks. An extension of the fundamental cost function of Eq. (2.26) to real-valued utilities is possible by weighting the samples in dependence from their utility or, better, utility gain

$$C = \sum_t k[J(\mathbf{z}_{t+1}) - J(\mathbf{z}_t)][\mathbf{v}_t - g(\mathbf{z}_t, \mathbf{w})]^2 \tag{2.29}$$

where k is a positive and monotonically growing function of utility gain. [Using the set S_1 in Eq. (2.26) is equivalent to trivial weighting of the samples with unity utility by one and those with zero utility by zero.] The reinforcement offset of Williams [155] can be viewed as an implicit weighting, where the gradient change is proportional to the difference between the reinforcement signal and the reinforcement baseline of a neuron. In this case, the weighting can become negative for samples with utility below the threshold, which amounts to using penalty in addition to reward. It is difficult to determine the weighting that ensures convergence to globally optimal performance. One of the reasons for this difficulty results from the lack of knowledge of maximum utility (or utility gain) in every point of state space. This maximum utility will inevitably vary with the state space point (some states have higher utility than others even with a globally optimal action network), and it cannot be figured out other than by finding the optimal action network. But using suboptimal actions for learning will lead to (what else?) learning suboptimal actions.

A useful principle is to adapt the weighting to the current performance of the action network by weighting positively only actions that have *higher utility* than the current action network can reach. This principle can be implemented iteratively, and it can improve the current performance. However, it is difficult to implement in more than a statistical manner: Whether an action is better than another for a *given state* can be determined only after both have been tried under identical conditions!

Although these difficulties remained unsolved, various principles for incremental weight adaptation have been proposed and successfully applied to problems of moderate size [e.g., 12, 6].

Note 2.4 The exploration by random perturbations of control actions is frequently viewed as a selectional type of learning; it is substantially different from the instructional one [12] of learning schemes that use the error gradient. This is suggested by the analytical gradient computation usually connected with the schemes based on closed-loop optimization (e.g., backpropagation through time) and popular terms such as "error signal."

From the optimization viewpoint, the difference between both schemes is less essential. Making n random uncorrelated perturbations in directions $\mathbf{d}_i$, $i = 1, \ldots$, provides, together with the evaluation of utility function increase ΔJ_i after these perturbations, complete information about the n-dimensional gradient. It defines a system of n equations

$$\Delta J_j = \nabla J \mathbf{d}_i \tag{2.30}$$

for n components of gradient ∇J. It can thus be viewed as a form of numeric gradient computation.

2.4.2 Training the Critic Network

The role of the critic network is to assess strategic utility. In the additive case [148], the strategic utility for a given initial state and a given sequence of actions is a sum of terms

$$\sum_{t=0}^{\infty} U(\mathbf{z}_t) \tag{2.31}$$

where $\mathbf{z}_t$ is recursively defined by the plant state equations (2.22). The infinite series is to be substituted by a finite one not only because it cannot be practically collected (which could be emulated by a recursive definition). The following two reasons are more important:

1. Even if the series converges for some action sequence (corresponding to stable controllers), it may diverge for suboptimal action sequences produced by still suboptimal action networks in early training phases.

2. The utility function may not fulfill the condition that its optimum is zero—then, the sum cannot converge. This is the case, for example, if some goal state has the (highest) utility zero but noise prevents the system from accurate stabilization at this point.

The defined role of the critic network leads to the following fundamental cost function:

The fundamental cost function for critic network learning measures the dissimilarity between the sum of utilities of future states and the critic network output, both for a given state.

Formally, for a finite horizon T, the exact fundamental cost function is

$$\sum_t \left[\left(\sum_{s=t}^{t+T-1} U(\mathbf{z}_s) \right) - J(\mathbf{z}_t, \mathbf{w}) \right]^2 \quad (2.32)$$

Note 2.5 In contrast to the closed-loop optimization scheme and to the training of action network in critic-based systems, the fundamental cost function (2.23) is *not* related to the utility or cost function of the control task. Rather, it is the error function for the approximation of a function, in this case of the utility function.

A recursive approximation of Eq. (2.32),

$$\sum_t \{[J(\mathbf{z}_{t+1}, \mathbf{w}) + U(\mathbf{z}_t)] - J(\mathbf{z}_t, \mathbf{w})\}^2 = \sum_t [J(\mathbf{z}_{t+1}, \mathbf{w}) - J(\mathbf{z}_t, \mathbf{w}) + U(\mathbf{z}_t)]^2 \quad (2.33)$$

has been proposed by Werbos [147, 148]. This approximation is based on substituting the estimated utility

$$J(\mathbf{z}_{t+1}, \mathbf{w}) \quad (2.34)$$

for the true utility

$$\sum_{s=t+1}^{t+T-1} U(\mathbf{z}_s) \quad (2.35)$$

Both representations have the same minimum, supposing the following:

The data are not noisy, so exact approximation (i.e., zero square error) of the utility function is possible

The system is stationary, so no learning delay biases the estimation

For incremental learning, the recursive definition (2.33) provides the following two benefits: First, it requires less overhead through avoiding the computations of the exact sum of partial utilities from stored T samples. Second, the error correction signals (usually based on error gradient) are available immediately, and not only after T samples. For batch schemes, the exact fundamental cost function is at least equally useful. The sums of partial utilities can be precomputed. The task is then to obtain the standard approximation of input/output pairs.

The definition (2.32) contains explicitly the sequence $\mathbf{z}_s, s = t, \ldots, t + T - 1$. Except for the initial state $\mathbf{z}_t$, these utility-determining states depend on the action sequences $\mathbf{u}, s = t, \ldots, t + T - 2$. With the definition (2.33), there is an explicit dependence only on the first action $\mathbf{u}_t$, the others being hidden in the assumption of equality of estimated and true utility for $\mathbf{z}_{t+1}$. In both cases, the action sequence has been considered to be given. Besides the nonoperational definition of the *optimum sequence* (which is not known at the time the critic is used), the sequence can only refer to a closed loop that includes a fixed action network. Obviously, both networks are strongly interdependent. This shifts our focus to the question of how to train both networks at a time.

2.4.3 Training a Critic-Based Controller

In Sections 2.4.1 and 2.4.2, the task of training a critic-based controller has been separated into two subtasks:

1. Training the action network using the critic network for determining the goal of training
2. Training the critic network using the action network for generating the action sequence that determines the strategic utility of a state

The straightforward way is by minimization of a weighted sum of an action network fundamental cost function, say, Eq. (2.29), and a critic network fundamental cost function, such as Eq. (2.33). Taking into account that the critic network in Eq. (2.29) depends on network parameters $\mathbf{w}_J$ and that the next state $\mathbf{z}_{t+1}$ in Eq. (2.33) depends on the action network via $\mathbf{z}_{t+1} = f[\mathbf{z}_t, g(\mathbf{z}_t, \mathbf{w}_g)]$, the composite fundamental cost function is

$$\sum_t \left[a_1 k[J(\mathbf{z}_{t+1}) - J(\mathbf{z}_t)][\mathbf{v}_t - g(\mathbf{z}_t, \mathbf{w}_g)]^2 + a_2 \left(J\{f[\mathbf{z}_t, g(\mathbf{z}_t, \mathbf{w}_g)], \mathbf{w}_J\} - J(\mathbf{z}_t, \mathbf{w}_J) + U(\mathbf{z}_t) \right)^2 \right] \quad (2.36)$$

Obviously, the gradient of the critic-network parameter vector $\mathbf{w}_J$ and the gradient of the action-network parameter vector $\mathbf{w}_g$ depend on both terms and can be exactly computed only simultaneously. However, this procedure is impractical if the model mapping f is not known. This is usually the case in applications where critic systems are useful; otherwise, the more straightforward closed-loop optimization approach is preferable, as argued above.

Thus, a usual approximation of the optimization of Eq. (2.36) is iterative:

1. Collecting data D_a about utility gains through various actions for a fixed critic network
2. Training the action network by minimizing an action-network fundamental cost function, such as Eq. (2.29), on training set D_a
3. Collecting data D_c about strategic utility of individual states for action sequences of this action network
4. Training the critic network by minimizing a critic network fundamental cost function, as shown in Eq. (2.32) or (2.33), on training set D_c

Using an incremental variant of this iterative batch scheme is certainly computationally attractive because of simplified bookkeeping and history storage overheads. It also provides (seemingly) a solution for the problem of reference between both networks, that is, the reason for the necessity of iterations in the batch scheme: It is the built-in forgetting of simple incremental schemes with a fixed learning rate. The influence of past data disappears with time. So the current version of, say, the critic network is based on a version of the action network that is not arbitrarily old. (But it is also not based on the current action network because the current critic network has emerged incrementally, based on the past data, too.)

On the other hand, this task is a functional approximation of a time-varying function. There are contradicting requirements that the approximation by each network must be based on as much data as possible to cover densely the state space and that the data should be forgotten as quickly as possible. So we can refer to the current, and not past, version of the other network.

Although the incremental form of the task has the same global optimum as the iterative batch form, a computing practitioner may have a hard time believing all this would converge! However, sophisticated critic systems have been tested on real-world examples and delivered good results. An example is the aircraft-landing system of Prokhorov et al. [118].

2.4.4 Critics, Dynamic Programming, and Lyapunov Functions

It has been pointed out by Werbos [144, 148], Watkins [142], and others, that the critic concept is closely related to dynamic programming. Dynamic programming is a general optimizing principle, developed by Bellmann [13], that is applicable to problems of the following form: Let $\mathbf{z}_t$ be a state vector and $\mathbf{u}_t$ be an action vector. Let the behavior of the system be described by the matrix difference equation

$$\mathbf{z}_{t+1} = f(\mathbf{z}_t, \mathbf{u}_t) \tag{2.37}$$

The meaning of index t is quite general in dynamic programming theory. In our case, it is interpreted as a time step index. Let the cost function be formulated in a backward-recursive form

$$C_t = g(C_{t+1}, \mathbf{z}_t, \mathbf{u}_t), \quad t < T \qquad (2.38)$$

with a predefined final value of C_T. The most common special case of Eq. (2.38) is the sum of individual cost-per-time-unit terms

$$C_t = \sum_{s=t}^{T} c(\mathbf{z}_s, \mathbf{u}_s) \qquad (2.39)$$

whose relationship to Eq. (2.39) can be seen if written as

$$C_t = C_{t+1} + c(\mathbf{z}_t, \mathbf{u}_t) \qquad (2.40)$$

The cumulative cost function C_t obviously depends on the system state $\mathbf{z}_t$, the action sequence $\mathbf{u}_t, \ldots, \mathbf{u}_{T-1}$ and time t.

The idea of dynamic programming consists in defining the minimum cumulative cost function C_t^*. The definition is recursive by setting $\mathbf{u}_t$ to a value $\mathbf{u}_t^*$ that minimizes the costs in and after the tth period:

$$C_t^* = g(C_{t+1}^*, \mathbf{z}_t, \mathbf{u}_t^*), \quad t < T \qquad (2.41)$$

Through fixing the action sequence to the optimal one, C_t^* depends only on the state $\mathbf{z}_t$ and time t. Setting T to infinity, dependence of C^* on time vanishes. More precisely, under some conditions imposed on Eq. (2.38), a function $C_{-\infty}^*(\mathbf{z})$ exists to which $C_t^*(\mathbf{z})$ converges with $t \to -\infty$. A so-defined function $C^*(\mathbf{z})$, in the context of the additive cost function (2.39), has been called the *strategic utility function* by Werbos [148]. This name reflects the importance of this function for the evaluation of an action: $C^*(\mathbf{z})$ represents the utility of state $\mathbf{z}$ including its consequences in the context of an optimal action sequence.

Dynamic programming provides an exact optimizing algorithm for a very broad class of control tasks, including the following:

Accounting for stochastic aspects such as noise or plant variability by including corresponding mean or worst-case terms in the function g of Eq. (2.41)

Including exploratory actions to improve the plant model (e.g., dual control)

However, a solution of any individual recursive step (2.41) requires finding an optimum action $\mathbf{u}$ as a function of state $\mathbf{z}$. In most nontrivial problems, an analytical solution is impossible. The last resort is then a numerical solution obtained by the following procedure:

1. Discretizing the n-dimensional state space, say, to a n-dimensional grid
2. Interpolating $C^*(\mathbf{z})$ within the grid hypercubes to receive values of C^* for all possible actions $\mathbf{u}$ (since action on a state in a grid node produces a state that is generally not on any grid node)
3. Selecting the optimal action

For a grid with k nodes along each dimension, n state variables, and m action variables for each of k^n grid nodes, k^m actions are to be tried. So for each time step, $k^{(n+m)}$ one-step model simulations are performed. This procedure is clearly feasible only for systems with very rough grids and few state variables. Even then, grid interpolation may lead to controllers of humble quality.

Computational expense for a straightforward numerical application of dynamic programming supplies a good reason for the search after approximations. It is here that critic systems enter the stage: Approximating C^* is exactly the goal of training a critic.

Let us now briefly explain another concept related to critic systems, the concept of a Lyapunov function. The theory of Lyapunov functions has been developed with the goal of receiving a tool for investigation of stability of nonlinear systems. For time invariant systems, one of the basic theorems of Lyapunov stability theory states the following. Let us have a system whose behavior can be described by a vector differential equation

$$\dot{\mathbf{z}} = f(\mathbf{z}) \qquad (2.42)$$

Furthermore, let a positive definite function $J(\mathbf{z})$ exist with a minimum at point $\mathbf{z}_0$ and the property

$$\frac{dJ}{d\mathbf{z}} f(\mathbf{z}) < 0 \qquad (2.43)$$

Then system (2.42) is asymptotically stable at $\mathbf{z}_0$; that is, the state $\mathbf{z}$ will converge to the state $\mathbf{z}_0$ with an arbitrary precision.

The proof idea is based on the fact that a positive-definite function has a single minimum. Since

$$\frac{dJ}{dt} = \frac{dJ}{d\mathbf{z}} \frac{d\mathbf{z}}{dt} = \frac{dJ}{d\mathbf{z}} f(\mathbf{z}) \qquad (2.44)$$

Eq. (2.43) guarantees that the value of J for a given system trajectory will diminish with time. For a function J with a single minimum at $\mathbf{z}_0$, the trajectory with diminishing values of J cannot but converge to $\mathbf{z}_0$.

32 FUNDAMENTAL APPROACHES TO NEUROCONTROL

This theoretical result can be used in several ways. In control theory and practice, the following two are usual: (1) For a given plant and a given controller (which altogether define a closed-loop system of the form (2.42)), stability can be proved via construction of a Lyapunov function. (2) For a given plant, a controller (or a class of controllers) can be sought such that a Lyapunov function exists for the closed-loop system (2.42). If the latter approach is followed, a constructive method for formulating a Lyapunov function is needed. One possibility is to use dynamic programming [79]. Suppose that the following conditions are satisfied:

1. The fixed point state z_0 (usually corresponding to a prescribed reference state if the system is a regulated closed loop) is accessible for every z.
2. The cost function expressing some kind of penalty for distance from the equilibrium point is of additive form (2.39) with nonnegative function c.
3. The costs c_t in time period t approach zero if z_t and u_t go to zero.

Then the dynamic programming procedure described above will produce two functions:

1. The function $u^*(z)$ represents the optimal controller
2. The positive definite function $C^*(z)$ represents the costs or utility of state z

The function C^* is a (discrete) Lyapunov function since the accessibility of state z_0 together with positive costs c_t imply that C^* is diminishing with time, which is equivalent to the continuous condition (2.43).

For every given stable controller, a similar procedure can be applied to receive stability proof. In this case, the function C^* resulting from actions u^* of the optimal controller is substituted by the function C^+ resulting from actions u^+ of the given controller.

The inverse relationship cannot be guaranteed. For an arbitrary given utility function (or strategic utility function, or the critic itself, in terms of critic terminology), it is not always possible to find a controller such that the closed loop is monotonically descending the utility function. (This also does not mean that there is no stable controller—the descent may work with another Lyapunov function.)

It can be seen that the Lyapunov function is a certain relaxation of dynamic programming concepts: As can be seen from the procedure of Kreisselmeier and Birkhölzer [79], it can be constructed not only for the optimal controller but for every stable controller. Also, the Lyapunov function is not unique; every positive-definite function that the closed loop descends in time does the job.

So it may be simpler to find a Lyapunov function than the (frequently unique) strategic utility function incorporated in the critic. So far, besides the computationally extremely expensive numerical approach, no general procedure for finding a Lyapunov function (no matter whether before or after finding the stable controller) has been found. This demonstrates how difficult the enterprise of learning critic systems is.

In summary, the principal magic of critic-based neurocontrol learning seems to be in solving a lot of difficult problems implicitly in the critic construction, including

Modeling the plant
Establishing the relationship between plant states and user's preferences
Determining how far the utility of a state is influenced by its late consequences
Evaluating the performance of a particular controller in time

This is a problem that holds some promise—it would be nice to find a fast and general algorithm for critic learning. Too bad it is so difficult!

2.5 SUMMARY OF FUNDAMENTAL APPROACHES TO NEUROCONTROL

Both template learning and plant inversion learning are approaches with special niches. For problems in their specific domains, they provide good results even with simple training methods such as backpropagation. For general control problems, the principal choice is between critic methods, usually coupled with reinforcement learning, and an approach based on closed-loop optimization.

The main points for comparison are the following:

1. Critic methods do not require explicit formulation of the relationship between the plant state and the cost or utility function. On the other hand, because it can be postulated that such a relationship must exist for the utility-based optimization to be feasible, the closed-loop optimization procedures based on identified models might be extended by identifying the utility function (or utility observer if the relationship can be expected to be dynamic) from pairs (measured state, assigned utility). I am not aware of research in this direction, but this task would be a relatively straightforward application of standard learning algorithms for functional approximation.

2. Critic methods seem to be substantially nearer to the way humans (except for control engineers) address control and decision problems, and thus may be combined more easily with knowledge-based approaches such as fuzzy control.

3. If we substitute "control engineers" for "humans" in the previous sentence, the approach based on identified models is better—it is based on a framework originally formulated within control theory.

4. The sequence "identifying the model" → "designing controller" is easier to check for the quality of intermediate results than the inseparable pair critic–controller. The correspondence between the simulation of the identified model and the measured time series can be evaluated in relatively straightforward ways by both numeric and graphic methods. A controller trained with a good model of reality can be evaluated off line by extensive simulations and thus has higher acceptance than one received from a black box.

5. Critic methods are better suited to online operation, although offline batch processing is possible. Online adaptation with the closed-loop optimization procedure is less elegant (but converges better). Closed-loop optimization methods fully show their advantages in batch operations.

6. Computational efficiency for complex nonlinear tasks is difficult to compare because there are no general results for either approach. A rough assessment can be made in the following way. The critic approach requires determining two parametrized functions of the measured state vector: the critic and the controller. Also the identification-based approach amounts to determining two parametrized functions of the measured state vector: the model and the controller. A similar number of function calls are necessary to reach a good solution by a similar optimizing method. The critic-based approach implies two types of obstacles to straightforward optimization:

a. As argued above, the critic approach requires *simultaneous* optimization of both networks. Consequently, the size of the optimization problem doubles. This can increase the computational expenses, since the complexity of global optimization is exponential in size. By contrast, the closed-loop optimization approach allows the optimization of the plant model and that of controller to be done in sequence.

b. Although simultaneous optimization of both networks is crucial for the critic approach, there is no obvious fundamental cost function for this simultaneous optimization. Functions of the type in Eq. (2.36) cannot be used because they require the plant model to be known in advance. A practical solution is to consider one network as fixed during the training. This results in a *sequence* of pairs of optimization tasks, instead of a single pair as is the case with closed-loop optimization.

In other words, the closed-loop optimization approach benefits from a formulation of identification and controller designs as separate optimization tasks with clearly specified cost functions. Then, sophisticated efficient optimization methods, such as those listed in Chapter 4, can be applied. For critic-based methods, such a straightforward approach is not applicable. In Chapter 9, methods are presented that assess both the controller and the Lyapunov function (by analogy to the critic above), using the same plant model as identified above. This can become a big obstacle in solving problems of realistic complexity.

In summary, optimization of a closed loop is a highly developed approach. It is currently the preferred choice for industrial applications. This is why the remaining chapters of this book focus on the optimization of a closed loop. However, some topics are also relevant to critic systems. They are, in particular, the neural network types for functional approximation (Chapter 3), controller structure (Chapter 6), and stability (Chapter 9).

3

NEURAL NETWORKS FOR CONTROL

Fundamental control tasks, such as feedback control designs, plant identification, and the training of a critic, are all functional approximation tasks. The functional approximations sought are the optimal feedback law, the plant model that is the most consistent with measured data, and the best approximation of the strategic utility function.

In Chapter 2, several frameworks were presented with functional approximation procedures and optimization goals (the basic cost function). In addition to the optimization method, a space of parametrized functions is needed. In this functional space, instances (characterized by certain parametric values) are sought that represent the best solution of the functional approximation task, that is, the minimum of the fundamental cost function.

Various types of neural networks developed in the past define such functional spaces. Individual neurocontrol learning algorithms are frequently explicit or disguised analogies of classical control design such as optimal control or numerical Lyapunov function based design methods. However, it is the representation of functions by neural networks that defines the field of neurocontrol in the broad sense and distinguishes neurocontrol from classical control.

There are several candidate neural network types for neurocontrol. The following sections discuss the advantages and disadvantages using the following criteria:

1. Type of nonlinearity used
2. Symmetric or asymmetric networks

3. Recursive or nonrecursive networks
4. Networks representing the approximated mapping directly or indirectly

3.1 NEURAL NETWORKS USING VARIOUS TYPES OF NONLINEARITIES

3.1.1 Linear Neural Networks

Let us start our discussion on nonlinearities with the case of no nonlinearities. A network without nonlinearities has a simple single-layer structure consisting of a set of input units and a set of output units. The activation of each output unit is equal to the weighted sum of the activations of the input units:

$$\mathbf{z} = \mathbf{W}\mathbf{x} \tag{3.1}$$

Formula (3.1) documents the trivial truth that linear neural networks can represent only linear controllers and linear plant models. There are plenty of mathematically established controller design algorithms based on algebraic operations. The prototype of these algorithms is the design of linear–quadratic optimal Ricatti controllers, a method that is sufficiently formalized to be implemented as a piece of software.

In software, these design algorithms are much faster than any neurocontrol learning algorithm can be. So, at first glance, there is no reason to consider a linear neurocontroller at all. Very roughly, this is true. However, let us at least list the aspects of control design that are not completely covered by any general linear classical design algorithm:

1. *Nonquadratic Optimization.* Ricatti controllers are optimal for any quadratic cost function. There is no general nonquadratic analogy to them. The immediate question is whether nonquadratic criteria are useful. The answer is frequently yes. In many practical tasks, the priorities between partial goals are different in different states. For example, comfort (i.e., low acceleration) may have highest priority in autonomous vehicles, but in the neighborhood of road margins, security considerations override comfort requirements. It may be very difficult to formulate such preferences in quadratic terms.

2. *Constraints.* The linear–quadratic approach assumes no bounds on state variables and control action. This assumption is frequently violated in practice. Introduction of hard limits on an action variable to a Ricatti controller may lead to a still working, but suboptimal, controller.

3. *Robustness.* Similar problems are encountered if robustness against non-Gaussian noise or plant parameter variations is required.

In all these situations, there is still no reason to expect that a linear controller is as good as a more general nonlinear one. On the other hand, linear controllers come along with $m \times n$ parameters (m being the number of controller actions and n the number of state variables). This moderate dimension of the parameter space may just make the optimization task computationally tractable even for complex problems. In any case, such a linear neurocontroller can serve as a benchmark for a nonlinear one.

3.1.2 Nonlinearities of Sigmoid Type

Nonlinearities of sigmoid type are probably the most widespread neural network nonlinearities. They have been made popular by Rumelhart et al. [126]. The network using this type of nonlinearities is frequently referred to as *multilayer perceptron*.

The sigmoid nonlinearity is described by the function

$$s(x) = \frac{1}{1 + e^{-x}} \qquad (3.2)$$

This function is plotted in Figure 3.1.

A modification of the sigmoid that is symmetric around the coordinate origin is

$$r(x) = \frac{2}{1 + e^{-x}} - 1 \qquad (3.3)$$

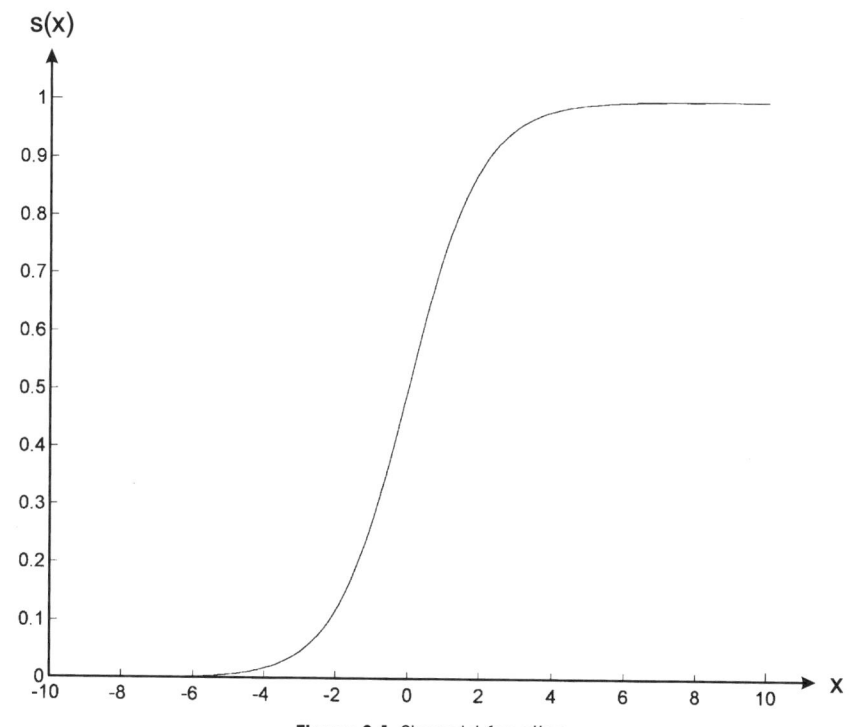

Figure 3.1 Sigmoid function.

40 NEURAL NETWORKS FOR CONTROL

The symmetric sigmoid function is plotted in Figure 3.2. Obviously, the only difference from Figure 3.1 is the range on the vertical axis. It has the property

$$r(-x) = -r(x) \tag{3.4}$$

as can be seen by

$$\begin{aligned}
r(-x) &= \frac{2}{1+e^x} - 1 \\
&= \frac{2e^{-x}}{e^{-x}+1} - 1 \\
&= \frac{2e^{-x} - e^{-x} - 1}{e^{-x}+1} \\
&= \frac{e^{-x}+1-2}{e^{-x}+1} \\
&= -\left(\frac{2}{1+e^{-x}} - 1\right) \\
&= -r(x)
\end{aligned} \tag{3.5}$$

Figure 3.2 Symmetric sigmoid function.

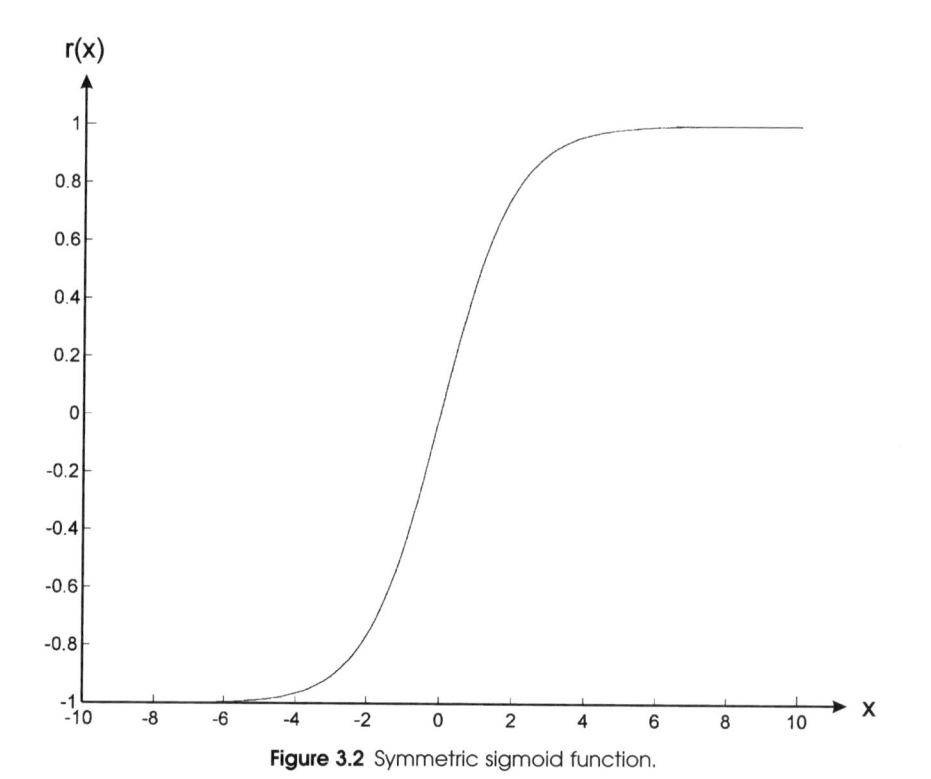

The argument x of a sigmoid unit is usually a weighted sum of the activations of the previous network layer. Generally, this sum is biased by a further constant, the *threshold* or *bias*.

A two-layer perceptron (a feedforward network with a single hidden layer, to be precise) with sigmoid hidden units as defined in Eq. (3.2) yields a vector function $\mathbf{z} = F(\mathbf{x})$ of the following form:

$$z_k = w_{k0} + \sum_j w_{kj} s\left(v_{j0} + \sum_i v_{ji} x_i \right) \tag{3.6}$$

Defining the vector sigmoid functions

$$S(\mathbf{x}) = [s(x_1), \ldots, s(x_n)]^T \tag{3.7}$$

and

$$R(\mathbf{x}) = [r(x_1), \ldots, r(x_n)]^T \tag{3.8}$$

we can express Eq. (3.6) in matrix notation:

$$\mathbf{z} = \mathbf{w}_0 + \mathbf{W} S(\mathbf{v}_0 + \mathbf{V}\mathbf{x}) \tag{3.9}$$

If symmetric hidden units with zero thresholds are used, the whole network is symmetric:

$$\begin{aligned} F(-\mathbf{x}) &= \mathbf{W} R[\mathbf{V}(-\vec{x})] \\ &= -\mathbf{W} R(\mathbf{V}\vec{x}) \\ &= -F(\mathbf{x}) \end{aligned} \tag{3.10}$$

The sigmoid function has three properties that seem to make it particularly interesting for control:

1. The sigmoid function is a monotonically increasing function. This property is intuitively attractive for control because common sense suggests that the controller action should be stronger for larger deviations between the goal state and the measured state (a clearly monotonic behavior). For most plants, this conjecture is correct.

2. There is a nearly linear region around $x = 0$. It is known that the optimal controller is linear for linear–quadratic problems. So, if we are lucky enough to encounter a nearly linear–quadratic problem, the three-layer perceptron with as many sigmoid hidden units as input units is expressive enough to represent an arbitrary linear mapping of the form

$$\mathbf{y} = \mathbf{V}\mathbf{x} \tag{3.11}$$

with arbitrary precision. This can be shown in the following way: The first three derivatives of the symmetric sigmoid function are

$$r' = \frac{1}{2}(1 - r^2) \tag{3.12}$$

$$r'' = \frac{r}{2}(1 - r^2) \tag{3.13}$$

$$r''' = \frac{1}{4}(1 - r^4) \tag{3.14}$$

Since $r''(0)$ is zero, the Taylor expansion of r at $x = 0$ is

$$r(x) = \frac{x}{2} + R_3 \tag{3.15}$$

with Lagrange residue

$$R_3 = \frac{x^3}{3!} \frac{1 - r^4(\rho x)}{4} = \frac{x^3 [1 - r^4(\rho x)]}{4!}, \qquad \rho \in (0, 1) \tag{3.16}$$

The residue is bounded by

$$|R_3| = \frac{|x|^3}{4!} \tag{3.17}$$

The difference between $2r(x)$ and x is

$$2r(x) - x = 2\left(\frac{x}{2} + R_3\right) - x = 2R_3 \tag{3.18}$$

So, an identity function $f(x) = x$ is approximated by $2r(x)$ with precision of order three with $x \to 0$. Consequently, if $\max_j |V_j x| = d$ and $\mathbf{I}$ is a unit matrix, the perceptron of the form

$$F(\mathbf{x}) = \frac{d}{\sqrt[3]{120\epsilon}} \mathbf{I} R\left(\frac{\sqrt[3]{120\epsilon}}{d} \mathbf{V} \mathbf{x}\right) \tag{3.19}$$

approximates the linear mapping (3.11) with precision ϵ.

It should be noted that the same order precision can be reached by any other odd function with the nonvanishing first derivative at the origin.

3. The sigmoid function is also able to approximate a hard-limit function

$$l(x) = \begin{cases} -1 & \text{if } x < 0 \\ 1 & \text{otherwise} \end{cases} \tag{3.20}$$

with arbitrary precision by rescaling the argument x [i.e., by using $r(cx)$ with large c]. In this case, the convergence is exponential, as can be seen by

$$r(cx) - 1 = \left(\frac{2}{1+e^{cx}} - 1\right) - 1 = \frac{2e^{-cx}}{1+e^{cx}} \qquad (3.21)$$

An activation function closely related to the symmetric sigmoid unit is the arctan function. Some researchers favor using the arctan function because its derivative

$$\arctan'(x) = \frac{1}{1+x^2} \qquad (3.22)$$

is larger than that of the sigmoid function for x far from zero. This is favorable for the convergence of simple learning methods.

Theoretically, functional approximation by a multilayer perceptron is based on such results as those of Cybenko [23], Funahashi [39], and Carrol and Dickinson [21], who state that a multilayer perceptron possesses the capability of approximating any continuous function with arbitrary precision. Bounds on the number of hidden units have not, so far, been found. This author's experience has not yet contradicted his own rather unscientific rule that sets the number of hidden units to $\min[\max(m,n), 6]$, where m and n are the number of perceptron inputs and outputs, respectively. The great attraction of this rule is that it limits the number of network parameters to a complexity order of $O[(m+n)^2]$.

3.1.3 Radial Basis Functions

Another popular class of activation functions are the so-called radial basis functions (RBFs). These are functions of the form

$$b(\mathbf{x}) = h(\|\mathbf{x} - \mathbf{c}\|) = h\left[\sqrt{\sum_i (x_i - c_i)^2}\right] \qquad (3.23)$$

where $\|\cdot\|$ is the Euclidian norm, $\mathbf{c}$ is a vector representing the center of a RBF, and h is a continuous function. The most usual choice of h is the Gaussian

$$h(p) = e^{-p^2/v} \qquad (3.24)$$

An important asset of RBFs is their close relationship to approximation and regularization theory. In particular, given K pairs $(\mathbf{x}_k, \mathbf{z}_k), k = 1, \ldots, K$, there is a linear function of K RBFs of the form $h(\|\mathbf{x} - \mathbf{x}_k\|)$ that exactly fits the pairs $(\mathbf{x}_k, \mathbf{z}_k)$. This function can be written as

$$G(\mathbf{x}) = \mathbf{H}^{-1}\mathbf{Z}\mathbf{b}(\mathbf{x}) \qquad (3.25)$$

where $K \times K$ matrix $\mathbf{H} = \left[h(\|\mathbf{x}_i - \mathbf{x}_j\|)\right]$, $i, j = 1, \ldots, K$, $\mathbf{Z}$ is a matrix with rows $\mathbf{z}_k$, and $\mathbf{b}$ is the vector of $h(\|\mathbf{x} - \mathbf{x}_k\|)$. A particularly attractive feature is the possibility of imposing, in addition to the requirement of fitting the functional to the measurements, a smoothness requirement with a continuous weight λ that leads to a modification of the approximation function (3.25):

$$G(\mathbf{x}) = (\mathbf{H} - \lambda \mathbf{I})^{-1} \mathbf{Z} \mathbf{b}(\mathbf{x}) \qquad (3.26)$$

In terms of RBF neural networks, for a hidden layer with as many RBF units as measurements, output layer weights can be determined that ensure either exact fitting or best fitting to measurements for a given weight of smoothness constraints. In both cases, the output layer weights can be computed analytically, a rare comfort for neural networks.

Unfortunately, the straightforward application of this approach suffers from a major shortcoming. In most cases, it is impossible to take as many RBF units as measurements. In applications where measurements are a part of the problem specification (e.g., classification applications), it is only a matter of quantity, since there are usually many more measurements than we wish (and can afford) to have as hidden units. Even worse, in controller design applications, the measurements are not explicitly known; looking for them is part of the problem.

A usual help is to let the measurements be represented by a set of *templates*, points in the domain of the sought function that are assumed to cover all important cases. This can be done either deliberately, by specifying the templates, or adaptively, by letting the system find such templates that only important parts of the domain are covered. The latter approach is an elegant one. The methods for this approach are called *adaptive resonance*, developed by Grossberg [43], and *learning vector quantization*, developed by Kohonen [76], and they have further been refined by many researchers. However, template adaptation is nonlinear optimization with potentially many optima. Moreover, it cannot be approached separately from the optimization of the output layer. What is important and frequent in the controller input domain is known only after a good controller has been found, though the hidden layer templates are a part of this controller. In summary, the comfort of analytical optimization of the RBF network output layer can scarcely be enjoyed in the controller design practice.

There is also another benefit of the fact that an RBF network has an output linear in coefficients of the output layer. Narendra [102] formulated a (likely the first and maybe the only) *globally stable adaptive rule* for nonlinear plant identification with an RBF neural network. Like the approximation theorems cited above, this adaptive rule assumes a fixed RBF hidden layer. Also, in this case the question of determining the parameters of RBF units remains open.

There are some further problems concerning RBF units. One of them is that of the scaling of input values. The activation of units of the form shown by Eq. (3.23) depend on the Euclidean distance between the actual input $\mathbf{x}$ and the template **c**: the smaller the distance, the higher the activation. The common-sense idea of RBF-based interpolation is that of weighting RBF templates by their proximity to the point to be interpolated. Euclidean distance is a good measure of proximity only if all input variables have approximately the same range.

To illustrate this point, let us imagine the following situation. A function of two arguments $f(x, y)$ is to be interpolated, with x taking values from $\langle 0, 1 \rangle$ and y taking values from $\langle 0, 100 \rangle$. One possibility is to take measurements from a grid with an equal number of nodes along each coordinate, say, with $\Delta x = 0.1$ and $\Delta y = 10$. This makes 121 grid points altogether. Let us now compare Euclidean distances of points $A = (0.50, 53)$ and $B = (0.53, 50)$ from individual measurement points on the grid. Point B can be easily assigned nearest templates: its nearest neighbor is $(0.50, 50)$ with Euclidean distance 0.03, distinctly nearer than the second nearest neighbor $(0.60, 50)$, with a 133% larger Euclidean distance of 0.07. By contrast, all 11 points $(\cdot, 50)$ have Euclidean distances less than 30.9, that is, within a 3% difference from the nearest neighbor $(0.50, 50)$. In other words, the position of point A on the x axis is hardly accounted for if the approximation template choice is based on Euclidean distance.

Another possibility, of course, is to take a grid equidistant along both coordinates. However, in our case, this would increase the number of grid points excessively. For a resolution of 0.1 units on both coordinates, 12,100 grid points are necessary.

A seemingly obvious remedy to this problem is to rescale all the variables to a normed range (e.g., $\langle 0, 1 \rangle$). However, in many cases (and particularly in cases of control), this merely alleviates the problem—the range of values that are the most relevant for the task is not identical with the range of possible values.

Another face of the same problem concerns irrelevant variables. In a sigmoid unit, an irrelevant variable can simply receive a zero weight and thus be out of consideration. To do the same for RBF units would amount to scaling the irrelevant variable to infinity—its value would not influence the activation of the unit. However, determining the relevance of a variable is part of learning. This suggests learning the scale as a free parameter of the RBF unit. Instead of Eq. (3.23),

$$b(\mathbf{x}) = h\left[\sqrt{\sum_i w_i(x_i - c_i)^2}\right] \quad (3.27)$$

is used.

Even the form (3.27) fails to capture linear dependences between variables. There is no way for a single RBF of the type shown in Eq. (3.27) to represent a good approximation of a linear function of the input variables such as $x + y$.

To provide sufficient expressions for such cases, a generalized RBF of the form

$$b(\mathbf{x}) = h\left[\sqrt{\sum_i \sum_j w_{ij}(x_i - c_i)(x_j - c_j)}\right] \quad (3.28)$$

can be used, with symmetric weights $w_{ij} = w_{ji}$. The special case of diagonal weight matrix is equivalent to Eq. (3.27). However, the expressive power of this type of RBF unit is at the expense of having, for n input variables, $n(n+1)/2 + n = n(n+3)/2$ free parameters per unit. For the basic RBF unit of the form shown in Eq. (3.28) which has n free parameters, the variable scaling shown in Eq. (3.27) seems to be a good compromise.

An important property of networks made of RBF-type units is the locality of representation [10, 102, 121]. This means that each unit is responsible for a relatively small subset of the network input space. The function h can be chosen so that these subsets for individual RBF units overlap only weakly. This is advantageous for incremental learning: What is learned about one subset (e.g., working point in control applications) is not easily overridden by information from another subset. (Unfortunately, such an appropriate choice of function h would have to account for input space dimensionality, number of RBF units, as well as scaling and distribution of values of individual variables.) These concepts in applications other than control have been intensively studied by Grossberg [43–47]. Other neural networks with strong partitioning of input space have been proposed by Kohonen [76] and Albus [3, 4]. There are also learning algorithms directly based on the partitioned input space: The BOXES learning system of Michie and Chambers [97], context learning automata [106], table lookup critic elements [12], Q-learning [142, 161], and the local basis/influence function networks of Baker and Farrell [10]. All of these benefit from the fact that there are no interferences during learning if the input space is partitioned.

Note 3.1 Of course, the locality of RBF-based representation is reached only *as long as* the input subspace in which individual RBF units are strongly activated do not overlap essentially. The overlapping case is that of *coarse coding*: If many units cooperate on the representation of the same information, the representation is (1) robust against failure of individual units and (2) more accurate than would be the case with direct representation by individual units, since individual inaccuracies are traded off statistically. Anyway, coarse coding and locality are contradicting properties, more of one of them means less of the other if the total number of units remains the same.

To some degree, locality also contradicts another property: the representational economy. Representational economy is improved if information is stored in a form as general as possible. The locality of RBF networks has its price: The number of RBF units necessary to represent a function of n arguments grows exponentially with n, as long as locality is retained. This is intuitively clear by realizing that locality is reached by partitioning the state space. A partitioning of a given resolution clearly leads to exponential growth of the number of partitions.

3.1.4 Higher-Order Units

A possible alternative to sigmoid units are higher order units (also called sigma–pi units) [e.g., see Giles and Maxwell, 42] of the form

$$f\left(\sum_k w_k \prod_{i \in S_l} x_i\right) \quad (3.29)$$

with $f(u)$ being a nonlinearity, summing over all subsets S_l (possibly limited-size subsets) of the set of the unit's inputs.

Higher-order units are very expressive and are certainly potentially capable of capturing the symmetry relationships discussed in the next subsection. However, their main shortcoming is the vast number of parameters that have to be learned. Even if we consider only second-order units, the number of their parameters grows with $O(n^2)$, where n is the number of inputs.

3.2 SYMMETRIC OR ASYMMETRIC NETWORKS

As long as neural networks are applied to pattern recognition and classification, their symmetric properties are of no particular importance. It is in the nature of the classification task that the relationship between input and output does not imply any symmetries, except for some special patterns in visual pattern recognition. Inputs are features with positive or negative values. Outputs are positive-class scores. Changing the task by adding constants to features or to class scores influences neither the theoretical nor the practical solvability of the classification problem considered. This is why activation functions without explicit symmetry representation are widely used as in the sigmoid function with a nonzero threshold.

By contrast, in neurocontrol, symmetry is much more essential. It is vital for some outputs to be odd functions of certain inputs. For example, control action usually has to be an odd function of the deviation of the controlled variable. With classical neural units, this relationship can be implemented in two ways: First, it can be enforced by taking a symmetric sigmoid function with threshold zero. The problem is that in nonlinear control and system identification, the odd-function relationship is not appropriate for all variables on which the output depends, such as variables determining the working point of the plant. Another possibility is to take the usual asymmetric sigmoids and rely on a learning procedure to discover the necessary symmetries. But this is not advisable—the learning algorithm will frequently fail to cope with this additional (and superfluous) burden. It has been confirmed by our extensive computational experiments that such failures are inevitable in nontrivial problems.

The symmetry requirements of plant model, controller, and critic networks are discussed in Sections 3.2.1 to 3.2.3, respectively, and formalized in Section 3.2.4. The symmetry problem can be addressed either on the level of a whole network or on individual nonlinear neurons. An example of the former approach is the gain vector network presented in Section 3.2.5. The latter approach is illustrated by the construction of a special activation function, the modulated sigmoid function [58], which is described in Section 3.2.6.

3.2.1 Symmetry in Plant Models

A classical linear plant model has the continuous time form

$$\dot{\mathbf{z}} = \mathbf{A}\mathbf{z} + \mathbf{B}\mathbf{u} \tag{3.30}$$

This model is symmetric around zero:

$$-\dot{\mathbf{z}} = \mathbf{A}(-\mathbf{z}) + \mathbf{B}(-\mathbf{u}) \tag{3.31}$$

In particular, for $\mathbf{z} = \mathbf{u} = \mathbf{0}$, $\dot{\mathbf{z}} = \mathbf{0}$; that is, the equilibirum point of the plant is zero. This model form is justified by most relationships among physical variables: For example, for zero force, there is zero acceleration, and so on.

A nonlinear generalization of this model is

$$\dot{\mathbf{z}} = f(\mathbf{z}, \mathbf{u}) \tag{3.32}$$

The equilibrium-point requirement probably is valid also for nonlinear systems. It can be formulated as

$$\mathbf{0} = f(\mathbf{0}, \mathbf{0}) \tag{3.33}$$

The symmetry requirement of the form

$$f(-\mathbf{z}, -\mathbf{u}) = -f(\mathbf{z}, \mathbf{u}) \tag{3.34}$$

is frequently valid approximately. It is justified by the linearization of the function around the origin. This is not the case for some classes of plants, such as diffusion systems.

If the plant behavior depends on a set of exogenous parameters **p** (temperature, age, etc.), this dependence is usually of an asymmetric nature. The equilibrium-point requirement is modified then so that

$$f(-\mathbf{z}, -\mathbf{u}, \mathbf{p}) = -f(\mathbf{z}, \mathbf{u}, \mathbf{p}) \tag{3.35}$$

is true for every **p**.

3.2.2 Symmetry in Control Applications

Also, classical linear controllers are usually defined in a symmetric way. The general form of a state feedback controller is

$$\mathbf{u} = -\mathbf{R}\mathbf{z} + \mathbf{W}\mathbf{w} \tag{3.36}$$

with **w** being the goal state. This controller guarantees zero action if a zero goal state has been reached. This controller equilibrium point defines an equilibrium point of a closed loop. More special configurations of controllers can define closed-loop equilibrium points for all goal states. An example is the dynamic controller of the form

$$\dot{\mathbf{u}} = \mathbf{R}(\mathbf{w} - \mathbf{z}) \tag{3.37}$$

which has an equilibrium point at, and is symmetric around, the goal state **w**. Nonlinear analogies are

$$\mathbf{u} = g\left(\left[\mathbf{z}^T, \mathbf{w}^T\right]^T\right) \tag{3.38}$$

and

$$\dot{\mathbf{u}} = g(\mathbf{w} - \mathbf{z}) \tag{3.39}$$

respectively. The equilibrium-point requirement for both can be formulated as

$$g(\mathbf{0}) = \mathbf{0} \tag{3.40}$$

and the symmetry requirement as

$$g(\mathbf{z}) = -g(-\mathbf{z}) \tag{3.41}$$

Also a nonlinear controller can be made dependent on exogenous parameters **p**. (This is an analogy to the gain scheduling approach to nonlinear control.) Then, the equilibrium-point requirement is

$$g(\mathbf{0}, \mathbf{p}) = \mathbf{0} \tag{3.42}$$

and the symmetry requirement is

$$g(\mathbf{z}, \mathbf{p}) = -g(-\mathbf{z}, \mathbf{p}) \tag{3.43}$$

For example, the lateral control of a vehicle [see 94] may receive as input the deviation of the vehicle course from the reference course. If this deviation is zero, the steering system is expected to do nothing; that is, the controller output (e.g., change of steering angle) must be zero. Suppose now that the steering is optimized to keep the lateral acceleration of the vehicle (caused by the control actions) independent from the forward velocity of the vehicle. Such an optimal nonlinear controller has to modify its gain for various forward velocities. The forward velocity thus becomes additional controller input. However, the controller output (the change of steering angle) should not be symmetric with regard to the forward velocity. In particular, the requirement that the controller output zero only if the forward velocity is zero would lead to a nonstable controller: The velocity is always positive in a moving vehicle.

3.2.3 Critic Networks

Critic networks are expected to approximate a function that evaluates the strategic utility of individual points in the state space. In the greatest number of cases, there is a point or a connected region with the highest utility. Outside of this region, the utility gradually decreases with increasing "distance" from the highest utility region. Here, the definition of distance must be adequate to the problem. This framework (with utility being substituted for by energy that has its minimum in the highest utility point) is formalized in the concept of a nonnegative–definite Lyapunov function, whose relationship to the critic systems was discussed in Section 2.4.1. The simplest example is a linear system that stabilizes at a goal state. Then, the negative utility can be expressed as a positive–definite quadratic Lyapunov function.

Suppose that the highest utility state $\mathbf{w}$ is known and its utility is J_0. This is frequently the case in control tasks—it is usually the state in which also the immediate utility U of Section 2.4 is the highest. Then, the critic can be sought in the form

$$J(\mathbf{w} - \mathbf{z}) \tag{3.44}$$

with the condition that

$$J(\mathbf{0}) = J_0 \tag{3.45}$$

Then, by transforming the utility values by

$$J' = J - J_0 \tag{3.46}$$

we obtain the equilibrium-point requirement

$$J'(\mathbf{0}) = 0 \tag{3.47}$$

The nonnegative–definiteness of the critic function is certainly difficult to learn from data. It may be useful to guarantee it by definition, such as

$$J'' = J'^2(\mathbf{w} - \mathbf{z}) \tag{3.48}$$

or

$$J'' = |J'(\mathbf{w} - \mathbf{z})| \tag{3.49}$$

Then, a network such as a single-output multilayer perceptron represents the mapping J'. The result is made positive–definite by transformation of the output value according to Eqs. (3.48) or (3.49).

If the equillibrium-point requirement (3.47) does not concern exogenous parameters $\mathbf{p}$, it can be reformulated as

$$J'(\mathbf{0}, \mathbf{p}) = 0 \tag{3.50}$$

for every $\mathbf{p}$.

3.2.4 Summary of Symmetry Requirements

It has been demonstrated that it is frequently useful to be able to predefine equilibrium points or symmetries of mappings represented by neural networks. If $\mathbf{x}$ are the inputs of the network

$$\mathbf{z} = f(\mathbf{x}, \mathbf{y}) \tag{3.51}$$

with regard to which f is to have a equilibrium point at, or be symmetric around, $\mathbf{0}$, and $\mathbf{y}$ are the inputs that are not included into the equilibrium point or symmetry definition, two requirements can be imposed:

1. The equilibrium-point requirement

    $$f(\mathbf{0}, \mathbf{y}) = \mathbf{0} \tag{3.52}$$

    for all $\mathbf{y}$.
2. The symmetry requirement (subsuming the equilibrium-point requirement)

    $$f(\mathbf{x}, \mathbf{y}) = -f(-\mathbf{x}, \mathbf{y}) \tag{3.53}$$

    for all $\mathbf{y}$.

Note 3.2 In many cases, the equilibrium-point requirement is strict, but the symmetry requirement is only approximate. Then, if a symmetric requirement is postulated and if the symmetric inputs **z** are duplicated in the asymmetric group

$$\mathbf{z} = f\left(\mathbf{x}, \begin{bmatrix}\mathbf{x}\\\mathbf{y}\end{bmatrix}\right) \tag{3.54}$$

then the weights on the network connections with asymmetric duplicate of **x** will account only for the *deviation from symmetry* (if these weights were zero, the symmetry is perfect). This trick may represent a substantial contribution to the training economy.

3.2.5 Gain Vector Networks

The principle of what is called here the gain vector network is very simple and has been used frequently in the past [e.g., by Alippi et al., 5]. The output of a neural network such as the multilayer perceptron is interpreted not as the next state but as the gain coefficients. It is an instance of the neural network's output being model or controller parameters [10]. For example, if a nonlinear dependence of n-dimensional vector **z** on m-dimensional vector **x**

$$\mathbf{z} = f(\mathbf{x}) \tag{3.55}$$

is to be represented, it can be done without loss of generality

$$\mathbf{z} = f'(\mathbf{x})\mathbf{x} \tag{3.56}$$

where f' is an $n \times m$ matrix (or a vector of length nm). The large number of network outputs is the price for the essential advantage that first-order information is encoded in output layer threshold alone, and it is only the higher-order information that passes from the input through the hidden layer. This makes the optimization of this network more tractable: A good starting point is defined by the thresholds as long as the weights between the hidden layer and the output layer are small.

This arrangement makes it immediately possible to formulate symmetric and asymmetric dependences. The mapping

$$\mathbf{z} = f'(\mathbf{y})\mathbf{x} \tag{3.57}$$

is symmetric for **x** and asymmetric for **y**.

The configuration for capturing the deviation from symmetry discussed in Note 3.2 is

$$\mathbf{z} = f'\left(\begin{bmatrix}\mathbf{x}\\\mathbf{y}\end{bmatrix}\right)\mathbf{x} \tag{3.58}$$

Some gain vector network structures are illustrated in the case studies of Chapters 12 to 15 (e.g., in Fig. 12.2).

3.2.6 Modulated Sigmoid Unit

A perceptron network with a single layer of sigmoid hidden units, linear output units, and zero thresholds corresponds to the mapping

$$z_k = \sum_j w_{kj} s\left(\sum_i v_{ji} x_i\right) \tag{3.59}$$

Suppose that the inputs are split into **x** contained in the symmetry or equilibrium-point requirements and **y** not contained there. Furthermore, suppose a modified sigmoid function with two scalar arguments, symmetric and asymmetric; that is, it has the form $s(x, y)$. The perceptron mapping can then be written as

$$z_k = \sum_j w_{kj} s\left(\sum_i u_{ji} x_i, \sum_i v_{ji} y_i\right) \tag{3.60}$$

If a function $s(x, y)$ can be found such that

$$s(0, y) = 0 \tag{3.61}$$

or

$$s(-x, y) = -s(x, y) \tag{3.62}$$

for any y, the network (3.60) would satisfy requirements (3.52) or (3.53), respectively.

A unit with such properties is the *modulated sigmoid unit* [58] of the form

$$s(x, y) = \frac{1}{1 + e^{-x-y-\Theta}} - \frac{1}{1 + e^{x-y-\Theta}} \tag{3.63}$$

Θ is the unit's threshold. This unit is called the *modulated sigmoid unit*. For zero asymmetric input and zero threshold, the unit is equal to the usual symmetric sigmoid unit:

$$s(x, 0) = \frac{1}{1 + e^{-x}} - \frac{1}{1 + e^x} = \frac{1}{1 + e^{-x}} - \frac{e^{-x}}{e^{-x} + 1}$$

$$= \frac{1 - e^{-x}}{1 + e^{-x}} = \frac{2 - (1 + e^{-x})}{1 + e^{-x}}$$

$$= \frac{2}{1 + e^{-x}} - 1$$

The validity of Eqs. (3.61) and (3.62) can be seen in the following way:

1. *The output of the modulated sigmoid unit (3.63) is zero if and only if the symmetric input x is zero.* If the symmetric input x is zero, then the unit's output is obviously

$$s(0, y) = \frac{1}{1 + e^{-y-\Theta}} - \frac{1}{1 + e^{-y-\Theta}} = 0 \qquad (3.64)$$

Suppose now that the unit's output is zero. Then

$$\frac{1}{1 + e^{-x-y-\Theta}} = \frac{1}{1 + e^{x-y-\Theta}} \qquad (3.65)$$

and thus

$$-x = x \qquad (3.66)$$

which can be true only for $x = 0$.

2. *For symmetric input $-x$ and asymmetric input y, the output of modulated sigmoid unit (3.63) is $-s(x,y)$.* Setting the symmetric input to $-x$, we get

$$\begin{aligned}
s(-x, y) &= \frac{1}{1 + e^{-(-x)-y-\Theta}} - \frac{1}{1 + e^{(-x)-y-\Theta}} \\
&= \frac{1}{1 + e^{x-y-\Theta}} - \frac{1}{1 + e^{-x-y-\Theta}} \\
&= -\left(\frac{1}{1 + e^{-x-y-\Theta}} - \frac{1}{1 + e^{x-y-\Theta}}\right) \\
&= -s(x, y)
\end{aligned}$$

The charts of the modulated sigmoid unit showing the dependence of function value on symmetric input x for asymmetric input y equal to 1, 3, and 5 are given in Figure 3.3.

The configuration for capturing the deviation from symmetry discussed in Note 3.2 is

$$z_k = \sum_j w_{kj} \left(\sum_i u_{ji} x_i, \sum_{i_x} v_{x_{ji_x}} x_{i_x} + \sum_{i_y} v_{y_{ji_y}} y_{i_y} \right) \qquad (3.67)$$

3.3 FEEDFORWARD AND FEEDBACK NETWORKS

Neurocontrol is concerned with dynamic systems; a concept typical for dynamic systems is that the state is necessary to characterize the behavior of the system in addition to the input and the output. The state of a system is

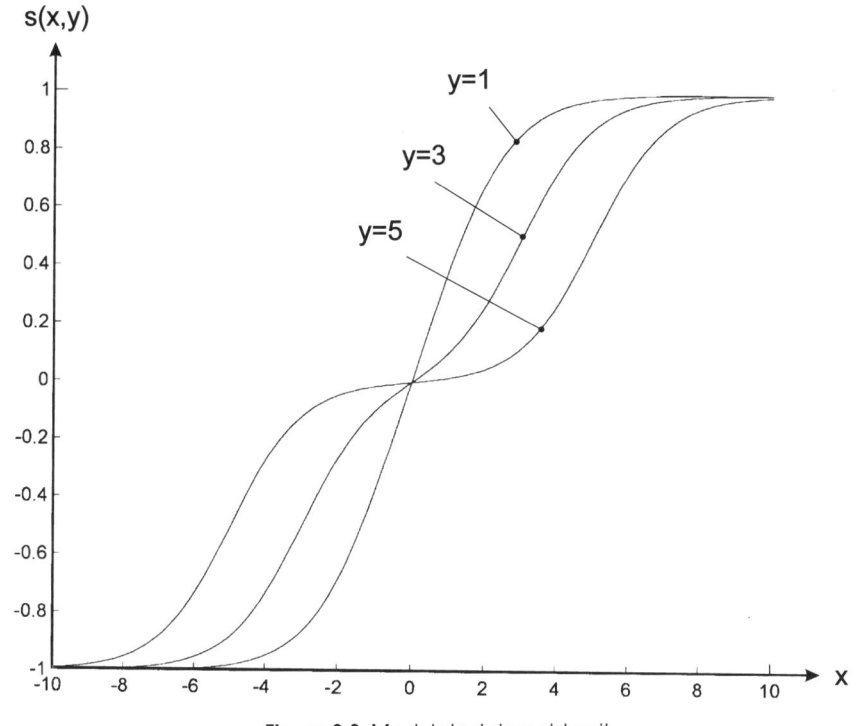

Figure 3.3 Modulated sigmoid unit.

jointly determined, in discrete time terms, by the system input and the past state. This makes clear that *state feedback* is inherent to dynamic systems.

The fact that feedback is an inherent feature of dynamic systems makes clear that neural networks representing such systems have to provide means for representing feedback. This lead to the idea [e.g., 157] of extending neural networks used in neurocontrol by internal/feedback connections. In contrast to the earlier interest on stable states or attractors of feedback networks [2, 51], neurocontrol focuses on their transient behavior. An alternative concept [e.g., 109] is based on taking a standard feedforward network and implementing external feedback by connecting some output nodes with some input nodes.

The existence of these approaches suggests the question:

Is external feedback sufficient to represent all systems, or does internal feedback add something to the expressiveness?

Let us investigate what representation is necessary for nonlinear control in dynamic plant models. It will be tacitly postulated that a multilayer perceptron with a single hidden layer can represent the arbitrary mapping required for modeling. The general form of the nonlinear discrete-time dynamic model is

$$\mathbf{z}_{t+1} = f(\mathbf{z}_t, \mathbf{u}_t) \tag{3.68}$$

with measurement vector

$$\mathbf{y}_{t+1} = h(\mathbf{z}_{t+1}) \tag{3.69}$$

The plant is of nth order if the state vector $\mathbf{z}$ is an n-dimensional vector. A network structure necessary for representing this model is given in Figure 3.4. In addition to the input layer representing the pair $(\mathbf{u}_t, \mathbf{z}_t)$, the state layer representing the $\mathbf{z}_{t+1}$, and the output layer representing $\mathbf{y}_{t+1}$, there are two hidden layers $\mathbf{h}_z$ and $\mathbf{h}_y$ necessary for the representation of nonlinear mappings f and h. Only both hidden layers contain sigmoid nonlinearities. There are feedback connections from the state layer to the input layer corresponding to passing one sampling period. The feedback comes into action only after a complete forward pass through all layers has taken place.

The network of Figure 3.4 contains internal feedback connections. However, the output (3.69) can be written, with help of (3.68), as

$$\mathbf{y}_{t+1} = h[f(\mathbf{z}_t, \mathbf{u}_t)] \tag{3.70}$$

so that a joint system description by

$$\begin{bmatrix} \mathbf{z}_{t+1} \\ \mathbf{y}_{t+1} \end{bmatrix} = f_{fh}(\mathbf{z}_t, \mathbf{u}_t) \tag{3.71}$$

can be substituted for Eqs. (3.68) and (3.69). This mapping corresponds to the *canonical form* of a discrete-time recurrent neural network formulated by Nerrand et al. [108]. Since there is only one nonlinear mapping to be represented now, the network structure is substantially simplified (see Figure 3.5).

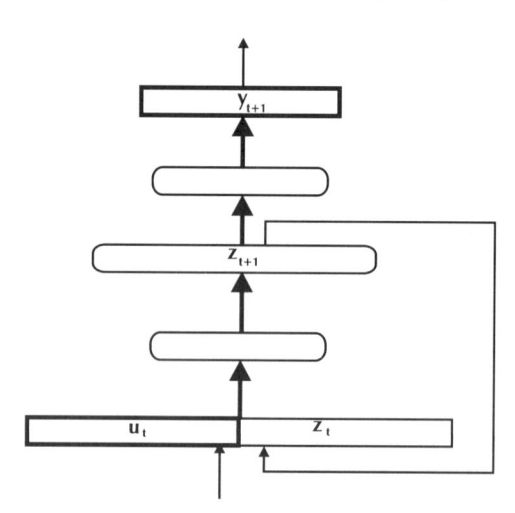

Figure 3.4 Neural network representation of the classical form of a general nonlinear plant.

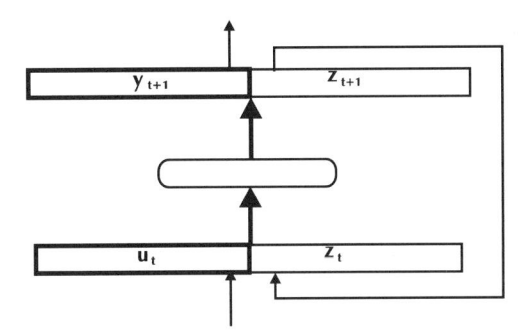

Figure 3.5 Canonical neural network representation of a general nonlinear plant.

The output layer now also contains units corresponding to nonmeasurable state variables z_{t+1}.

The canonical form (3.71) gives the answer to the question concerning internal versus external feedback:

External feedback is sufficient to represent all dynamic systems.

Note 3.3 The sufficiency of external feedback does not mean that representations of Figures 3.4 and 3.5 are equally economical. A rough comparison might be by the number of input/output dependences that have to be represented. In the general case, the representation by (3.68) and (3.69) implies n dependences on $n+p$ variables plus q dependences on n variables, while that by (3.68) implies $n+q$ dependences on $n+p$ variables. The latter case is clearly a less economical representation. The reason is simply that it contains the mapping (3.68) twice. In terms of causality structure, it ignores the fact that y_{t+1} depends on z_t and u_t only via z_{t+1}.

Note 3.4 In many cases, the output variables are simply a selection of state variables (or can be redefined so that this is the case). Then, the state vector **z** can be split into the measurable variables **y** and nonmeasurable variables **x**. Then, the external feedback mapping (3.68) represents the system completely.

Note 3.5 The basis of the representations presented are the universal approximation theorems [e.g., 23] stating that any continuous mapping can be represented by a multilayer perceptron with a single hidden layer. It is assumed that there is a state vector of *fixed length* that characterizes the system. This vector is the input to the particular perceptron. It is then the task of (an unspecified number of) hidden units to implement the mapping. However, in recurrent operations, there is no difference between hidden units, on the one hand, and input/output units that represent nonmeasurable states, on the other hand. In fact, one is free to view the recurrent perceptron as a nonlayered recurrent network of (at least) the same expressive power. This is consistent with the theorem that a recurrent network has the computational capability of a Turing machine.

This justifies some popular recurrent network architectures that are, formally, subsets of those of Figures 3.4 and 3.5. The recurrent network of Elman [26] follows the basic structure of Figure 3.4 motivated by Eqs. (3.68) and (3.69) but does not contain the hidden layers necessary for representation of general mappings f and h. Another such network has been proposed by Jordan [67].

Besides the state description by Eqs. (3.68) and (3.69), an observable plant can be expressed in the input/output form

$$\mathbf{y}_{t+1} = f(\mathbf{y}_t, \ldots, \mathbf{y}_{t-d_o}, \mathbf{u}_t, \ldots, \mathbf{u}_{t-d_o}) \qquad (3.72)$$

with $d_o \leq n$ being the observability index. The network structure for input/output representation is given in Figure 3.6 (for a more precise structure, see the elastomer test bench plant model of Fig. 12.2). Triangles below the input layer represent storing past values of $\mathbf{u}$ and $\mathbf{y}$ and their expansion to the extended input vector. The main difference from the structure of the canonical network of Figure 3.5 is that the external feedback in the input/output representation concerns only external, that is, measurable, variables. By contrast, canonical state representation feeds back the values of nonmeasurable state variables. To distinguish these cases, the terms *purely external feedback* and *formally external feedback* are used for the latter case in what follows.

The question to pose now is the following:

Is purely external feedback sufficient to represent all systems, or does formally external feedback add something to the expressiveness?

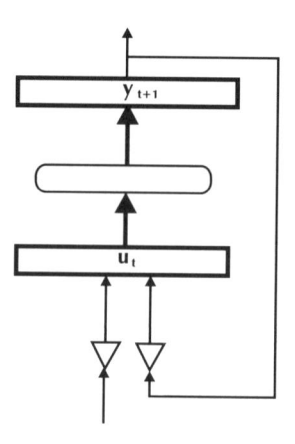

Figure 3.6 Input/output neural network representation of an observable nonlinear plant.

In principle, the answer to this question is simple. Since the input/output representation applies only to observable systems, the answer can be only that the formally external feedback is more expressive. However, let us briefly discuss two further aspects that are important from the practical viewpoint:

1. Is the class of systems that cannot be represented by purely external feedback, but can be represented with help of formally external feedback, interesting for applications?
2. Is there an efficient way to really identify the plants from this class?

The assumption of observability is usual in classical control theory. For nonlinear systems, observability concepts are typically based on the linearization around a working point [86]. (More general concepts are discussed, for example, by Kwakernak and Siwan [82].) In continuous state spaces, linearization concepts work in most cases. However, systems containing noninvertible mappings may require input/output sequences of a length exceeding the order of the system. For example, in the plant described by Srinivasan et al. [134], where

$$z_{t+1} = az_t + bu_t$$
$$y_t = z_t^2 \tag{3.73}$$

the input/output description of the form

$$y_{t+1} = f(y_t, u_t) \tag{3.74}$$

cannot be found because of noninvertible mapping $y_t = z_t^2$ that does not allow us to compute the state x_t, which can be either $\sqrt{y_t}$ or $-\sqrt{y_t}$. The dichotomy can be resolved only by observing input and output at time $t - 1$. Then, the equation

$$z_t - az_{t-1} = bu_{t-1} \tag{3.75}$$

allows (for $a \neq 1$) the consistent one to be selected out of four combinations of signs in front of z_t and z_{t-1}. Similar systems with bifurcations can arise, for example, in diffusion processes. There are also wide classes, such as chaotic systems [130], for which practical numeric observability strongly depends on the time horizon (e.g., entering the observability definitions by sampling the period's length).

For more exotic systems in control practice, such as state machines containing Boolean functions (an example is cited by Williams [155]), observability in a linearized sense cannot be applied. The reconstruction of the state from measurements is not connected with the usual concept of system order. For example, if the state is characterized by two Boolean variables z_1 and z_2, and the only measurement available is $y = z_1 \wedge z_2$, z_1 cannot be reconstructed as long as z_2 is "false." Another example is the system with the state z unity if an event has occurred ever before and zero if not. Then, the output series necessary to identify the state is of unbounded length. Williams [155] has called such systems "systems with strongly hidden states."

Even if the triggering events are guaranteed to occur in finite time so that the input/output sequence necessary for the state identification is bounded, the state representation is much more economical: A single state does the work of a finite, but maybe very long, observation series. (Such systems can be viewed as being of degenerate order, from the input/output point of view.)

Although Boolean systems may appear too distant from control practice, many realistic processes containing saturations (e.g., the wastewater treatment plant of Chapter 15) behave nearly Boolean in some working points.

Note 3.6 The input/output representation using time-delayed inputs and outputs has also been criticized [26] for capturing time (i.e., subsequent measurements) through a spatial structure (i.e., parallel inputs into the network). For continuous inputs, this is not justified if it is taken into account that delayed values can be substituted, without loss of expressiveness, by *temporal derivatives*. For example, an input/output model represented by mapping with arguments $\mathbf{y}_{t-i}$ and $\mathbf{u}_{t-i}$ for $i = 0, \ldots, n-1$ is equivalent to the model using (numerical) derivatives $\mathbf{y}_t^{(t-i)}$ and $\mathbf{u}_t^{(t-i)}$ for $i = 0, \ldots, n-1$. This representation does not require an (assumedly unbiological) shift operator. Instead, derivative signals, possibly coming from derivative sensors (e.g., edge detectors in visual processing), are used.

The cost for representational superiority of state representation with nonmeasurable states is the difficulty to extract the model from data. The task to be solved is to identify the model together with the nonmeasurable state trajectory. This is difficult even if the model is linear. Suppose that the model is of the form

$$\begin{bmatrix} \mathbf{y}_{t+1} \\ \mathbf{x}_{t+1} \end{bmatrix} = \mathbf{A} \begin{bmatrix} \mathbf{y}_t \\ \mathbf{x}_t \end{bmatrix} + \mathbf{B}\mathbf{u} \qquad (3.76)$$

with p action variables $\mathbf{u}$, q measurable variables $\mathbf{y}$, and $n-q$ unmeasurable variables $\mathbf{x}$. Suppose also that the measurements are exact. If both $\mathbf{y}$ and $\mathbf{x}$ can be measured, there will be, with data from T sampling periods, Tn linear equations with $n(n+p)$ variables (unknown coefficients of matrices $\mathbf{A}$ and $\mathbf{B}$). After $T = n+p$ sampling periods, the solutions are exactly determined. For unmeasurable $\mathbf{x}$, there is the same number of equations, but $n(n+p) + (T+1)(n-q)$ unknown variables [including $(T+1)$ vectors $\mathbf{x}$ of hidden states]. But now, the equations are nonlinear, containing products of unknown matrix coefficients with unknown states.

A particular problem is the initial state. The network units corresponding to an unmeasurable state must be set to some values. These values can be assessed only if the state variables are known to exist but are unmeasurable. Even then, the harmful effects of the initial state can be surmounted only by dynamic optimization over a relatively long horizon.

It is difficult to assess the chance for success in identifying a model with hidden states for problems of realistic size. An example is presented in Chapter 15. The wastewater treatment plant has been identified using the scheme shown in Figure 3.5 with unmeasured state variables for "biological oxygen demand." The conditions for success have been relatively advantageous:

Unmeasured states are definitely known to exist in concrete form.

Reasonable initial states can be assessed by domain experts.

There are only four nonmeasurable state variables from a total of 15 state variables.

The trajectories of these discovered states have not been dissimilar from what was expected, but some of them have diverged over time. After introducing additional stabilization measures, the identified model exhibited reasonable trajectories for the discovered states. However, the information gain by these additional states has been almost imperceptible—the prediction error was approximately the same as for the model omitting biological oxygen demand and postulating an observability index of two instead of one. So, my congratulations in advance for the future researcher who finds a working algorithm to identify models with hidden states (even "weakly" hidden ones).

Finally, it has to be noted that in addition to the representation of plant models, recurrent networks can also be used to represent dynamic controllers. Dynamic controllers are useful if the input/output representation of a plant is used or if robustness against systematic disturbances or nonidentified plant behavior is required. The purely external feedback is usually sufficient. For more on such controller structures, see Chapter 6.

3.3.1 Direct and Indirect Representation of Mappings

The usual way to represent a mapping of vector $\mathbf{x}$ to vector $\mathbf{z}$ by a neural network is by identifying $\mathbf{x}$ with an input layer of the network and $\mathbf{z}$ with an output layer. However, it is not the only representation possible. Let us observe the Taylor expansion of the mapping of $\mathbf{x}$ to the ith component of $\mathbf{z}$:

$$\begin{aligned} f(\mathbf{x}) &\approx f(\mathbf{x}_0) + \nabla f_{\mathbf{x}_0}(\mathbf{x} - \mathbf{x}_0)^T + \tfrac{1}{2}\left[(\mathbf{x} - \mathbf{x}_0)\mathbf{H}_{\mathbf{x}_0} + \ldots\right](\mathbf{x} - \mathbf{x}_0)^T \\ &= f(\mathbf{x}_0) + g(\mathbf{x}_0, \mathbf{x} - \mathbf{x}_0)(\mathbf{x} - \mathbf{x}_0)^T \end{aligned} \quad (3.77)$$

where ∇f_{x_0} is the gradient of function f at x_0 and $\mathbf{H}_{x_0}$ is the Hessian matrix at x_0.

A neural network can represent both the mapping f and the mapping g. In the latter case, the neural network would have to be embedded into the problem to represent the mapping f. For mappings f with property

$$f(\mathbf{x}_0) = \mathbf{0} \qquad (3.78)$$

the term $f(\mathbf{x}_0)$ can be omitted and Eq. (3.77) is the gain vector representation of Section 3.2.5.

What are the arguments for using one or another representation? The direct representation is certainly the more economical in space, at least with a fixed number of hidden units. If mapping f is from R^m to R^n, the neural network involved has m input units and n output units. With indirect representation via mapping g, neural network has $2m$ input units and mn output units.

However, there is a difference in the difficulty of mapping to be represented. If the function f is quadratic, so is its direct representation, but in the case of the indirect representation, the neural network can represent only a linear mapping. If the function f is cubic, the direct representation is cubic and the indirect representation is quadratic.

In other words, the direct method implies representing difficult relationships with the help of a few parameters, whereas the indirect method represents simple relationships with the help of many parameters. The way biological neural networks address the complexity problem seems to be the many-parameter approach. From the viewpoint of representing the problem so that it is easy to solve with the help of numerical methods, the second method seems to be the better one.

4

OPTIMIZATION METHODS FOR NEUROCONTROL

The biological roots of neural networks are responsible for the widespread use of the term *learning* to describe the process during which the network parameters are changed to improve the performance of the neural-network-based system. Another frequently used term, *training*, is preferred if the process of parameter tuning is directed to deliberately selected situations. Both terms have a common mathematical equivalent, *optimization*. Because neurocontrollers are usually designed on digital computers with the help of numerical mathematics, it is helpful to exploit the mathematical connection as extensively as possible.

Chapter 2 has reviewed the most important approaches to neurocontrol, particularly with regard to optimization. Each fundamental approach was assigned a fundamental cost function, that is, a function whose explicit or implicit minimization is characteristic for the approach. This chapter will discuss the numerical methods adequate to do the minimization job. The large range of optimization methods does not allow this chapter to be more than a review. Nevertheless, the review hints as to what class of methods can be expected to lead to best results. For professional implementation, the literature must be consulted or good commercial products applied.

Optimization methods for continuous problems (which are all the neurocontrol approaches listed in Chapter 2) can be roughly classified as either local or global. Local methods, which are the topic of Section 4.1, can find only the minimum convex function, that is, the function with a single minimum. For nonconvex functions, local methods find the minimum basin of attraction (i.e., the convex region), which is the starting point for optimization. In contrast, global methods, surveyed in Section 4.2, claim to be able to find (at least in probability) a global minimum of some class of nonconvex functions.

Since learning tasks for nonlinear neural networks, such as multilayer perceptrons, are known to have multiple minima, local methods are, in fact, inadequate. However, there are several reasons why they are popular:

> They are efficient even for large problems with thousands of free parameters.
>
> They are less difficult to implement than global methods.
>
> The global optimization claim of global methods is strictly justified only for optimization problems of modest size, usually below the size of neurocontrol tasks of industrial relevance.
>
> They can be modified to make their locality less strict.
>
> They can be transformed, to some extent, to incremental algorithms.

However, their practicality does not change their inadequacy. The author has consistently witnessed the superiority of global methods over local ones. The difference between the two has been frequently as large as that between a success and a complete failure.

The last item on the list of arguments in favor of local methods introduces an important auxiliary topic of this present chapter. The extent to which sophisticated optimization methods can be applied depends essentially on whether a batch or an incremental computational scheme is used. This extent is clearly biased in favor of batch schemes. A batch scheme provides a directly computable function value, and frequently also a directly computable cost function gradient. This is all that most optimization algorithms require. So virtually *any* widespread optimization algorithm can be used with batch schemes.

This is not the case for incremental schemes with successive presentation of individual measurements. The cost function value and gradient enter the computation only statistically and under assumptions that cannot be verified in practice. So the comparison between batch and incremental schemes can be paraphrased as "What gets lost with incremental computation?" This is the title of Section 4.4.

4.1 LOCAL OPTIMIZATION METHODS

Local optimization is a mathematical concept with a rich range of efficient implementations. The discussion of this section is inspired by Chapter 10 in the famous book *Numerical Recipes in C* by Press et al. [117]. Likewise here the two components of local optimization are treated separately:

1. Determine the direction of the next step
2. Determine the length of the step or line search.

Implementation details of all local optimization algorithms described here also depend on Press et al. [117]. The following discussion presents only the basic principles necessary to assess the appropriateness of particular algorithms for neurocontrol tasks.

4.1.1 Line Search

Line search can be viewed as a special algorithm for determining optimum step size. Step size is equivalent to the learning rate in simple neural network learning algorithms. The main difference between the concept of line search and that of learning rate is that the learning rate is usually constant or changes independently from the current parameter vector, whereas line search finds a particular optimal step length for each current parameter vector.

Technically, line search is looking for a local optimum in a specified direction. Separating the line search algorithm from the search direction algorithm is justified by a particular property of one-dimensional optimization: That is, a local optimum in one dimension can be *bracketed* by two boundary points and an auxiliary point. If there is an auxiliary point b between two boundary points a and c such that $C(b) < C(a)$ and $C(b) < C(c)$, then the existence of a local minimum between a and c is guaranteed. This makes clear the outstanding position of a one-dimensional search space; in more than one dimension, no ordering is defined, and thus it cannot be determined whether a point is between two others or not.

Press et al. [117] presented several algorithms for efficient search after a local minimum bracketed by three points. The simplest is the golden section search algorithm. This algorithm is robust and independent from any assumptions about the cost function form. It consists of iteratively performing the following steps:

1. A new point x between a and c is generated according to the rule

$$x = \begin{cases} b + G(c-b) & \text{if } c-b > b-a \\ b - G(b-a) & \text{otherwise} \end{cases} \quad (4.1)$$

   with $G = (3 - \sqrt{5})/2 \approx 0.38197$ being the golden section constant. In other words, the larger of segments (a,b) and (b,c) is partitioned by the golden section, with the smaller subsegment being adjacent to b.

2. From the point set a, b, c, x, the point with the minimum cost function value is substituted for the former point b, and its left and right neighbors are substituted for former a and c, respectively. For example, if $a < b < x < c$ and $C(b) = \min[C(a), C(b), C(x), C(c)]$, then the original triple (a, b, c) is substituted by (a, b, x).

The use of the golden section constant guarantees superlinear convergence of the algorithm; that is, additional orders of accuracy are gained at less than linear computational expense, in terms of cost function evaluations.

Press et al. [117] also presented algorithms exploiting (1) the quadratic approximation of cost function and (2) derivative information. They make the search more efficient in special circumstances. An algorithm for finding the initial triple (a, b, c) is presented in [117], too.

4.1.2 Search Direction

The methods for determining search direction are usually classified by their order. The order of a method is determined by the order of approximation of the cost function. If the cost function is approximated by a linear function

$$C(\mathbf{x}) \approx c - \mathbf{bx} \qquad (4.2)$$

the method is of first order. Second-order methods are based on the quadratic approximation

$$C(\mathbf{x}) \approx c - \mathbf{bx} + \tfrac{1}{2}\mathbf{xAx} \qquad (4.3)$$

In both cases, the approximation is usually based on a Taylor expansion around a point $\mathbf{x}_0$:

$$C(\mathbf{x}) = C(\mathbf{x}_0) + \sum_i \frac{\partial C}{\partial x_i} x_i + \frac{1}{2} \sum_i \sum_j \frac{\partial^2 C}{\partial x_i \partial x_j} x_i x_j + \ldots \qquad (4.4)$$

Then, the scalar c of Eqs. (4.2) and (4.3) is equal to the value of cost function C at point $\mathbf{x}_0$, the vector $\mathbf{b}$ is equal to the negatively taken gradient ∇C at point $\mathbf{x}_0$, and the matrix $\mathbf{A}$ to the Hessian matrix at point $\mathbf{x}_0$.

First-Order Methods The simplest way to determine the search direction is to set it equal to a negatively taken gradient. The rationale for using a gradient is that an infinitely small step in the direction and size of vector $\mathbf{y}$ leads to a change in the cost function equal to

$$\Delta C(\mathbf{x}) = \nabla C \mathbf{y} \qquad (4.5)$$

If the step is $-\alpha \nabla C$, the cost function change is

$$\Delta C(\mathbf{x}) = -\alpha \nabla C \nabla C = -\alpha \|\nabla C\| \qquad (4.6)$$

This cost function change is always negative or zero, since the gradient norm $\|\nabla C\|$ is nonnegative. (Zero change results only for the zero gradient.)

There are two basic possibilities to determine the size of search step in the gradient direction:

1. Very small step size α
2. Step size determined by line search

The idea behind very small step size is to make steps so small that the gradient direction does not change substantially. This guarantees monotonical descent. Since there is no reasonable way to define "small step" with just first-order information, the usual method has been trial and error. There have been many accounts on how to schedule or adapt step size. (As stated in Section 4.1.1, the usual synonym for step size in neural network literature is *learning rate*.) Some proposals use a gradually decreasing step size as learning progresses, but this cannot be justified without special assumptions about the cost function. (Intuitively, the idea of decreasing steps as one is approaching the minimum works also without diminishing the step size—the gradient norm itself is decreasing and converging to zero for smooth functions.) The smaller the step size, the longer is the search. So the use of step size so small that the gradient can be assumed not to change leads to prohibitively long optimization times.

The approach of determining the step size by line search is, unfortunately, only a partial remedy for gradient descent. The reason for this is, paradoxically, that the line search finds with high precision the point at which the cost function is minimum in a given search direction. Suppose that the search direction (determined by the gradient computed in the previous step) is **y**. At the point found by the line search, the minimum along this direction is reached. This means that the change of C at this point, that is, $\Delta C(\mathbf{x})$ of Eq. (4.5), is zero:

$$\nabla C \mathbf{y} = 0 \qquad (4.7)$$

But this equation implies that the new gradient and the old search direction are orthogonal. Because the new search direction is equal to the new gradient with reverse sign, the new and the old search directions are mutually orthogonal.

The enforced orthogonality of subsequent search directions is harmful for minimization. Suppose that the convex basin of attraction has the form of a valley that is long along one direction but narrow along another. The steepest gradient will not point to the minimum but rather to the bottom of the valley. The next orthogonal search direction will be similarly hopeless. The search directions may then alternate transversally across the valley, approaching the minimum very slowly.

The shortcomings of first-order methods arise from the oversimplified model of the cost function. The first-order model is a linear model, and a linear function has no minimum! This makes it impossible to locate the minimum in a predictive way.

From this viewpoint, the step to second-order methods is not just the step from one to two. The second-order approximation is a quadratic that has a unique minimum; that enables inferences to be made about the probable location of the minimum and so the search is more efficient.

Second-Order Methods Not Using Gradient There is a well-established tradition in the neural network community to connect learning with the determination of gradient. However, there are efficient optimization methods that do not require gradient computation. They comprise not only relatively exotic approaches without guaranteed polynomial convergence, such as downhill simplex [107], simulated annealing [96], and the various evolutionary algorithms but also local optimization algorithms with superlinear convergence.

A family of these algorithms is based on creating a set of search directions that have the advantageous property that the next line searches do not affect the results reached along the line searches of previous directions. Such search directions are called *conjugate directions*. To formulate this requirement formally, suppose that the cost function has the form shown in Eq. (4.3). Then, the gradient at point **x** is

$$\nabla C(\mathbf{x}) = \mathbf{A}\mathbf{x} - \mathbf{b} \tag{4.8}$$

Suppose further that the previous search direction has been **u** and that the line search has found the minimum at point $\mathbf{x}_0$. So, at $\mathbf{x}_0$, the cost function change in direction **u** is zero:

$$\nabla C(\mathbf{x}_0)\mathbf{u} = (\mathbf{A}\mathbf{x}_0 - \mathbf{b})^T \mathbf{u} = 0 \tag{4.9}$$

If the next search direction **v** is chosen to preserve what has been reached before, it has to satisfy

$$\nabla C(\mathbf{x}_0 + \delta \mathbf{v})\mathbf{u} = [\mathbf{A}(\mathbf{x}_0 + \delta \mathbf{v}) - \mathbf{b}]^T \mathbf{u} = 0 = (\mathbf{A}\mathbf{x}_0 - \mathbf{b})^T \mathbf{u} \tag{4.10}$$

or

$$\mathbf{u}^T \mathbf{A} \mathbf{v} = 0 \tag{4.11}$$

The direction sets with such properties can be constructed both with and without gradient computation.

One such algorithm is the Powell algorithm (see [18 or 117]). It starts with an arbitrary set of n linearly independent directions $\mathbf{u}_i$, for example, n unit basis vectors $\mathbf{e}_i$ and an initial state $\mathbf{x}_0$. Then, the following steps are iteratively repeated for $k = 1, \ldots$:

1. Make successive line searches in directions $\mathbf{u}_i$, $i = 1, \ldots, n$. The first line search starts in point $\mathbf{z}_0 = \mathbf{x}_{k-1}$. Each next line search starts in the minimum $\mathbf{z}_{i-1}$ found by the previous one.
2. Perform line search in the direction $\mathbf{z}_n - \mathbf{z}_0$. The minimum found by this line search is denoted $\mathbf{x}_k$.
3. Shift the search direction vectors in the direction set according to $\mathbf{u}_i \leftarrow \mathbf{u}_{i+1}$ (leaving off $\mathbf{u}_1$). Set $\mathbf{u}_n \leftarrow \mathbf{z}_n - \mathbf{x}_0$. In other words, put $\mathbf{z}_n - \mathbf{z}_0$ to a FIFO (first-in–first-out) queue, leaving off $\mathbf{u}_1$.

During the iterative loop, the original nonconjugate directions are substituted by conjugate ones. For exact quadratic forms, the algorithm finds the exact minimum after n iterations, that is, after $n(n+1)$. It requires $O(n^2)$ memory, mainly for storing n search directions.

For cost functions that are not exactly quadratic, more than n iterations are necessary. Then, the the vectors of the direction set may become mutually linearly dependent and so constrain the search to a subspace of the parameter space. The simplest remedy is to reinitialize the direction vectors after every n iterations. More sophisticated procedures are discussed by Press et al. [117] and Brent [18].

Second-Order Methods Using a Gradient A sequence of conjugate directions satisfying the condition (4.11) can be generated also with help of gradient computation. This is the case with the conjugate gradient family of algorithms.

One version of a conjugate gradient algorithm [117] starts with an arbitrary pair of vectors $\mathbf{h}_0 = \mathbf{g}_0$ and an initial parameter vector $\mathbf{x}_0$. Then, the following steps are iteratively performed for $k = 1, \ldots$:

1. Perform the line search from $\mathbf{x}_{k-1}$ in the direction $\mathbf{h}_{k-1}$. The optimum is denoted by $\mathbf{x}_k$.
2. Compute the gradient $\mathbf{g}_k$ at the point $\mathbf{x}_k$.
3. Compute the next search direction $\mathbf{h}_k$ by

$$\mathbf{h}_k = \mathbf{g}_k + \frac{\mathbf{g}_k^T \mathbf{g}_k}{\mathbf{g}_{k-1}^T \mathbf{g}_{k-1}} \mathbf{h}_{k-1} \tag{4.12}$$

known as the Fletcher–Reeves variant, and by

$$\mathbf{h}_k = \mathbf{g}_k + \frac{(\mathbf{g}_k - \mathbf{g}_{k-1})^T \mathbf{g}_k}{\mathbf{g}_{k-1}^T \mathbf{g}_{k-1}} \mathbf{h}_{k-1} \tag{4.13}$$

known as Polak–Ribiere variant.

The conjugacy is guaranteed by the line search property that the next gradient is orthogonal to the direction of line search. This means that line search is a prerequisite for the operation of the conjugate gradient algorithm.

The conjugate gradient algorithm reaches the minimum of exact quadratic form in n iterations, that is, after computing n line searches and n gradients. Its main property is that it requires only $O(n)$ memory—the new and old versions of vectors $\mathbf{u}_k$, $\mathbf{g}_k$, and $\mathbf{x}_k$.

Another group of algorithms is based on the idea of computing the exact minimum of the quadratic function shown in Eq. (4.3). Suppose that the current parameter vector is $\mathbf{x}_0$. The gradient at this point is

$$\nabla C(\mathbf{x}_0) = \mathbf{A}\mathbf{x}_0 - \mathbf{b} \qquad (4.14)$$

The gradient at the cost function minimum $\mathbf{x}_{\min}$ is

$$\nabla C(\mathbf{x}_{\min}) = \mathbf{A}\mathbf{x}_{\min} - \mathbf{b} = 0 \qquad (4.15)$$

so that

$$\mathbf{A}\mathbf{x}_{\min} = \mathbf{b} \qquad (4.16)$$

If the gradient $\nabla C(\mathbf{x}_0)$ and the Hessian matrix A are known, the exact minimum $\mathbf{x}_{\min}$ can be computed directly from Eqs. (4.14) and (4.16):

$$\mathbf{x}_{\min} = \mathbf{x}_0 - \mathbf{A}^{-1}\nabla C(\mathbf{x}_0) \qquad (4.17)$$

This is the formula for solving Eq. (4.16) by the Newton method.

While the gradient should be computed directly, the inverse of the Hessian matrix $\mathbf{A}^{-1}$ is approximated by the sequence of matrices $\mathbf{H}_k$. This is the main idea of variable-metric methods. Because of their relationship to the Newton method, they are also called *quasi-Newton* methods. Some versions of variable-metric methods are discussed by Press et al. [117], Dennis and Schnabel [24], and Růžička and Kober [128]. Like the conjugate gradient algorithms, variable-metric methods reach the minimum of exact quadratic form in n iterations, that is, after computing n line searches and n gradients. They require $O(n^2)$ memory, essentially for the matrices $\mathbf{H}_k$.

4.1.3 Comparison of Local Optimization Methods

For first-order methods, there are no estimates of convergence even for exact quadratic functions. In other words, they can converge arbitrarily slowly.

Second-order methods can be compared by the number of cost function calls that are necessary to reach a minimum exact quadratic function. Powell's method, which does not use the gradient, requires $n(n+1)$ line searches. Second-order methods that use the gradient, that is, the conjugate gradient and variable metric methods, require n gradient computations and n line searches. On average, a line search needs a constant number of cost function calls, say, K.

To assess the expense for gradient computation, we must distinguish between analytic and numeric gradients. Analytic gradients can be computed if exact analytic formulas are available. Then, its expense is usually in a relatively fixed relationship (e.g., some factor L) to the cost function computation. If backpropagation formulas are used, the expense for an analytic gradient computation is comparable to that for forward computation, as long as all forward-pass information (hidden unit inputs and activations) is stored. If this storage is incomplete, the factor L may amount to 5 to 10.

If analytic formulas are not available or are not used for some reason, the numeric gradient can be computed. The elements of the gradient vector are

$$g_i = \frac{C(x_1, \ldots, x_i + \delta_i, \ldots, x_n) - C(x_1, \ldots x_i, \ldots, x_n)}{\delta_i} \quad (4.18)$$

(the asymmetric version), or

$$g_i = \frac{C(x_1, \ldots, x_i + \delta_i, \ldots, x_n) - C(x_1, \ldots, x_i - \delta_i, \ldots, x_n)}{2\delta_i} \quad (4.19)$$

(the symmetric version). The symmetric version is more correct and should be used particularly for large δ_i (e.g., for the filtering described in Section 4.2.1). For small δ_i, the asymmetric version is usually sufficient. If filtering is not part of the goal, the computing precision must be accounted for with appropriate δ_i [117]. If the accuracy of computing the function C (which cannot exceed the accuracy of the data type given but can approach it) is denoted as ϵ, the truncation and round-off errors are minimum for

$$\delta_i = \sqrt{\epsilon \frac{C}{C^{(2)}}} \quad (4.20)$$

with asymmetric formula (4.18), and

$$\delta_i = \sqrt[3]{\epsilon \frac{C}{C^{(3)}}} \quad (4.21)$$

with symmetric formula (4.19), and C, $C^{(2)}$, and $C^{(3)}$ are, respectively, the zero-th, the estimate of the second, and the estimate of the third derivatives of function C at the point where the numeric derivatives are computed. Of course, $C^{(2)}$ and $C^{(3)}$ must be substituted by estimates. To compute the numeric gradient, $n+1$ or $2n$ cost function calls are necessary for asymmetric and symmetric formulas, respectively.

With these definition, the following numbers of cost function calls are required:

$Kn(n + 1)$ for Powell's method
$(K + L)n$ for methods using the analytic gradient
$(n + 1)n + Kn$ for methods using the asymmetric numeric gradient
$2n^2 + Kn$ for methods using the asymmetric numeric gradient

From this viewpoint, the most efficient methods are those using the analytic gradient, followed by those using the numeric gradient, with the most expensive being the Powell's method.

However, further analysis may lead to different conclusions. The first concerns the numerical character of the analytical gradient computation. The computation of analytical gradients, as well as that of the cost function call, is subject to imprecision. Besides multiplication and addition errors, often a source of inaccuracy is the derivative of numerically computed functions such as the sigmoid

$$s(x) = \frac{1}{1 + e^{-x}} \qquad (4.22)$$

The relationship

$$\frac{ds}{dx} = s(1 - s) \qquad (4.23)$$

is true only analytically—numerical deviations may be substantial. Accumulating such errors over recursive computations with larger horizons, as is necessary for plant identification and controller training, may lead to inconsistencies between the computations of cost function gradient, on one hand, and those of cost function values, on the other hand. Then, for example, the determination of search direction may be inconsistent with the results of the line search. This is disastrous for all methods because the consistency assumption is the very foundation of their underlying principles (e.g., see the conjugate gradient algorithm of Section 4.1.2).

For nontrivial tasks with recursion horizons larger than ten, this author has seen the analytical gradient pointed in a direction opposite to that of the numeric one [which is fully consistent with the cost function value computation, within some accuracy bounds given by the size of δ_i in Eq. (4.18) or Eq. (4.19)]. This leads to a complete collapse of convergence of even sophisticated higher-order methods.

An example of such computational experiments with plant identification is provided in the case studies of Chapters 12 and 15. The elastomere test bench plant model of Chapter 12 uses a multilayer perceptron (as a gain vector network) with four input nodes, four hidden nodes (in a single hidden layer), and four output nodes—a total of 40 parameters (2 × 16 weights and 2 × 4 thresholds), which is a relatively small model. The plant model of the wastewater treatment plant of Chapter 15 consists of six multilayer perceptrons, each having the same number of input, hidden, and output nodes, namely five, four, three, six, six, and four for the individual perceptrons. This results in a total of 310 plant model parameters—a relatively large model.

Optimizations of both plant models for various multistep forecast horizons (see Section 5.4.1) have been performed using the optimization methods of the neurocontrol package TREG mentioned in Chapter 11. The results are given in Table 4.1, which presents (1) the optima over the training sets, as found by the global optimization method, and (2) the precision of correspondence between the worst (i.e., the worst-corresponding) element of the analytical gradient, on one hand, and the numeric gradient, on the other, at an identical starting point.

Observing the optima reached, it is obvious that the analytic gradient results deteriorate with an increasing forecast horizon. This is the case for both applications, but the deterioration sets are for the smaller horizon in the large application, that is, the wastewater treatment plant. This is surprising since the optimization resources have been limited by the number of cost function calls. Thus the analytic-gradient-based procedure has a clear advantage; the numeric gradient computation consumes a number of cost function calls equal to the number of free parameters, while analytic gradient does not. In other words, the analytic-gradient-based procedure has performed a substantially larger number of iterations than that based on the numeric gradient and found much worse results. Inspection of optimization protocols has shown that the analytic-gradient-based procedure tries many more local optimizations, but none of the local optimizations converge nearly as well as is the case with the numeric gradient.

Additional insight can be gained from observing the differences between the analytic and the numeric gradient. The worst correspondences in digits is given

Table 4.1 Plant model identification using analytic and numeric gradients

Plant	Gradient	Horizon			
		1	10	100	300
		Optima reached			
Test bench	Analytic	0.043	0.062	0.251	0.562
Test bench	Numeric	0.045	0.057	0.126	0.189
Wastewater plant	Analytic	0.031	0.237	0.523	0.752
Wastewater plant	Numeric	0.031	0.042	0.121	0.118
		Worst precision (digits)			
Test bench	—	8	4	2	0
Wastewater plant	—	7	2	0	0

in Table 4.1. Both vectors have been almost identical for small forecast horizons. This shows that the analytical gradient formulas (based on backpropagation through time) have been correctly derived for the plant models and network types involved. With increasing forecast horizon, the correspondence deteriorates, becoming disastrous for horizons of 100 and more; then, even the sign of individual gradient elements is frequently different.

Because of advantageous properties of plant models received with longer forecast horizons (see Section 5.4.1), it is no exaggeration that

inconsistencies resulting from the inaccuracy of analytic gradient computation make the analytic gradient a bad choice for neurocontrol tasks of any realistic size.

Note 4.1 It is important to point out that this statement about preferability of the numeric gradient is valid only in the given neurocontrol context, with cost functions depending on numerically approximated exponential nonlinearities and on the behavior of dynamic systems (which tend to react exponentially to changes of plant parameters). For smooth convex functions such as quadratic functions of cost function parameters, the analytic gradient is frequently superior—the numeric gradient based on finite perturbations is only an approximation and its computation is expensive.

Note 4.2 Many researchers have reported receiving good results with the analytic gradient, using algorithms such as backpropagation through time [134] or dynamic backpropagation [105]. This success seems to depend on the specific problem, but also, as shown above, on general characteristics such as problem size and forecast horizon.

For not especially large problems, a compromise can certainly be found for the forecast horizon, one that is small enough that the optimization procedure converges even with the analytic horizon and large enough that the plant model corresponds sufficiently to reality. The sensitivity of simple first-order algorithms such as the steepest-gradient procedure with a fixed step length is certainly lower than that of higher-order algorithms (though they may not converge at all). However, those practitioners who have not directly compared the results received with the help of an analytic and numerical gradient may not be aware of the precision loss.

Another shortcoming of the gradient is its strict locality. The local optimization algorithms have been designed for smooth convex functions, so they exhibit guaranteed convergence properties. In practice, and particularly in neurocontrol practice, they are applied to nonconvex, and frequently nonsmooth, cost functions with many local minima. If the cost functions have large, deep, and rough basins of attraction with small disturbances generating many small, shallow local minima, it is useful not to make inferences about appropriate search directions from points on the surface (as is the case with the gradient, particularly the analytic gradient) but rather from past solution trajectories. The latter is the case in Powell's method which computes the new search direction from the result of the n previous line searches.

This may be the reason why Powell's method frequently is better than the gradient methods in robustness and thus also in the quality of the final solution.

4.2 GLOBAL OPTIMIZATION METHODS

An extremum of a completely arbitrary function of real-valued arguments is not computable because it requires a combinatorial search in infinite spaces. This is why some global optimization approaches make additional explicit or implicit assumptions about the function or are satisfied with trying to allocate the available resources so that the probability of finding the optimum is as high as possible.

The presentation of the methods in this section focuses on how they simplify the nontractable general global optimization task. This allows the reader to assess whether a particular method will successfully tackle the problem.

4.2.1 Generalizations of Local Search

One group of approaches can be viewed as a generalization of local search. It is implicitly assumed that the function to be minimized is roughly convex, but its surface is modulated by disturbances of high spatial frequencies and small amplitudes (Figure 4.1).

A straightforward way to address the roughly convex function minimization is to filter out the high-frequency disturbance and make gradient descent on this smoothed function. Suppose that the basis convex function is ax^2 and that the disturbance is $b\sin(cx)$:

$$ax^2 - b\cos(cx) \qquad (4.24)$$

The gradient of this function is

$$2ax + bc\sin(cx) \qquad (4.25)$$

The simple gradient descent converges to the global minimum only as long as

$$|bc| < 2|ax| \qquad (4.26)$$

Whether this is satisfied depends not only on the frequency and amplitude of the disturbance but also on the slope of the underlying convex function. For underlying functions of order higher than one, the nearer the actual point is to the minimum at $x = 0$, the less likely it is that condition (4.26) is satisfied. If the

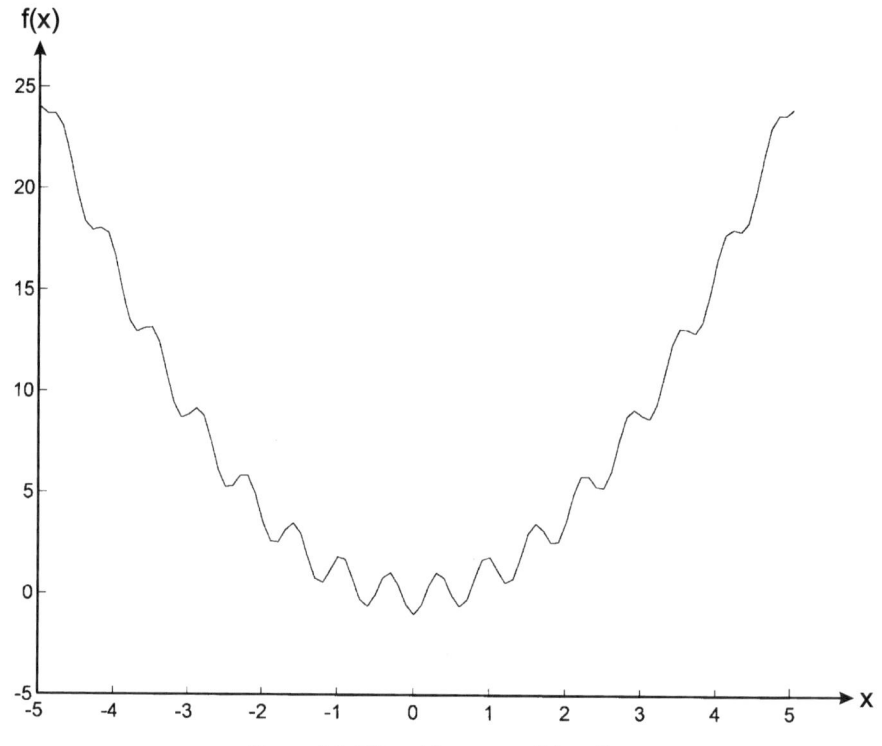

Figure 4.1 "Roughly convex" function.

high-frequency disturbance is reduced with the help of filtering by a factor $d < 1$, the condition (4.26) becomes $|bcd| < |2ax|$. For a given frequency c, this is satisfied with higher amplitudes b than before filtering.

There are two simple ways to embed the filtering into the gradient computation. The first is based on the use of the well-known momentum term. For the ∇f_k gradient at step k, the momentum-based smoothed gradient is recursively computed as

$$\nabla_s f_{k+1} = (1 - \alpha)\nabla_s f_k + \alpha \nabla f_k \qquad (4.27)$$

with $\alpha \ll 1$. This is obviously a first-order low-pass filter.

Another possibility is to compute numeric gradient

$$\nabla_d f(x) = \begin{bmatrix} \dfrac{(x_1 + h_1) - (x_1 - h_1)}{2h_1} \\ \vdots \\ \dfrac{(x_n + h_n) - (x_n - h_n)}{2h_n} \end{bmatrix} \qquad (4.28)$$

at point **x** with help of relatively large steps h_i. For the function (4.24), the (scalar) gradient is

$$\frac{\{a(x+h)^2 - b\cos[c(x+h)]\} - \{a(x-h)^2 - b\cos[c(x-h)]\}}{2h}$$

$$= \frac{4axh}{2h} - \frac{b}{2h}\{\cos[c(x+h)] - \cos[c(x-h)]\}$$

$$= 2ax - \frac{b}{2h}\{\cos[c(x+h)] - \cos[c(x-h)]\} \tag{4.29}$$

The condition (4.26) is then substituted by

$$\left|\frac{b}{2h}\right| < 2|ax| \tag{4.30}$$

This is more advantageous than Eq. (4.26) whenever $2h > 1/c$. In other words, for h substantially larger than the period of the disturbance, this is a low-pass filter.

This simple discussion makes clear the shortcomings of both methods based on filtering.

- They are efficient only if the frequency and amplitude of the disturbance and the slope of the underlying convex function can be approximately assessed and used to determine appropriate values of momentum weight α or numeric gradient step h, respectively.
- Their effect will usually deteriorate in the neighborhood of the minimum.

Additionally, the momentum term method exhibits another shortcoming. The low-pass filter (4.27) represents a first-order dynamic term with a certain response time depending on its time constant, which is proportional to the reciprocal of the weight α. Because the momentum-based gradient is usually computed from the sequence of successive points x_k received by successive downhill steps, the smoothed gradient, in fact, refers to some earlier point x_{k-d}. So the gradient method with a fixed-length step may overshoot the minimum of the underlying convex function.

The momentum term is difficult to use with higher-order local optimization methods that do not always make simple sequences of downhill steps. In addition, the overshooting property of the momentum-based gradient may produce contradicting information about the second-order approximation of the function (two step sequences approaching the minimum from different directions will locate the minimum at different spots).

Nevertheless, despite these caveats, the momentum term and the long-step numeric gradient are, if used with care, easy to compute in upgrading local optimization algorithms. In both cases, increasing the locality in the course of minimization by increasing the coefficient α or by decreasing the gradient computation step-step-size h will alleviate the problems mentioned. It corresponds to the commonsense idea of successively refining the search.

Roughly convex functions are a special case of the hierarchic functional structure with few local minima on the coarse level (with low frequency and high amplitude) and many local minima on the fine level (with high frequency but low amplitude).

Random search methods make use of the idea that for coarse improvements either long steps or long sequences of steps with a nonmonotonic decrease of the function value are necessary. Such a coarse optimization phase is followed by a fine optimization phase with short steps and sequences of monotonic decreases.

A popular and theoretically well-founded class is that of *annealing algorithms*, proposed by Metropolis et al. [96], Kirkpatrick et al. [74], and Kirkparick [75]. Theoretical foundations of annealing and their relationship to the generalization of Markov chains, random Markov fields, have been investigated by Geman and Geman [41]. Annealing algorithms were originally designed for discrete optimization. On the discrete state space, a neighborhood structure is defined. In each point of the state space, or *node*, the set of neighbors of this node corresponds to the set of possible steps in the next search iteration. The next step from this set is selected probabilistically, depending on the cost function values of the individual nodes. The probability is proportional to

$$p(\mathbf{x})\ e^{-\frac{C(\mathbf{x})}{T}} \qquad (4.31)$$

where T is the temperature parameter. The temperature parameter allows the control of the proportion of acceptance probabilities between neighbors with low- and high-cost function values, that is, between downhill and uphill steps. The downhill step with the highest probability (4.31) represents the steepest descent. To reach state space regions that are far away from the current point and to which there is no descending path, it is necessary to make many uphill steps. To this end, the temperature T must be set to high values. By contrast, to reach the local minimum of the attractor in which the current point is, the most efficient way is to make only steepest descent downhill steps. This is the case with very low temperature T. For the hierarchic function structure, the appropriate procedure is to start with a high temperature to find a region of low-cost function values and to proceed with decreasing temperature to find the local minimum. The theorem of Geman and Geman states that under certain conditions the global optimum is reached with a probability one if an annealing schedule with logarithmically decreasing temperature and certain starting temperature is reached. The hierarchic structure is no precondition of the theorem. However, the logarithmic annealing schedule under which the general case of the global minimum can be found is too slow for practical application. On the other hand, it has been successfully applied to practical optimization problems.

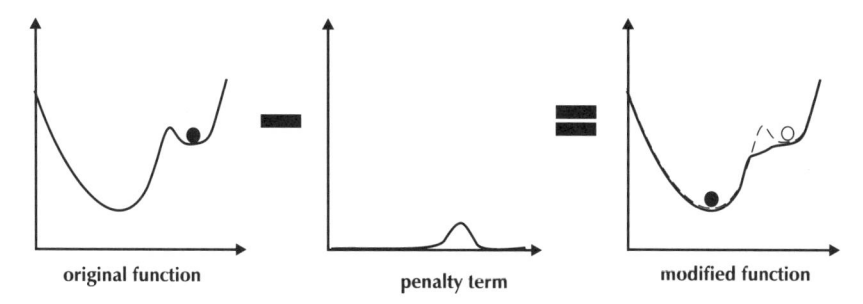

Figure 4.2 Deforming the cost function with the help of a penalty term.

To solve optimization problems in real-valued settings, the neighborhood definition has to be replaced by a set of possible steps of various directions and lengths. The algorithm then consists of the following steps:

1. Evaluating the cost function for the states $\mathbf{x}_i$ resulting from possible steps.
2. Determining the probabilities

$$p(\mathbf{x}_i) = \frac{e^{-C(\mathbf{x}_i)/T}}{\sum_i e^{-C(\mathbf{x}_i)/T}} \quad (4.32)$$

for the given T.
3. Selecting a next state in random from the distribution (4.32).

The temperature T is successively decreased.

Another approach addresses the problem of getting stuck in local minima in the most direct way. After reaching such a minimum, the cost function is deformed with the help of a penalty term [40] in the neighborhood of the local minimum so that the minimum is canceled (Figure 4.2). Conceptually related are also the tunneling methods [87]. Local deformations have only local impact. For this reason this class of methods can succeed only if the local minima are viewed as disturbances of an otherwise tractable function. This justifies the classification of these methods as generalizations of local search.

4.2.2 Approaches Based on Complete or Restricted Covering of the Cost Function Domain

An assumption weaker than those of the roughly convex function or the hierarchic function structure is that of known bounded curvature (i.e., the second derivative). Under this assumption, the function can be partitioned into a set of discrete regions, for example, a grid of certain density. On this grid, combinatorial search can be performed. An efficiency increase is possible by excluding regions that, with given curvature bounds, cannot contain a solution better than one already found. (This is analogous to the branch-and-bound approach of discrete optimization in operations research.) The key to the efficiency of the method is obviously in determining nontrivial bounds. The first problem is that the bounds are not known. Even worse, if the bounds were known, they may be so high that the search would be too expensive to be feasible. This would then reflect the complexity inherent in the problem that cannot be surmounted by any algorithmic sophistication. These reasons prevent the application of this method with theoretically guaranteed results to neurocontrol as well as to most other practical applications.

If the exhaustive search is not feasible, it is logical to try to cover the cost function domain as far as possible with given resources and to use some local optimization method. Then, it is sufficient if every interesting local basin of attraction is hit by a starting point. For n-dimensional domains, the trivial coverage by all combinations of only two values per dimension leads to generating 2^n vectors. With these two values normalized to a uniform difference between both, these combinations correspond to the corners of an n-dimensional hypercube (Fig. 4.3).

Since each of these vectors is then usually used as a starting point for a local optimization, this is not feasible for n greater than about 20, and this number is where simple neurocontrol problems just start.

Yet sparser coverage (for normalized variables) is by $2n$ vectors x_i, $i = 1, \ldots, 2n$, with elements

$$\begin{aligned} x_{ij} &= 1 \quad \text{for } i = 2j - 1 \\ &= -1 \quad \text{for } i = 2j \\ &= 0 \quad \text{otherwise} \end{aligned} \quad (4.33)$$

This corresponds to starting points in the centers of the hypercube sides (Fig. 4.4).

Besides the simple solutions of Figures 4.3 and 4.4, sophisticated schemes can be implemented corresponding to m points found on the n-dimensional hypersphere that are equidistant in some sense. A popular alternative is to generate m random samples from some (usually uniform) distribution.

Optimal Use of Computational Resources The simplest version of the methods based on incomplete coverage of the cost function domain performs a local optimization search for all generated starting points. The motivation for incomplete coverage comes from computational resource limitations. The fact that the resource expense is proportional not only to the number of samples but also to the cost of individual local optimization suggests that economical use of local optimization would gain resources for taking more samples and thus for covering the domain more densely.

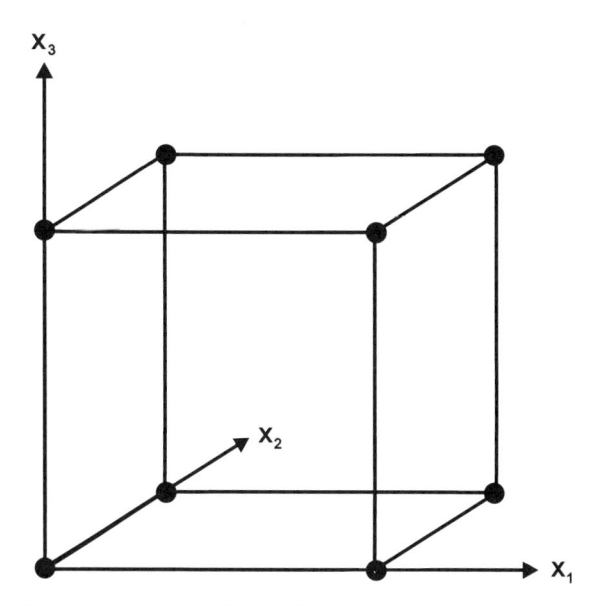

Figure 4.3 Starting points at the corners of the hypercube.

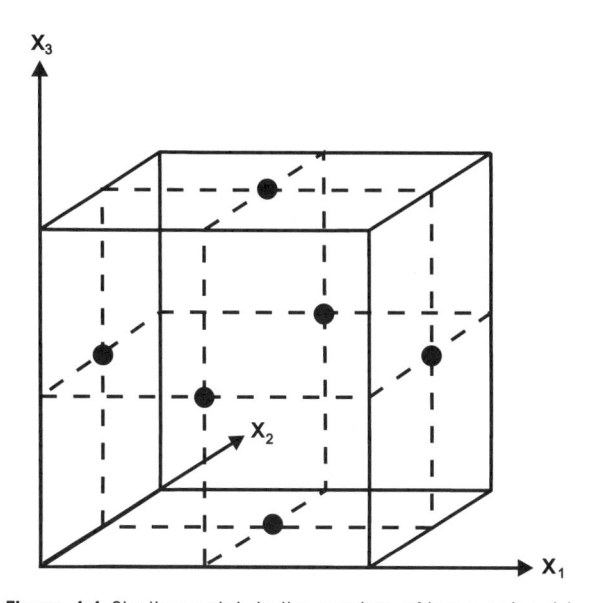

Figure 4.4 Starting points in the centers of hypercube sides.

The simplest implementation of this idea is by applying local optimization to the single best sample (single-start methods) or to the best subset of samples (multi-start methods, e.g., Bayesian reduced multistart method of Boender and Rinnooy Kan [16]).

More sophisticated approaches take the spatial relationships between samples into account. The *clustering methods* [17] try to find clusters of samples that belong to the same attractor by means of statistical clustering methods. *Multilevel methods* [123] reach this by searching for sample chains with increasing cost function values. The local optimization is then applied to the best sample of each cluster or chain. The assumption underlying the operation of this family of methods is obviously that the number of distinct attractors is small enough that most of them are hit by at least one random sample, with the full potential of resource economy being exploited only if there are multiple samples in most attractors.

Another method of resource allocation is implemented in *evolutionary algorithms* [33, 50, 131]. In these algorithms inspired by genetic evolution, the set of samples is called population. The local search, called mutation, is usually random. The resources allocation is optimized by selection: The samples with lower cost function value are subject to more intensive local search by cloning them (or enabling them to have more offspring). Unsuccessful samples die out.

In addition to mutation, evolutionary algorithms frequently use another operator, the crossover. This operator combines features of two successful samples. Implicit here is the assumption about the form of the cost function: It is assumed that the contributions of individual elements of the cost-function argument vector are, at least to some degree, *decomposable*. The simplest example of such a decomposable cost function is one composed of additive terms, each depending only on a single argument:

$$C(\mathbf{x}) = \sum_i f_i(x_i) \tag{4.34}$$

Some particularly sophisticated approaches are based on statistical modeling of the cost function [22], [137]. Their commitment to *average rationality* (in the Bayesian sense) is an approach to optimizing resource allocation. Their complexity has so far prevented their broader application.

4.2.3 Computational Experience

Růžička and Kober [128] have performed a unique comparative study of application of various global optimization methods to neurocontrol. They tested the following methods:

1. A single-start random sample method (SS).
2. A Bayesian reduced multistart method (BRMS) after Boender and Rinnooy Kan [16].

3. A combined tunneling function and random search method (TFRS), freely modeled after Levy and Montalvo [87].
4. Clustering methods, among others the density clustering method [16].
5. Various variants of multilevel methods, for example, the modified multilevel single-linkage method (MMLSL), developed by Rinnooy Kan and Timmer [123].

All methods have been tested on four neurocontrol tasks with 20 to 36 free parameters. The neurocontrol approach has been closed-loop optimization, with neurocontroller represented by multilayer perceptron. All four examples indicated similar ranks of individual methods. An excerpt of the results for the most complex test example is given in Table 4.2.

It is important to note that the results reached by any of these global optimization methods are clearly superior to those attained by local optimization. This can be seen by realizing that the single-start method (SS) consists, roughly, in generating 2416 random samples and performing local optimization from the best of them. Even then, the results of this method are by far the worst of all methods tested.

4.3 HOW TO MAKE OPTIMIZATION EASIER

Second-order local optimization algorithms (and also global optimization algorithms using the local ones as subprocedures) are designed for smooth functions for which the quadratic function is a good approximation, at least in the neighborhood of minimum.

There is no reason to assume that optimization tasks arising in neurocontrol have these properties. On the contrary, there are at least two reasons to assume the opposite:

1. It is a property of unstable dynamic systems that a small error may be amplified exponentially. A small change in a parameter (e.g., a weight of the neurocontroller) may lead to a small error, which can be

Table 4.2 Results for selected global optimization methods

Method	Minimum Reached	Cost Function Calls	Gradient Evaluations
SS	0.102	2416	0
BRMS	0.081	99120	0
TFRS	0.075	2866	2124
DC	0.081	2922	85598
MMLSL	0.069	2133	85892

exponentially amplified by the system dynamics and finally result in an exponential increase in cost function that depends on the closed-loop behavior. Although the goal of controller optimization is to produce a stable system, large regions of parameter space in which the optimization procedure is probably operating most of the time may correspond to unstable systems.
2. The nonlinearities in neural networks, such as sigmoid or Gaussian radial basis functions, contain exponential terms.

Nevertheless, the optimization algorithms are astonishingly robust and frequently cope with these violations of mathematical assumptions. However, the task should not be made more difficult than necessary. There are some recommendations that should be considered if they are not contradictory to the basic objectives of the particular control problem. The first of them are directly related to second-order methods assumptions. The cost functions for plant identification and controller training should be smooth (or at least continuous) and not too far from second order, that is, from quadratic functions. For example, the cost function for controller training described by Eq. (2.18), in which is shown the quality criterion for closed-loop behavior, should not contain powers of state and action variables higher than, say, four.

Note 4.3 The term *order* used in connection with cost function denotes the order of a polynomial; that is, it is the highest power appearing in the cost function. There is no direct relationship to the order of the plant model or of the neurocontroller to the polynomial order of plant model nonlinearities.

Note 4.4 The fact that the cost function is a polynomial of, say, second order in plant output variables or controller actions does not strictly imply that it is a second-order polynomial in plant model parameters such as neural network weights, and thus the second-order assumptions underlying the local optimization algorithms are satisfied.

For unstable plants or closed loops (even for linear ones), the plant output after some time interval may be a rapidly diverging (e.g., exponential) function of the plant parameters.

The second recommendation has its roots in the numeric precision of optimization. The contributions of individual subgoals to the value of cost should be of the same order of magnitude. For example, if the cost function for plant identification consists of the sum-of-square errors of all state variables, the sum-of-square errors are to be weighted so that they have similar values. Otherwise, small and unimportant improvements in the error of an overscaled variable may overshadow large improvements of underscaled ones. Then, the improvements in underscaled variables may be prevented by the numerical accuracy of the algorithms.

The third recommendation concerns the formulation of the neurocontrol problem as an optimization problem. The optimization algorithms should not be misused to discover trivial and well-known facts. If the controller or the plant is known to be symmetric, the symmetry should be fixed in the network structure (see Section 3.2). If they are known to be almost symmetric, the difference from symmetry should be identified separately. If two plant components are known to be homomorphic (i.e., with different inputs and outputs but identical behavior, described by identical, although unknown, differential equations), both should be represented by the same neural network. If a plant or controller is known to be linear, it should be represented by a linear network. If it is assumed to be nearly linear, the linear behavior and the difference from linearity should be represented separately (and possibly optimized successively).

4.4 WHAT GETS LOST WITH INCREMENTAL COMPUTATION?

Batch learning is a model of computation in which the data used for computing the cost function are collected *before the optimization procedure is started*.

This statement should not be confused with the question of whether the data collected to define the cost function are sufficient for this cost function to be a good approximation of some ideal cost function. For example, the cost function for plant identification may be defined as the square of the forecast errors of a neural network plant model, summed over a given set of, say, ten thousand sample measurements. This cost function is only an approximation of the ideal cost function consisting of the integral of squared forecast errors integrated over the entire state space. This approximate or pragmatic cost function can still be evaluated anytime with a deterministic outcome that depends only on the values of cost function parameters, for example, on the values of network weights.

By contrast, in incremental learning, the data used for computing the cost function are collected only during the optimization procedure. This can work only under certain assumptions about the relationship between the data continually presented and the complete set of data necessary for the computation of the (possibly approximate) cost function.

Let us consider first a slightly more general than usual incremental processing in which the data are collected during optimization but evaluated only after a certain (possibly large) number of samples and not after every sample. The difference between this scheme and batch processing is that whereas batch processing always uses the same data set, the incremental scheme always works with new data as they are coming in. For example, in batch processing, the gradient is computed from a fixed set of data, whereas in generalized incremental computing, it is computed the first time from, say, samples $\mathbf{d}_t$ for $t = 1, \ldots, 1000$, the second time from those with $t = 1001, \ldots, 2000$, and so on. Let us first investigate under which conditions such an incremental scheme can be used in evaluating a cost function that is a good approximation of the cost function defined for the batch-processing case.

The usual assumptions are that the model is time-invariant during the data collection process and that the data sampled are distributed in a way that is consistent with the distribution required by the cost function. For example, if the ideal cost function for plant identification is the integral of equally weighted squared forecast errors integrated over the entire state space, the distribution consistent with this function would be the uniform distribution of data samples over the state space. If this assumption is at least approximately satisfied, an additive cost function can be incrementally computed by summing the partial cost functions (e.g., squared errors for individual forecasts based on single samples) over a time horizon sufficiently long for the law of large numbers to be applicable. A cost function computed in this way would have the property of having an almost deterministic value that depends only on the values of the cost function parameters. More exactly, the value of this cost function would vary depending on the size of the sampled data set being summed, and the variance would be arbitrarily small for an arbitrarily large sample (except for artificial cases that a smart statistician would be able to present readily). If the variance is negligible, the cost function computed by this procedure can be used in any optimization algorithm applicable to the cost function computed in batch.

The importance of a small variance in the optimization algorithm can be illustrated by a numerical gradient computation. The ith element of the gradient vector is computed with the help of the formula

$$\frac{\partial C}{\partial x_i} \approx \frac{C(x_1, \ldots, x_i + \delta_i, \ldots, x_n) - C(x_1, \ldots, x_i, \ldots, x_n)}{\delta_i} \quad (4.35)$$

If the cost function is replaced by its statistical estimate C_e with variance D, then the difference $C_e(x_1, \ldots, x_i + \delta_i, \ldots, x_n) - C_e(x_1, \ldots, x_i, \ldots, x_n)$ has the variance $2D$. Then the relative error of the gradient element is

$$\frac{\sqrt{2D}}{C(x_1, \ldots, x_i + \delta_i, \ldots, x_n) - C(x_1, \ldots, x_i, \ldots, x_n)} \quad (4.36)$$

Since $C(x_1, \ldots, x_i + \delta_i, \ldots, x_n) - C(x_1, \ldots, x_i, \ldots, x_n)$ is usually a very small number, and the acceptable variance D is very small, too, Eq. (4.35) cannot be useful in computations. The variance of cost function is similarly harmful to line search, where the stopping rules depend on precision, and to virtually all other algorithms.

It is important to point out that there are different requirements for determining the appropriate number of samples for batch and incremental computation.

For Batch Computation of Cost Function The number of samples is determined by the density of coverage of state space required for the cost function to be a sufficiently good approximation of ideal cost function, as discussed above. With a given sample set, the cost function is completely deterministic, no matter how small the sample set.

For Incremental Computation of Cost Function The sample set size has to guarantee, in addition to the state space coverage, a variance of cost function small enough so that this cost function can be viewed as deterministic to the extent sufficient for application of a particular optimization method. This sample set size may be orders of magnitude larger than what is sufficient for state space coverage (and sometimes also orders of magnitude larger than what is computable in reasonable time).

This discussion makes clear that incremental learning with sample sets large enough to guarantee a deterministic evaluation of cost function is not practical. Also it has no advantage compared to batch optimization because it is not genuinely adaptive in real time.

Therefore, practical incremental schemes not only collect the data incrementally, but they also make parameter changes after every sample, or after a number of samples that are usually far from being sufficient to deliver deterministic cost function values. Then, the cost function, in fact, is stochastic, with more or less essential variance of the cost function value for a given parameter vector.

From a theoretical point of view, this is the case for stochastic approximation. The stochastic approximation theory [25, 73, 124] is directly applicable to solving stochastic equations. Because local optimization of a smooth function C can be viewed as a task of solving the equation

$$\nabla C(\mathbf{x}) = 0 \qquad (4.37)$$

the stochastic approximation theory can be applied here in a straightforward manner. In commonsense terms, if the cost function parameters (i.e., network weights) are to converge to stable fixed values, the first-order gradient method must be used, with the step size decreasing slowly enough to be able to pass the distance to the error minimum and rapidly enough to become zero, that is, to no longer react to the stochastic perturbation in the current sample once the close neighborhood of the minimum is reached.

One possible set of mathematical conditions imposed on the step size δ_t is

$$\sum_t^\infty \delta_t = \infty \\ \sum_t^\infty \delta_t^2 < \infty \qquad (4.38)$$

The stochastic approximation theory says nothing about the speed of convergence and is thus inappropriate for literal implementation by mortal users without further, problem-specific considerations. However, it makes obvious the point that in an incremental setting, the step size cannot be determined by considering the optimal convergence speed, for it is strictly bounded by stability requirements of stochastic approximation theory. (A specific theorem concerning the bounded learning rates for neural networks, related but not explicitly referring to stochastic approximation theory, has been formulated by Srinivasan et al. [134].)

To summarize, genuine incremental schemes are not open to controlled cost function experiments (i.e., evaluating the cost function for arbitrary parameter values under otherwise identical conditions). The only theoretically founded approach is to perform gradient learning in time with small and diminishing step sizes.

What are the drawbacks in choosing this optimization method? The brief list as the follows:

1. *No Line Search Can Be Performed.* Line search can be viewed as a procedure for determining the optimum step in a given direction: that is, the step for which the cost function has a minimum (and a zero derivative) along this direction. Frequently, this conflicts with the stochastic approximation schedule.
2. *Second-Order Methods for Requiring Line Search Cannot Be Used.* This concerns particularly the conjugate gradient methods and Powell's algorithm. These algorithms are not able keep the search directions conjugate without line search.
3. *Numerical Gradient Cannot Be Used.* All that numerical gradient requires is to perform exactly n or $2n$ controlled experiments.
4. *Most Global Optimization Methods Cannot Be Used.* Except for simulated annealing and evolutionary algorithms, global optimization algorithms require the cost function to be evaluated at various points of the state space with the results compared. For simulated annealing and evolutionary algorithms, or, more generally, for all random search methods, this is not the case. However, hardly anything can be said about the convergence of these (themselves not particularly efficient for continuous optimization) algorithms in the incremental mode of operation.

After these negatives, we need some good news: The first-order gradient descent is not the last resort for incremental learning. The *Kalman training algorithm* proposed by Singhal and Wu [133] and extended and applied to several neurocontrol problems by Puskorius and Feldkamp [120, 121] exploits second-order information by building a parameter covariance matrix, an analogy to the Hessian matrix in variable metric local optimization methods. The Kalman training algorithm has been formulated for a cost function of the form

$$C(\mathbf{w}) = \tfrac{1}{2}[\mathbf{d} - \mathbf{z}(\mathbf{w})]^T \mathbf{S}[\mathbf{d} - \mathbf{z}(\mathbf{w})] \qquad (4.39)$$

where **w** is the parameter vector and **d** is the vector of reference values. This cost function is formally quadratic but, in reality, nonquadratic because of arbitrary nonlinear function **z(w)**. It is general enough for a variety of plant identification, tracking, and stabilization tasks with **w** being the parameter vector of the plant model, of the neurocontroller, or of both.

A parameter update iteration of the Kalman training algorithm is the following:

$$\mathbf{A}_t = \alpha(\mathbf{S}^{-1} + \alpha \mathbf{H}_t^T \mathbf{P}_t \mathbf{H}_t)^{-1}$$
$$\mathbf{K}_t = \mathbf{P}_t \mathbf{H}_t \mathbf{A}_t$$
$$\mathbf{w}_{t+1} = \mathbf{w}_t + \mathbf{K}_t[\mathbf{d} - \mathbf{z}(\mathbf{w}_t)]$$
$$\mathbf{P}_{t+1} = \mathbf{P}_t - \mathbf{K}_t \mathbf{H}_t^T \mathbf{P}_t + \mathbf{Q}_t$$

(4.40)

where $\mathbf{H}_t$ is the matrix of partial derivatives $h_{ij} = \partial z_j / \partial w_i$, $\mathbf{A}_t$ is an artificial noise matrix for avoiding degeneracy, and α is the learning rate. With noncorrelated weights equally scaled with regard to their influence on the cost function, the matrix $\mathbf{P}_t$ is a unit matrix. For a small learning rate α, the matrix **A** is equal to $\alpha \mathbf{S}$. The parameter update rule from Eq. (4.40) then becomes

$$\mathbf{w}_{t+1} = \mathbf{w}_t + \alpha \mathbf{H}_t \mathbf{S}[\mathbf{d} - \mathbf{z}(\mathbf{w}_t)] = \mathbf{w}_t - \alpha \nabla C_{\mathbf{w}_t} \quad (4.41)$$

that is, the usual first-order gradient formula. The learning rate α of the Kalman filter can be varied over time to be consistent with stochastic approximation conditions.

Even with sophisticated methods such as the Kalman training algorithms, the expectations about incremental training cannot be entirely sanguine. For the identification of strongly nonlinear plants of realistic size and difficulty, I have found that tens of thousands of sampled measurements may be necessary. The cost function for measuring the errors for all samples to be evaluated is of the order of 10^5 times if global optimization methods are used (and they should be used if the plant identification is done from scratch, rather than adapting a dependably good model). So altogether, approximately 10^9 samples must be processed! Since incremental methods are substantially less efficient, due to the problem of applying line search and global optimization, we can hardly expect to do with less data. Even a 1-ms sampling rate would call for hundreds to thousands of hours of plant operation before the result is reached.

Of course, for low-dimensional plants with nearly linear characteristics, this expense is drastically reduced. For linear univariable plants, 10 to 20 samples are sufficient for identification. The reason is that linearity allows *extrapolation*, that is, inference from an arbitrary state space point to an arbitrary other point. But for nonlinear plants, the domain for incremental learning of industrial scope will probably remain the fine-tuning of neurocontrollers originally optimized in a batch scheme. This is recommended also by researchers otherwise committed to purely incremental adaptive schemes [e.g., 104].

Note 4.5 In neurocontrol optimization applications such as plant identification or control design, the individual data samples come from various points of state space. It is advantageous if the pieces of information about one state space point or regions are stored separately from those about another point or region. Formally, the cost function can be viewed then as a sum of partial cost functions over all regions indexed by r:

$$C\left(\begin{bmatrix} \mathbf{w}_1 \\ \cdot \\ \cdot \\ \cdot \\ \mathbf{w}_n \end{bmatrix}\right) = \sum_r C_r(\mathbf{w}_r) \qquad (4.42)$$

with each partial cost function C_r having its own parameter vector $\mathbf{w}_r$ (which does not overlap with the parameter vectors of other regions). Then, the convergence to the optimum can be treated separately for each state space region. To a certain degree, this is a property of RBF networks, or more generally, the local representation based on partitioning of input space, discussed in Section 3.1.3.

Note 4.6 The discussion of properties of incremental optimization would be much more difficult if a time-variant environment were considered. Then, the effects of forgetting, or, more exactly, the interferences among forgetting, parameter space covering, and convergence of stochastic approximation have to be taken into account. These topics are briefly addressed in Chapter 8.

Note 4.7 In incremental schemes, the instability of the adaptation, in the sense of divergence from the optimum, is a serious problem. A discussion of these aspects is given in Chapter 5.

Note 4.8 In this section, the incremental schemes have been defined as those for which data are collected during training. Using so-defined incremental schemes is justified for improving performance during the plant operation.

Additionally, incremental schemes with fixed training sets are occasionally used. Such schemes do not exhibit properties advantageous for adaptive learning parallel to the plant operation (or, more exactly, they are as appropriate for adaptive learning as are batch schemes; see Chapter 8). Possible justifications for using this type of incremental schemes concern algorithmic properties. The discussion in this section suggests that there are hardly any arguments in favor of such schemes. With the advantage of online adaptation being discarded, such schemes have too many deficiencies, in particular if compared with efficient second-order local algorithms or even with global optimization algorithms.

The most frequently cited arguments in favor of these schemes are the following:

1. The deviation of the gradient from its true (i.e., computed for the whole training set) value caused by its evaluation with the help of a single sample or a subset of training set introduces a certain *random component* into the search. This is true but will hardly ever be an efficient and well-founded way to randomize the search. If randomizing the search is a goal, more direct approaches (e.g., those discussed by Müller et al. [98] or Archetti [7]) are preferable.
2. It has been argued that a subset of a large training set may be sufficient for computing an acceptably accurate gradient. Then, the computational expense for gradient computation is reduced. However, if this is the case and can be proved or assessed, a complete optimization using only such a (carefully selected) subset is usually a better alternative because the convergence-deteriorating random disturbance caused by the variably selected subset is eliminated.

Sometimes computing experiments support the superiority of incremental schemes with a fixed training set. However, at least as far as I know, such incremental schemes are compared with simple first-order schemes and not with higher-order batch schemes, nor with global algorithms. Also the personal experience of the author has shown the overwhelming superiority of genuine batch schemes.

5

PLANT IDENTIFICATION

5.1 CLASSICAL VIEW OF IDENTIFICATION

Classical control theory can contribute to neural network plant identification in two ways. First, there are theoretically founded statements about general linear model structures that can be used, by analogy, in neural-network-based schemes. Second, classical identification algorithms are classified in a way that can be fruitful in showing how neural-network-based identification can proceed.

This section considers only linear approaches. Nonlinear approaches are treated together with neural-network-based ones in Sections 5.2 through 5.7.

5.1.1 Classical Model Structures

Discrete-Time Models In classical control theory, a discrete-time linear plant is described by a state equation of the form

$$\mathbf{z}_{t+1} = \mathbf{A}\mathbf{z}_t + \mathbf{B}\mathbf{u}_t \tag{5.1}$$

with $\mathbf{z}_t$ being the n-dimensional state of the plant at time t and $\mathbf{u}_t$ the p-dimensional action (or input) at time t. A model of the form (5.1) is referred to as the *state model*.

The state model can be used directly only if all state variables can be measured. Otherwise, a model must be constructed that works with measurable

94 PLANT IDENTIFICATION

output variables. The measurable q-dimensional output of a plant $\mathbf{y}_t$ is often supposed to be a linear function of state $\mathbf{z}_t$:

$$\mathbf{y}_t = \mathbf{C}\mathbf{z}_t \tag{5.2}$$

By recursive application of Eq. (5.1), we get

$$\mathbf{z}_{t+k} = \mathbf{A}^k \mathbf{z}_t + \sum_{i=0}^{k-1} \mathbf{A}^{k-1-i} \mathbf{B} \mathbf{u}_{t+i} \tag{5.3}$$

To reconstruct the state $\mathbf{z}_t$, which is not directly measurable, with the help of measurable outputs $\mathbf{y}_t$ and inputs $\mathbf{u}_t$, $n-1$ matrix equations received from Eq. (5.3) and (5.2) can be considered, that is,

$$\mathbf{y}_{t+k} = \mathbf{C}\mathbf{A}^k \mathbf{z}_t + \sum_{i=0}^{k-1} \mathbf{C}\mathbf{A}^{k-1-i} \mathbf{B} \mathbf{u}_{t+i} \tag{5.4}$$

for $k = 1, \ldots, n-1$. Together with Eq. (5.2) itself, this is a system of n linear matrix equations with n unknown variables z_i. These n matrix equations are, in fact, nq ordinary equations. They have a solution if the rank of the matrix of all coefficients associated with $\mathbf{z}_t$, that is, of matrix

$$\begin{bmatrix} \mathbf{C} \\ \mathbf{C}\mathbf{A} \\ \cdot \\ \cdot \\ \cdot \\ \mathbf{C}\mathbf{A}^{n-1} \end{bmatrix} \tag{5.5}$$

is n [e.g., 1]. This condition is called the *observability condition*. Frequently, the first $n_o - 1$, $n_o \leq n$, of Eqs. (5.4) are, together with Eq. (5.2), sufficient to determine $\mathbf{z}_t$. Integer number n_o is called the *observability index*. To still have a system of at least n equations, n_o must be such that $n_o q \geq n$. Solving this system of linear equations is equivalent to applying a linear operator. So the state $\mathbf{z}_t$ is a linear function of future outputs $\mathbf{y}_{t+i}$, $i = 0, \ldots, d_o - 1$ and inputs $\mathbf{u}_{t+j}$, $i = 0, \ldots, d_o - 2$:

$$\mathbf{z}_t = \sum_{i=0}^{n_o-1} \mathbf{E}_i \mathbf{y}_{t+i} + \sum_{j=0}^{n_o-2} \mathbf{F}_j \mathbf{u}_{t+j} \tag{5.6}$$

Using Eq. (5.3) with $t \leftarrow t - n_o$ and $k \leftarrow n_o$ and substituting Eq. (5.6) for $\mathbf{z}_{t-n_o}$, we get

CLASSICAL VIEW OF IDENTIFICATION

$$\begin{aligned}\mathbf{z}_t &= \mathbf{A}^{n_o}\mathbf{z}_{t-n_o} + \sum_{i=0}^{n_o-1}\mathbf{A}^{n_o-1-i}\mathbf{B}\mathbf{u}_{t-n_o+i}\\
&= \mathbf{A}^{n_o}\left(\sum_{h=0}^{n_o-1}\mathbf{E}_h\mathbf{y}_{t-n_o+h} + \sum_{j=0}^{n_o-2}\mathbf{F}_j\mathbf{u}_{t-n_o+j}\right) + \sum_{i=0}^{n_o-1}\mathbf{A}^{n_o-1-i}\mathbf{B}\mathbf{u}_{t-n_o+i}\\
&= \sum_{h=0}^{n_o-1}\mathbf{A}^{n_o}\mathbf{E}_h\mathbf{y}_{t-n_o+h} + \sum_{j=0}^{n_o-2}\mathbf{A}^{n_o}\mathbf{F}_j\mathbf{u}_{t-n_o+j} \sum_{i=0}^{n_o-1}\mathbf{A}^{n_o-1-i}\mathbf{B}\mathbf{u}_{t-n_o+i}\\
&= \sum_{i=0}^{n_o-1}\mathbf{G}_i\mathbf{y}_{t-1-i} + \sum_{j=0}^{n_o-1}\mathbf{H}_j\mathbf{u}_{t-1-j}\end{aligned}$$ (5.7)

But then, $\mathbf{y}_{t+1}$ can be written, with the help of Eqs. (5.2) and (5.7) and applied to $\mathbf{z}_{t+1}$, as

$$\begin{aligned}\mathbf{y}_{t+1} &= \mathbf{C}\mathbf{z}_{t+1}\\
&= \mathbf{C}\left(\sum_{i=0}^{n_o-1}\mathbf{G}_i\mathbf{y}_{t-i} + \sum_{j=0}^{n_o-1}\mathbf{H}_j\mathbf{u}_{t-j}\right)\end{aligned}$$ (5.8)

This representation can be simplified to

$$\mathbf{y}_{t+1} = \sum_{i=0}^{n_o-1}\mathbf{P}_i\mathbf{y}_{t-i} + \sum_{j=0}^{n_o-1}\mathbf{Q}_j\mathbf{u}_{t-j}$$ (5.9)

This is known as the *input–output plant model*.

Another model form pays particular attention to the temporal relationship between the inputs and the outputs of the model [104]. This alternative form of input-output model, which is based on relative degree, is frequently called the *decoupling model*. The output of a model of the form shown in Eqs. (5.1) and (5.2) at time $t+k$ can be written, with the help of Eq. (5.3), as

$$\mathbf{y}_t = \mathbf{C}\mathbf{z}_t = \mathbf{C}\mathbf{A}^k\mathbf{z}_{t-k} + \mathbf{C}\sum_{j=0}^{k-1}\mathbf{A}^j\mathbf{B}\mathbf{u}_{t-1-j}$$ (5.10)

For some plants, some of vectors $\mathbf{c}_i\mathbf{A}^j\mathbf{B}$ may be zero, where $\mathbf{c}_i$ is the *i*th row of matrix $\mathbf{C}$, that is, the row of C that describes the relationship between the state vector and the *i*th output variable. If this is the case, for some i and $j = 0,\ldots,r_i - 1$, the number r_i is called the *relative degree* of the *i*th output variable.

In discrete time, the meaning of relative degree is the following: Action $\mathbf{u}_t$ performed at time t has an observable effect on the *i*th measurement only after r_i sampling periods, that is, at time $t + r_i$. Before that time, the measurements are completely determined by the memory of the system materialized by the state vector $\mathbf{z}$. (This memory, of course, includes the effects of actions before time t.)

For observable and controllable plants, it is $1 \leq r_i \leq n$. Using, for each output variable, the ith row of the matrix equation (5.10) with k set individually to the corresponding r_i, we get a plant representation by

$$\mathbf{y}^*_{t+1} = \mathbf{A}^* \mathbf{z}_t + \mathbf{B}^* \mathbf{u}_t \tag{5.11}$$

with

$$\mathbf{y}^*_{t+1} = \begin{bmatrix} y_{1,t+r_1} \\ \cdot \\ \cdot \\ \cdot \\ y_{n,t+r_n} \end{bmatrix} \tag{5.12}$$

Substituting for $\mathbf{z}_t$ from Eq. (5.7) into Eq. (5.11), we get

$$\begin{aligned} \mathbf{y}^*_{t+1} &= \mathbf{A}^* \left(\sum_{i=0}^{n_o-1} \mathbf{G}_i \mathbf{y}_{t-1-i} + \sum_{j=0}^{n_o-1} \mathbf{H}_j \mathbf{u}_{t-1-j} \right) + \mathbf{B}^* \mathbf{u}_t \\ &= \sum_{i=0}^{n_o-1} \mathbf{P}_i \mathbf{y}_{t-i} + \sum_{j=0}^{n_o-1} \mathbf{Q}_j \mathbf{u}_{t-j} \end{aligned} \tag{5.13}$$

Continuous-Time Models The state model for continuous-time linear plant is described by a state equation of the form

$$\dot{\mathbf{z}} = \mathbf{A}\mathbf{z} + \mathbf{B}\mathbf{u} \tag{5.14}$$

with output equation

$$\mathbf{y} = \mathbf{C}\mathbf{z} \tag{5.15}$$

As we did for discrete time, an input–output model can be derived for continuous time. By recursive application of Eq. (5.14), that is, by successive differentiation, we get

$$\mathbf{z}^{(k)} = \mathbf{A}^k \mathbf{z} + \sum_{i=0}^{k-1} \mathbf{A}^{k-1-i} \mathbf{B} \mathbf{u}^{(i)} \tag{5.16}$$

The upper index (i) denotes the ith time derivative.

To reconstruct the state $\mathbf{z}$ with the help of measurable outputs $\mathbf{y}$ and inputs $\mathbf{u}$, $n-1$ matrix equations from Eqs. (5.15) and (5.16) can be considered such that

$$\mathbf{y}^{(k)} = \mathbf{CA}^k \mathbf{z} + \sum_{i=0}^{k-1} \mathbf{CA}^{k-1-i} \mathbf{Bu}^{(i)} \qquad (5.17)$$

for $k = 1, \ldots, n-1$. This constitutes a system of equations that can be solved if the observability contition identical with the discrete-time model is satisfied, that is, if matrix (5.5) has rank n [e.g., see 20, 31]. The observability index n_o, formally defined in a way analogous to the discrete-time case, determines what number of derivatives of outputs and inputs are required to reconstruct the state z:

$$\mathbf{z} = \sum_{i=0}^{n_o-1} \mathbf{E}_i \mathbf{y}^{(i)} + \sum_{j=0}^{n_o-2} \mathbf{F}_j \mathbf{u}^{(j)} \qquad (5.18)$$

Substituting Eq. (5.18) into (5.17) with $k \leftarrow n_o$, we get the input–output model:

$$\begin{aligned}\mathbf{y}^{(k)} &= \mathbf{CA}^k \left(\sum_{i=0}^{n_o-1} \mathbf{E}_i \mathbf{y}^{(i)} + \sum_{j=0}^{n_o-2} \mathbf{F}_j \mathbf{u}^{(j)} \right) + \sum_{i=0}^{k-1} \mathbf{CA}^{k-1-i} \mathbf{Bu}^{(i)} \\ &= \sum_{i=0}^{n_o-1} \mathbf{P}_i \mathbf{y}_{(i)} + \sum_{j=0}^{n_o-1} \mathbf{Q}_j \mathbf{u}_{(j)} \end{aligned} \qquad (5.19)$$

Also the decoupling model for continuous time [e.g., 20, 31] is formally similar to the discrete-time case. The concept of relative degree is based on higher-order derivatives rather than on delays. The relative degree of the ith output variable is its lowest-order time derivative that is directly influenced by the action vector. For example, for an integrator plant $\dot{y} = u$, the relative degree is one because the action affects the first derivative of y. For a double integrator plant $\ddot{y} = u$, the relative degree is two.

The decoupling model is

$$\mathbf{y}^* = \sum_{i=0}^{n_o-1} \mathbf{P}_i^* \mathbf{y}_{(i)} + \sum_{j=0}^{n_o-2} \mathbf{Q}_j^* \mathbf{u}_{(j)} \qquad (5.20)$$

with

$$\mathbf{y}^* = \begin{bmatrix} y_1^{(r_1)} \\ \vdots \\ y_n^{(r_n)} \end{bmatrix} \qquad (5.21)$$

5.1.2 Algorithms for Identification of Models from Data

Plant identification is a topic extensively treated in classical control. Algorithmic construction of a plant model has been studied since the 1960s. The bulk of the analytic results concerns linear models. Nevertheless, classical plant identification uses some *computational schemes* that can be adapted to neurocontrol in a straightforward way.

Classical identification procedures have been proposed for both time and frequency domains. Although the frequency domain procedures are useful (and popular) for linear systems, their applicability to nonlinear systems generally (and neural-network-based systems in particular) is limited; hardly any automatic nonlinear identification procedure has found broader application. By contrast, a time-domain view is equally applicable to linear and nonlinear problems (except for a substantially lower extent to which analytic solutions are available). So the most important analogies between classical and neural-network-based plant identification concern *time-domain parameter estimation algorithms*.

Individual approaches vary along several lines. The first thing to distinguish is between batch schemes, on one hand, and incremental schemes, on the other. Batch schemes consist of a certain number of collected measurements and then the model computed from this collection. The computation may be based on an explicit formula, as is the case for the least-squares approach with linear models [e.g., 8], or an optimization scheme, which is computationally more expensive but applicable to the general case. In both cases, the sought model parameters **A** and **B** are those that minimize for a model of the form

$$\mathbf{z}_{t+1} = \mathbf{A}\mathbf{d}_t + \mathbf{B}\mathbf{u}_t \tag{5.22}$$

the term

$$\sum_t f(\|\mathbf{z}_t - \mathbf{d}_t\|) \tag{5.23}$$

where $\mathbf{d}_t$ is the measurement vector at time t, the time index t varies over all measurements, and $f(s)$ is a growing function, usually s^2 or $|s|$.

Incremental schemes recompute the model after each new measurement. The recomputation algorithm may be based on an algebraic transformation of a batch algorithm (e.g., recursive least squares) [8], or on a self-contained rule (usually in the form of a differential or difference equation) designed with a special regard to stability or speed of convergence. The latter type of model has been studied by Narendra and Annaswamy [103]. Two possible model forms are

$$\dot{\mathbf{z}} = \mathbf{A}\mathbf{z} + \mathbf{B}\mathbf{u} \tag{5.24}$$

and

$$\dot{z} = A_m z + (A - A_m)d + Bu = A_m(z - d) + Ad + Bu \quad (5.25)$$

with a fixed matrix A_m. Although there is an incremental adaptive rule with guaranteed stability for model (5.25), it is not the case for model (5.24).

While the incremental methods are provided with the adaptive attribute, batch algorithms, if applied periodically, can equally be viewed as adaptive because they improve their performance by interaction with their environment via measurements. (For a discussion of definitions of adaptiveness, see Narendra and Annaswamy [103], and the references given there.)

Another classification line for approaches to plant identification concerns the choice between static schemes and dynamic schemes (also called series-parallel and parallel models, respectively, by Narendra [101], or model-error and output-error schemes, respectively, in classical parameter estimation). Here, the terms static and dynamic concern the way in which the the input to the model is computed. Scheme (5.22) uses state variable measurements d. By contrast, scheme (5.24), whose discrete analogy for small Δ_t would be

$$z_{t+1} = z_t + \Delta_t(Az_t + Bu_t) \quad (5.26)$$

uses past model outputs z and includes its own dynamics. The scheme (5.25), whose discrete analogy is

$$z_{t+1} = z_t + \Delta_t[A_m(z_t - d_t) + Ad_t + Bu_t] \quad (5.27)$$

uses its own history only together with the fixed gain matrix A_m and thus represents a compromise between dynamic and static approaches.

5.2 PLANT IDENTIFICATION BY NEURAL NETWORKS

Identification by neural networks exhibits a lot of analogies to the classical identification. Frequently, model structures as well as basic computational schemes are directly inspired by classical algorithms.

On the other hand, entering the nonlinear domain brings about some completely new aspects. Analytic solutions mostly cannot be formulated, and their place is taken by numerical ones. The lost theoretical foundation of analytic solutions is traded off by the freedom not to confine oneself to analytically tractable simplifications and ways of solution.

Identification has several facets, each of which provides a set of questions to be answered. The first question is what structure of the model is sufficiently general from the theoretical point of view to be able to represent a large family of real plants. The generality of this structure is particularly important in the neural-network-based plant identification: It is a method that is almost entirely based on inference from data rather than on knowledge that might be used for designing problem-specific structures. A complementary question is what restrictions to this structure are imposed by the fact that data on which the identification builds are gained with limited accuracy, and their choice is constrained by available sensors. Another related question is what neural network representation is advantageous from the numeric optimization viewpoint. These topics are discussed in Section 5.3.

Since neural network plant identification is solved by learning algorithms, which are, in fact, numeric optimization algorithms, a cost function for plant identification has to be formulated. Generally, the fundamental cost function for plant model identification measures the deviation between the behavior of the plant and that of the model. This leads to the question of how this similarity is to be measured, which is discussed in Section 5.4.

With the problems of model structure and similarity measure solved, plant identification can be viewed as a problem of minimizing the cost function. The optimization algorithm aspects, in particular, those related to incremental and batch computation, are addressed in Section 5.5.

The product of a successful optimization is a model that is similar to the plant; that is, its behavior is characterized by available data sampled from the plant. Whether in new situations this similarity will remain close depends on the extent to which the data are representative for the entire state space and simultaneously statistically relevant. This is discussed in Section 5.6.

Finally, the possibility of combining neural-network-based structures with analytic ones should not be ignored. If either the plant model or the controller are implemented as neural networks, and the other part analytically, one is still justified to call it a neurocontrol approach. This is also the case if optimization methods identical to those used with neural networks are applied to nonneural structures with free parameters. The case of identifying an analytic plant with free parameters is discussed briefly in Section 5.7.

5.3 MODEL STRUCTURE

5.3.1 What Theory Says

The most general form of discrete-time plant description is by the state equation

$$\mathbf{z}_{t+1} = f(\mathbf{z}_t, \mathbf{u}_t) \tag{5.28}$$

with f being a nonlinear vector mapping. Like linear systems, the generality of this form is based on the idea of a state vector $\mathbf{z}_t$ that incorporates complete information relevant for the future behavior of the plant.

Other representational alternatives are analogous to their linear counterparts. With plant output characterized by the nonlinear mapping

$$\mathbf{y}_t = h(\mathbf{z}_t) \tag{5.29}$$

the input–output model

$$\mathbf{y}_{t+1} = f(\mathbf{y}_t, \mathbf{y}_{t-1}, \ldots, \mathbf{y}_{t-n_o+1}, \mathbf{u}_t, \mathbf{u}_{t-1}, \ldots, \mathbf{u}_{t-n_o+1}) \tag{5.30}$$

and the decoupling model [104]

$$\mathbf{y}^*_{t+1} = f(\mathbf{y}_t, \mathbf{y}_{t-1}, \ldots, \mathbf{y}_{t-n_o+1}, \mathbf{u}_t, \mathbf{u}_{t-1}, \ldots, \mathbf{u}_{t-n_o+1}) \tag{5.31}$$

where $\mathbf{y}^*_{t+1}$, defined as in Eq. (5.12), can be substituted for the state model.

Note 5.1 Levin and Narendra [86] have formulated a local (i.e., valid in a certain neighborhood of the equilibrium) analogy to linear observability based on linearized representation with the help of Jacobian matrices. To prove global observability may be substantially more difficult. In particular, global observability does not necessarily imply that the observability index is less or equal to the model order. An example for this is the system described by Eq. (3.73) of Section 3.3. The state of this system can be reconstructed only by taking into account an input–output series of length exceeding the system order by one; that is, an additional observation is necessary to resolve the bifurcation caused by the square term in the output equation.

Continuous counterparts of these models are (1) the state model

$$\dot{\mathbf{z}} = f(\mathbf{z}, \mathbf{u}) \tag{5.32}$$

with output

$$\mathbf{y} = h(\mathbf{z}) \tag{5.33}$$

(2) the input–output model

$$\mathbf{y}^{(n_o)} = f(\mathbf{y}, \mathbf{y}^{(1)}, \ldots, \mathbf{y}^{(n_o-1)}, \mathbf{u}, \mathbf{u}^{(1)}, \ldots, \mathbf{u}^{(n_o-1)}) \tag{5.34}$$

and (3) the decoupling model

$$\mathbf{y}^* = f(\mathbf{y}, \mathbf{y}^{(1)}, \ldots, \mathbf{y}^{(n_o-1)}, \mathbf{u}, \mathbf{u}^{(1)}, \ldots, \mathbf{u}^{(n_o-1)}) \tag{5.35}$$

with $\mathbf{y}^*$ as in Eq. (5.21).

Now, we have a choice among six models. At first glance, the decision is simple: The plant identification by neural networks is usually based on discrete sampling, and the complete state is never known with certainty, so a discrete-time input–output model must be used. The discrete-time decoupling model can be taken if the relative degrees are known.

However, the matter is not so simple. There are some additional considerations that must be taken into account:

- The discrete-time model is not simply a discretized continuous-time model. Whereas the former is based on delays, the latter is based on derivatives. What the relationship is between a discrete-time model and a discretized continuous-time model must be clarified as well as which of these alternatives is more advantageous.
- The input–output model assumes that the plant state **z** is observable by means of a measurement vector **y**. What happens, and what should be done, if it is not?
- Can the knowledge of relative degree be acquired and used?

Comparison of Discrete-Time Models and Discretized Continuous-Time Models Usually discrete-time models are based on delayed variables whereas continuous-time models are formulated in terms of time derivatives. The relationship between the two models can be established with numeric derivatives. A numerical derivative is an approximation of the definition of derivative:

$$\frac{dx}{dt} \approx \frac{x(t + \Delta_t) - x(t)}{\Delta_t} \tag{5.36}$$

An exact derivative is defined by Eq. (5.36) for $\Delta_t \to 0$. The simplest possibility to approximate the derivative is by taking Eq. (5.36) with some small Δ_t.

Integration procedures can choose the step size Δ_t so that a specified accuracy of approximating the continuous differential equation is reached. For example, they may be used for simulating the state space model (5.32) with a neural network representing the mapping f. Using such methods is usually not very advantageous for the following reasons:

1. Sampled data, with the help of which the model is to be trained, are often equidistant. Simulating the model with variable step size would require synchronization of both time series at the time points of the samples.
2. The main reason for using variable step size is to guarantee a good approximation of a given continuous-time system. The goal of neural-network-based plant identification is to receive a model that forecasts the data sampled at (usually equidistant) discrete time points, no matter what the values between these time points are.
3. A plant model is usually used for neurocontroller training, and this neurocontroller is mostly used as a discrete-time controller. Training a discrete-time controller must be done in a discrete-time framework. Using a continuous model in such a discrete-time framework would bring about excessive algorithmic complexity.

4. If an analytic error gradient is used, special formulas for continuous model would have to be derived.

However, for very unstable plants, simulation with fixed step size may cause the error function to behave too irregularly, even in a close neighborhood of the optimum. Then, using a continuous model with sophisticated integration methods such as those of Runge–Kutta [e.g., 117] may improve the convergence of the optimization algorithm.

Another possibility is a rough but practical approximation of the continuous model based on formula (5.36) with Δ_t equal to the sampling period used in the data acquisition. Then, the kth derivative of vector $\mathbf{x}_t$ (standing for either $\mathbf{z}_t$ or $\mathbf{y}_t$) can be defined recursively as

$$\mathbf{x}_t^{(k)} = \frac{1}{\Delta_t}(\mathbf{x}_t^{(k-1)} - \mathbf{x}_{t-1}^{(k-1)}) \tag{5.37}$$

with $\mathbf{x}_t^{(0)} = \mathbf{x}_t$. Obviously, the kth derivative $\mathbf{x}_t^{(k)}$ can be expressed in terms of $\mathbf{x}_t, \ldots, \mathbf{x}_{t-k}$:

$$\mathbf{x}_t^{(k)} = \sum_{i=0}^{k} p_i \mathbf{x}_{t-i} \tag{5.38}$$

with some fixed scalar constants q_i. For example,

$$\mathbf{x}_t^{(2)} = \frac{1}{\Delta_t^2}(\mathbf{x}_t - 2\mathbf{x}_{t-1} + \mathbf{x}_{t-2}) \tag{5.39}$$

Inversely, $\mathbf{x}_{t+1}$ can be expressed with the help of $\mathbf{x}_{t+1}^{(k)}$ and past lower derivatives $\mathbf{x}_t^{(i)}$, $i < k$, as

$$\mathbf{x}_{t+1} = \Delta_t^k \mathbf{x}_{t+1}^{(k)} + \sum_{i=0}^{k-1} \Delta_t^i \mathbf{x}_t^{(i)} \tag{5.40}$$

This can be easily shown by mathematical induction. The relationship for $k = 0$ is trivial, and

$$\begin{aligned}\mathbf{x}_{t+1} &= \Delta_t^k \mathbf{x}_{t+1}^{(k)} + \sum_{i=0}^{k-1} \Delta_t^i \mathbf{x}_t^{(i)} \\ &= \Delta_t^k \mathbf{x}_{t+1}^{(k)} - \Delta_t^k \mathbf{x}_t^{(k)} + \Delta_t^k \mathbf{x}_t^{(k)} + \sum_{i=0}^{k-1} \Delta_t^i \mathbf{x}_t^{(i)} \\ &= \Delta_t^{k+1} \mathbf{x}_{t+1}^{(k+1)} + \sum_{i=0}^{k} \Delta_t^i \mathbf{x}_t^{(i)} \end{aligned} \tag{5.41}$$

104 PLANT IDENTIFICATION

Now, the relationship between discrete-time models and continuous-time models with numeric derivatives can be established. The continuous-time state model (5.32) with numeric derivatives can be written as

$$\dot{\mathbf{z}}_{t+1} = \frac{\mathbf{z}_{t+1} - \mathbf{z}_t}{\Delta_t} = f(\mathbf{z}_t, \mathbf{u}_t) \tag{5.42}$$

or

$$\mathbf{z}_{t+1} = \mathbf{z}_t + \Delta_t f(\mathbf{z}_t, \mathbf{u}_t) = f'(\mathbf{z}_t, \mathbf{u}_t) \tag{5.43}$$

which is identical with Eq. (5.28).

The continuous-time input–output model (5.34) with numeric derivatives

$$\mathbf{y}_{t+1}^{(n_o)} = f(\mathbf{y}_t, \mathbf{y}_t^{(1)}, \ldots, \mathbf{y}_t^{(n_o-1)}, \mathbf{u}_t, \mathbf{u}_t^{(1)}, \ldots, \mathbf{u}_t^{(n_c-1)}) \tag{5.44}$$

can be written, with the help of Eqs. (5.40) and (5.38) as

$$\begin{aligned}
\mathbf{y}_{t+1} &= \Delta_t^{n_o} \mathbf{y}_{t+1}^{(n_o)} + \sum_{i=0}^{n_o-1} \Delta_t^i \mathbf{y}_t^{(i)} \\
&= \Delta_t^{n_o} f(\mathbf{y}_t, \mathbf{y}_t^{(1)}, \ldots, \mathbf{y}_t^{(n_o-1)}, \mathbf{u}_t, \mathbf{u}_t^{(1)}, \ldots, \mathbf{u}_t^{(n_c-1)}) + \sum_{i=0}^{n_o-1} \Delta_t^i \mathbf{y}_t^{(i)} \\
&= f'(\mathbf{y}_t, \mathbf{y}_{t-1}, \ldots, \mathbf{y}_{t-n_o+1}, \mathbf{u}_t, \mathbf{u}_{t-1}, \ldots, \mathbf{u}_{t-n_o+1})
\end{aligned} \tag{5.45}$$

which is equivalent to Eq. (5.30).

Analogous transformation can be done between the decoupling models (5.35) and (5.31). This shows that

both model families, the discrete-time models and the continuous-time models with numeric derivatives, are equivalent.

A neural network can represent the mapping f of either Eq. (5.28) or Eq. (5.32).

If both model families are equivalent, are there reasons to prefer one or the other group? The advantage of discrete-time models with time delays is that the state at time $t+1$ is computed directly, for example, by the mapping f of Eq. (5.28). If this mapping is directly represented by a neural network, the next state is identical with the output of the network's output layer. Consequently, published formulas for analytic gradients, such as backpropagation formulas, can be applied in a straightforward way. By contrast, if the neural network represents the mapping f of continuous-time model (5.42) with numeric derivatives, the next state has to be computed by numeric integration of the form

$$\mathbf{z}_{t+1} = \mathbf{z}_t + \Delta_t \dot{\mathbf{z}}_t \tag{5.46}$$

Formula (5.43) had collapsed Eqs. (5.42) and (5.46) into one. For the input–output model (5.34), integration has to be performed n_o times. (The computational expense for this is negligible if compared with that for the forward pass of the neural network itself.) If the analytic gradient is used, its computation must account for this. The dynamic backpropagation of Narendra and Parthasarathy [105] is a good conceptual basis for such generalization. Nevertheless, spending some time with the derivation of correct formulas is inevitable.

On the other hand, the models using numerical derivatives are superior as far as convergence of optimization algorithms is concerned. The reason for this is that delayed values are strongly autocorrelated. That is, x_t is strongly correlated with x_{t-1}; the values of successive samples are very similar (at least if sampling periods are small enough to be consistent with the sampling theorem). By contrast, numeric derivatives of successive orders are approximately statistically independent. For example, let us observe the successively sampled values of variable x, x_{t-1}, and x_t. For simplicity, the mean value of x is assumed to be zero. The variance of x is V_x. The correlation coefficient between x_t and x_{t-1} can be assumed to be almost one:

$$r_t = r(x_t, x_{t-1}) = \frac{E[x_t x_{t-1}]}{\sqrt{E[x_1^2]E[x_2^2]}} = \frac{E[x_t x_{t-1}]}{V_x} \approx 1 \tag{5.47}$$

where $E[\cdot]$ is expected value.

The correlation between x_t and the first numerical derivative $x_t^{(1)} = (1/\Delta_t)(x_t - x_{t-1})$ is

$$\begin{aligned} r(x_t, x_t^{(1)}) &= r\left(x_t, \frac{1}{\Delta_t}(x_t - x_{t-1})\right) \\ &= \frac{E[x_t(x_t - x_{t-1})]/\Delta_t}{\sqrt{\mathrm{Var}[x_t]\mathrm{Var}[x_t - x_{t-1}]}/\Delta_t} \\ &= \frac{E[x_t^2] - E[x_t x_{t-1}]}{\sqrt{V_x \times 2V_x}} \\ &= \frac{V_x - r_t V_x}{V_x \sqrt{2}} \\ &= \frac{1 - r_t}{\sqrt{2}} \\ &\approx 0 \end{aligned} \tag{5.48}$$

The same relationship for the correlation between successive higher order derivatives can be inferred analogously. Now, one might ask: What has the correlation of delayed values and the independence of numeric derivatives to

106 PLANT IDENTIFICATION

do with the convergence of the optimization algorithm? Each neural network input is associated with a weight or another type of parameter. Similar (correlated) inputs suggest similar influences on the cost function value, and they appear roughly substitutable for each other. Consequently, the weights associated with similar inputs tend to develop in the same direction, that is, to become correlated.

Many optimization methods work substantially better if the inputs of the function involved are orthogonal or statistically independent. Otherwise, the cost function tends to have the form of a long, narrow valley. This is exactly where first-order gradient methods probably will fail (see Chapter 4).

Second-order methods cope with this problem to some degree, but it is better to avoid the problem than to cope with it! Because the convergence of optimization is a dominant factor for having and not having a success with neurocontrol, the following is recommended:

Use the model representation based on numeric derivatives of model variables whenever possible.

With optimizing methods using no gradient or numeric gradient, the efficiency increase is free. With the analytic gradient, it is necessary to derive the nonstandard gradient formulas. Even then, it should pay off.

Note 5.2 The conceptual structure of the plant model must be distinguished from the operational structure. The considerations of input and output orthogonality are relevant for the conceptual structure of the plant model represented by a neural network (no matter whether the representation is direct or in the form of the gain vector; see Section 3.3.1). On this level, statistical independence of inputs or outputs and thus the independence of network weights must be reached.

By contrast, the operational structure of the plant model involves the computation of a plant model action from sampled measurements. The measurements are almost always successively delayed values, except for cases where derivatives are measured directly by special sensors (e.g., by a distance sensor, velocity sensor, and an acceleration sensor). This is why the operational model usually must compute numeric derivatives from delayed values. The operational model output, that is, the next output vector, is then received by integration of the next higher-order derivative of output, produced by the conceptual neural network model.

Thus, a call of operational plant model consists of three steps:

1. Computing the required numeric derivatives from sampled measurements.
2. Evaluating the conceptual plant model containing numeric derivatives.
3. Integrating, if necessary, integrating the plant model output given in the form of a numeric derivative to the next action vector.

Comparison of State Models and Input–Output Models For plant identification methods that use explicit modeling, it makes sense to define nonmeasurable state variables. Their definition is then based on the knowledge that such variables enter the dynamic description of the system in a known way. For identification methods without explicit models, for instance, neural-network-based identification, the relationships that underlie system behavior are not known a priori and cannot be used for definitions of meaningful nonmeasurable variables. The opposite case is discussed briefly below.

First, we share a little philosophy. Let us compare a state model, for example, Eq. (5.32) with its input–output counterpart (5.34). For the computational scheme to be complete, the input–output model must be supplied with auxiliary integration equations such that

$$\mathbf{y}_{t+1}^{(k)} = \mathbf{y}_t^{(k)} + \Delta_t \mathbf{y}_{t+1}^{(k+1)} \tag{5.49}$$

for $k = 0, \ldots, n_o - 2$, and

$$\mathbf{u}_{t+1}^{(k)} = \mathbf{u}_t^{(k)} + \Delta_t \mathbf{u}_{t+1}^{(k+1)} \tag{5.50}$$

for $k = 0, \ldots, n_o - 2$. Together with

$$\mathbf{y}_{t+1}^{(n_o-1)} = \mathbf{y}_t^{(n_o-1)} + \Delta_t f(\mathbf{y}_t, \mathbf{y}_t^{(1)}, \ldots, \mathbf{y}_t^{(n_o-1)}, \mathbf{u}_t, \mathbf{u}_t^{(1)}, \ldots, \mathbf{u}_t^{(n_o-1)}) \tag{5.51}$$

this is a system of $qn_o + p(n_o - 1)$ difference equations for $qn_o + p(n_o - 1)$ variables $\mathbf{y}, \mathbf{y}^{(1)}, \ldots, \mathbf{y}^{(n_o-1)}, \mathbf{u}, \mathbf{u}^{(1)}, \ldots, \mathbf{u}^{(n_o-2)}$, with input $\mathbf{u}^{(n_o-1)}$. In other words, formally, this is a state model with a $qn_o + p(n_o - 1)$-dimensional state space. The difference between this extended state space and the state space is that the model described by Eqs. (5.49) to (5.51) may be redundant. The extent of redundancy can be determined only if the genuine order n is known. There are order assessment algorithms for linear systems, for example, those based on determinant ratios [e.g., 140], but their nonlinear counterparts would amount to applying them to every point of the state space, which can hardly be reduced to a pragmatic procedure. On the other hand, real plants are frequently of very high (or even infinite) order if all modes are taken into account. This is why the input–output model can be taken for a real state equation if no information is available about the composition of the state vector.

The commitment to measurable variables can lead to the following three situations:

1. The measurable variables are identical with the complete state.
2. The measurable variables are only a subset of the complete state, but the complete state is observable.
3. The state is not completely observable.

108 PLANT IDENTIFICATION

If you are lucky enough to encounter a plant for which the assumption is reasonable that all state variables are directly measurable, the state model can be used directly. A formally equivalent situation occurs if the observability index and relative degrees of all output variables are equal to one. Then, there is no difference between the input–output model and the state model with outputs taken for real states.

The case of the observable state indicates the use of an input–output model. How to select the value of observability index n_o is discussed in Section 5.3.2.

In last case above in which the state is not completely observable, there are three possibilities: First, we can install additional sensors and measure more. This is the most trivial and efficient way, and improvement is almost guaranteed.

Second, we can treat unobservable states as disturbances and hope that these disturbances will not overshadow the dynamics of the observable part of the plant. This will work if the unobservable states are associated with weak (or at least stable) modes of the plant. Most controllers in the world are designed with the help of incomplete models; that is, they ignore some states and many of them work. For nonlinear systems, the assumption of unmodeled dynamics being associated with stable modes can hardly be checked in another way than by trying to identify the model. If this assumption is violated, the forecast error of the model will remain large. To be more concrete, if the measurements are reasonably accurate (e.g., 12-bit accurate), and the squared error of the forecast *derivative* (using a continuous-time model with numeric derivatives) exceeds, say, 10% of the derivative variance, unmodeled dynamics will come into play. The presence of unmodeled dynamics is substantially easier to recognize if the correspondence of a whole trajectory is compared. Other terms for trajectory comparison are the *output error model* or *parallel model*. This configuration makes possible (see Section 5.4) divergance of the model from the measurement and thus makes inconsistencies explicit.

An obvious precondition for this kind of check for unmodeled dynamics is that there is no other reason for the model optimization to fail, such as a bad optimization method. For example, using a first-order method in incremental mode, one could fail to get a good model even if the state is completely measurable. Even if the target configuration is online adaptive, it is advisable to use, at least at this stage, batch processing and a global optimization method together with second-order local optimization and to avoid using an analytic gradient (see Chapter 4).

Finally, if the input–output model fails, and no additional measurements are possible, the situation becomes serious. The last resort is to attempt to identify strongly hidden states [157]. The word "attempt" is chosen deliberately: the chances for success are some orders of magnitude smaller than if one stays on the firm ground of observable plants and input–output models. This situation is discussed in Section 3.3. More precisely, the input–output model is extended by the additional hidden-state vector **x**. A proposed form of the extended model with numerical derivatives is the following:

$$\begin{aligned}\mathbf{y}_{t+1}^{(n_o)} &= f(\mathbf{y}_t, \mathbf{y}_t^{(1)}, \ldots, \mathbf{y}_t^{(n_o-1)}, \mathbf{x}_t, \mathbf{x}_t^{(1)}, \ldots, \mathbf{x}_t^{(n_o-1)}, \mathbf{u}_t, \mathbf{u}_t^{(1)}, \ldots, \mathbf{u}_t^{(n_o-1)}) \\ \mathbf{x}_{t+1}^{(n_o)} &= f(\mathbf{y}_t, \mathbf{y}_t^{(1)}, \ldots, \mathbf{y}_t^{(n_o-1)}, \mathbf{x}_t, \mathbf{x}_t^{(1)}, \ldots, \mathbf{x}_t^{(n_o-1)}, \mathbf{u}_t, \mathbf{u}_t^{(1)}, \ldots, \mathbf{u}_t^{(n_o-1)})\end{aligned} \quad (5.52)$$

The hidden state variables are treated in the same way as the output variables except that they do not appear in the error function: They have no counterparts in the measurements.

I will refrain from discussing at length whether it is better to consider a longer vector $\mathbf{x}$ or a short one (possibly a scalar) together with its first $n_o - 1$ derivatives as proposed in (5.52). An argument in favor of the latter alternative is that a shorter vector $\mathbf{x}$ is satisfied with a simpler specification of the initial state, if the initial state of the whole plant is approximately stationary. Then, in the scalar case, only a single specification of the initial state $\mathbf{x}_0$ is required, $\mathbf{x}_0^{(i)}, i > 0$, being set to zero. This may be advantageous because the initialization of hidden states is a formidable problem. In fact, the best principle to apply is to assign the hidden state a conceptual meaning and set an initial state consistent with this meaning. But this is only possible if there are variables in the plant that are not measurable but whose meaning is conceptually known. This is evident in the case study of the wastewater treatment plant in Chapter 15, where a meaningful variable, the contaminant substrate quantity, is not measurable online but can be estimated by expert knowledge or a laboratory test.

Value of Relative Degree The decoupling models (5.31) and (5.35) can be viewed as alternative forms of the input–output model. Before addressing the question of when they are preferrable to conventional input–output models, let us briefly compare the meaning of relative degree, a crucial concept for decoupling models in both discrete time and continuous time.

It is almost a rule that wherever a delay by i sampling periods appears in a discrete-time model, the ith derivative is found in a corresponding continuous-time model. So, it is no surprise that (1) the discrete-time relative degree is the delay of a given output variable on which the plant input has a direct influence, whereas (2) the continuous-time relative degree is the time derivative of a given output variable on which the plant input has a direct influence.

One aspect of this is intriguing. Suppose that the same plant is modeled in both continuous and discrete time. It is assumed that both models are qualitatively analogous. For example, state vectors have identical definitions, observability conditions are similar, and the order of both models is the same. With the continuous model, even if it is discretized by using numeric derivatives, the action $\mathbf{u}_t$ affects the r_ith derivative of variable $y_{i,t}$ immediately, that is, at time $t + 1$, the discrete model states that variable y_i is influenced only after a delay of r_i sampling periods. Moreover, discrete models with different sampling periods Δ_t either have to have different relative degrees r_i or be contradictory in postulating delays $\Delta_t \times r_i$ of different duration.

To understand the mechanism, let us observe a simple double-integrator plant with continuous-time model

$$\dot{x}_1 = u$$
$$\dot{x}_2 = x_1 \quad (5.53)$$
$$y = x_2$$

The relationship between input u and output y is

$$\ddot{y} = u \quad (5.54)$$

that is, the relative degree of y is two. A discrete-time model received by the simplest method, the Euler integration, is

$$x_{1,t+1} = x_{1,t} + \Delta_t u_t$$
$$x_{2,t+1} = x_{2,t} + \Delta_t x_{1,t} \quad (5.55)$$
$$y_t = x_{2,t}$$

with decoupling equation

$$\begin{aligned} y_{t+2} &= x_{2,t+2} \\ &= x_{2,t+1} + \Delta_t x_{1,t+1} \\ &= x_{2,t+1} + \Delta_t x_{1,t} + \Delta_t^2 u_t \end{aligned} \quad (5.56)$$

Relative degree of y is, once more, two.

Let us now discretize more exactly, for example, by taking the average from $x_{1,t+1}$ and $x_{1,t}$, instead of $x_{1,t}$ alone, in the second state equation:

$$\begin{aligned} x_{1,t+1} &= x_{1,t} + \Delta_t u_t \\ x_{2,t+1} &= x_{2,t} + \Delta_t \frac{x_{1,t+1} + x_{1,t}}{2} \\ &= x_{2,t} + \Delta_t \frac{2x_{1,t} + \Delta_t u_t}{2} \\ &= x_{2,t} + \Delta_t x_{1,t} + \Delta_t^2 \frac{u_t}{2} \\ y_t &= x_{2,t} \end{aligned} \quad (5.57)$$

with decoupling equation

$$\begin{aligned} y_{t+1} &= x_{2,t+1} \\ &= x_{2,t} + \Delta_t x_{1,t} + \Delta_t^2 \frac{u_t}{2} \end{aligned} \quad (5.58)$$

Now, the relative degree of y is one.

So the relative degree of discrete-time models received by discretizing the continuous models depends on the discretization method. If the continuous model contains no delays, it can be transformed by sufficiently accurate discretization method into a discrete-time method with relative degrees equal to one. In this sense, relative degree in discretized systems is a kind of artifact of the discretization method.

Is the relative degree in discrete-time systems only an illusion without real meaning? Definitely, it is not. Comparing the decoupling models (5.56) and (5.56), observe that, in both cases, y is affected by u with factor Δ_t^2. Generally, with relative degree r_i, the effect of input on the change of output would be proportional to $\Delta_t^{r_i}$. In other words, the effect of persistent input on an output with relative degree r_i grows with the r_ith power of time.

Note 5.3 The preceding discussion has been based on the assumption that a continuous-time model is primary. This is the case for most physical models because physical laws operate in continuous time. Of course, there are also systems for which the discrete-time description is primary. Then, the interpretation of relative degree as a delay reflects inherent properties of the plant.

With the interpretation of relative degree in mind, the appropriate choice between the conventional input–output model and the decoupling model can be discussed. The decoupling model has been formulated by Falb and Wolowich [27] as a part of a particular controller design method based on the idea of decoupling between individual controller actions and plant outputs. In neurocontrol, Narendra and Mukhopadhyay [104] have proposed several design algorithms based on analogous principles. The basic idea is the following: With the input–output model (5.30), the action that has a real influence on $\mathbf{y}_{t+1}$ may be any of $\mathbf{u}_{t-i}, i = 0, \ldots, n_o - 1$ (and it is not explicitly known which of them). By contrast, with the decoupling model (5.30), the effective action is guaranteed to be $\mathbf{u}_t$. Because $\mathbf{y}_{t-i}, i = 0, \ldots, n_o - 1$ and $\mathbf{u}_{t-i}, i = 1, \ldots, n_o - 1$ are constant past values, at time t, there is an explicit relationship between the cause represented by $\mathbf{u}_t$ and the effect represented by $\mathbf{y}_{t+1}^*$. If the vectors $\mathbf{u}$ and $\mathbf{y}$ are of the same dimension, this relationship may even be *invertible*; that is, the action for reaching a certain state can be directly figured out. For continuous-time models with numeric derivatives, this procedure can be also applied after some algebraic manipulation of the right-hand side of the model.

If the plant model is not intended for use with such a generalized inversion scheme but rather with closed-loop optimization, both the conventional input–output model and decoupling model can be used. The advantage of the input–output model is the statistical independence between its output $\mathbf{y}^{n_o}$ and each of its arguments $y^{(i)}, i < n_o$ (see above). (Of course, there is a dependence of the output on the *whole set* of its arguments; otherwise, no model would be identifiable.) The advantage of the decoupling model is that the r_ith derivative is the most directly influenced derivative by the model input $\mathbf{u}$ and that this relationship is relatively simple (and thus easy to find by learning from data). The most decisive point is probably that relative degrees are hardly possible to determine a priori, without performing the identification.

To summarize the discussion:

Input–output models are more universal and are thus the first choice. For special neurocontroller design techniques based on plant inversion, decoupling models are required. Their use is restricted by the necessity of assessment of relative degrees for all output variables. In any case, relative-degree definition based on derivatives is preferable.

5.3.2 What Works in Practice?

The results of Section 5.3.1 can be summarized into the recommendation to use the input–output model

$$\mathbf{y}_{t+1}^{(n_o)} = f(\mathbf{y}_t, \mathbf{y}_t^{(1)}, \ldots, \mathbf{y}_t^{(n_o-1)}, \mathbf{u}_t, \mathbf{u}_t^{(1)}, \ldots, \mathbf{u}_t^{(n_o-1)}) \tag{5.59}$$

with numeric derivatives computed according to

$$\mathbf{x}_t^{(k)} = \frac{1}{\Delta_t}(\mathbf{x}_t^{(k-1)} - \mathbf{x}_{t-1}^{(k-1)}) \tag{5.60}$$

whenever there is no particular reason to take something else.

The model (5.59) is completely specified only after the observability index n_o is determined. As is usual with nonlinear systems, the easiest way to determine this index is by trying what works and what does not work. If an observability index is set too small, there is a danger that what is observable in absolute terms is not observable with *this* particular index. Then, the problem of unmodeled dynamics, briefly mentioned in Section 5.3.1, is encountered once more.

However, the observability index problem can also be seen from another point of view. The question of what *should* be done can be substituted by the question of what *can* be done. A serious constraint that every control engineer has encountered is the *precision of sampled data*. The following discussion is concerned with the question of how this precision constrains the feasible observability index.

One example of noise source for all is the discretization error. Suppose that sine signals with maximum amplitude a_{max} and maximum frequency f_{max} are sampled. The sampling theorem requires a sampling period that is at most half of the period of the maximum frequency signal, that is, satisfying

$$\Delta_t < \frac{1}{2}\frac{1}{f_{max}} \tag{5.61}$$

This is frequently applied using a reserve factor c so that

$$\Delta_t = \frac{1}{2c}\frac{1}{f_{max}} \tag{5.62}$$

Numeric derivatives of sampled variable z are computed from consecutive sampled values. For example, the first derivative is

$$\dot{z}_t = \frac{z_t - z_{t-1}}{\Delta_t} = \frac{\Delta z}{\Delta_t} \tag{5.63}$$

Generally, the ith derivative is

$$z^{(i)} = \frac{\Delta^{(i)}}{\Delta_t^i} \tag{5.64}$$

with $\Delta^{(i)}$ being the ith-order difference. The ith derivative of the sine signal has the maximum amplitude

$$(2\pi f_{max})^i a \tag{5.65}$$

The maximum value of the ith-order difference (as used for the computation of the ith numerical derivative) is then

$$(2\pi f_{max})^i a \Delta_t^i \tag{5.66}$$

Using Eq. (5.62), this maximum value is

$$(2\pi f_{max})^i a \left(\frac{1}{2c}\frac{1}{f_{max}}\right)^i = a\left(\frac{\pi}{c}\right)^i \tag{5.67}$$

This means that for a sine signal with maximum frequency consistent (with given reserve) with the sampling theorem and amplitude a_{max}, the values of the ith order difference vary between $-a(\pi/c)^i$ and $a(\pi/c)^i$. For lower frequencies and amplitudes, this range is, of course, smaller.

Usual 12-bit analog–digital converters provide, if the sensor is optimally calibrated, 4096 different values. It is probable that most of the time values will stay within a fraction of this range, optimistically, that is, maybe a third. It is a general rule that sampling rates should be at most a tenth of those required by the sampling theorem; that is, the reserve factor c is about ten. So although usual signal values take on about $4096/3 \doteq 1400$ values, the first derivative takes only $1400\pi/10 \doteq 440$, and the second derivative about $1400(\pi/10)^2 \doteq 138$ different values in the maximally excited case. For a frequency that is a quarter of the maximum, the resolutions are approximately

1400, 110, 9. Using high-resolution analog–digital converters alleviates the discretization source of noise but not other sources such as white noise, which behave in a qualitatively similar way. A practical consequence is that numeric derivatives of order higher than two (and mostly also higher than one) have hardly any information content and should not be used in modeling.

As stated in Section 5.3.1, higher-order derivatives of a variable can be substituted by delayed values of the same variable. This approach is very popular in the neurocontrol community, probably because discrete sampling of measurements suggests a relationship to discrete systems theory rather than to continuous systems. Superficially, the impression may arise that because delayed values are as precise as nondelayed one, systems of arbitrarily high order may be identified. This is not, of course, the case. It is not the delayed variable itself that is important, but its contribution to the information about system trajectory. As pointed out in Section 5.3.1, an important difference between derivatives and delayed values is that although derivatives, including the zeroth derivative, of deterministic systems (or stochastic systems with uncorrelated noise) are mutually uncorrelated, delayed variables exhibit high autocorrelation. Correlated variables contain less information than uncorrelated ones. So the contribution of delayed variables is exactly the same as that of numeric derivatives; the latter are merely an orthogonalization of the former.

Under these practical restrictions, the determination of observability index can be reduced to:

Try one; if it fails, try two. With very precise sensors, three may be also worth trying.

In terms of concrete models, the first alternative to be attempted is

$$\dot{\mathbf{y}}_{t+1} = f(\mathbf{y}_t, \mathbf{u}_t) \tag{5.68}$$

which is identical with the state model. Its operational form (explicitly using only sampled values; see Note 5.2), which can be used directly in a simulation, is

$$\mathbf{y}_{t+1} = \mathbf{y}_t + \Delta_t f(\mathbf{y}_t, \mathbf{u}_t) \tag{5.69}$$

In this case, there is an implicit assumption of the complete state being measurable either directly or by algebraic operations on the measurements.

If the forecast error of the model (5.68) is too high, the second-order model

$$\ddot{\mathbf{y}}_{t+1} = f(\mathbf{y}_t, \dot{\mathbf{y}}_t, \mathbf{u}_t, \dot{\mathbf{u}}_t) \tag{5.70}$$

with operational form

$$\mathbf{y}_{t+1} = \mathbf{y}_t + \Delta_t \left[\frac{\mathbf{y}_t - \mathbf{y}_{t-1}}{\Delta_t} + f\left(\mathbf{y}_t, \frac{\mathbf{y}_t - \mathbf{y}_{t-1}}{\Delta_t}, \mathbf{u}_t, \frac{\mathbf{u}_t - \mathbf{u}_{t-1}}{\Delta_t}\right) \right]$$
$$= 2\mathbf{y}_t - \mathbf{y}_{t-1} + \Delta_t^2 f\left(\mathbf{y}_t, \frac{\mathbf{y}_t - \mathbf{y}_{t-1}}{\Delta_t}, \mathbf{u}_t, \frac{\mathbf{u}_t - \mathbf{u}_{t-1}}{\Delta_t}\right)$$
(5.71)

must be identified. The second-order model is worth accepting only if its forecast error is substantially lower than that of the first-order model.

The third-order model

$$\mathbf{y}_{t+1}^{(3)} = f(\mathbf{y}_t, \dot{\mathbf{y}}_t, \ddot{\mathbf{y}}_t, \mathbf{u}_t, \dot{\mathbf{u}}_t, \ddot{\mathbf{u}}_t)$$
(5.72)

can be attempted if the information content of the third derivative is substantial. This can be assessed by computing the second derivatives from sampled data and investigating the following:

Does a sufficient number of different values occur?
Are the different values distributed so that they provide interesting information most of the time?

The operationalization of the third-order model is analogous to the second-order model.

Note 5.4 With more instrumental expense, higher derivatives can be measured with the help of faster sampling and filtering or smoothing techniques. For example, if there are k different values of the second derivative with sampling period Δ_t, their number will increase to ck if the moving average of the last c sampled values is computed with sampling period Δ_t/c. This will introduce a delay of $\Delta_t/2$ into the measurement process. (Other types of recursive filters may introduce a time constant and increase the order of the plant instead of the delay.) In this way, higher-order derivatives may be exploited in input–output models. On the other hand, the delay or order increase will make the plant more difficult to control; it must be checked cautiously whether such additional dynamics are substantial for the given plant.

Note 5.5 Bringing the derivative-based models back to the operational form that uses delayed variables is no contradiction to the preference for using derivatives discussed above. The advantage of using orthogonal derivatives concerns the inputs and outputs of the mapping *directly represented by the neural networks*. Then, the relationship between orthogonality of inputs and output, on the one hand, and the orthogonality of associated network weights

116 PLANT IDENTIFICATION

is established. Using such a neural network in a different framework does not change anything in this principle.

5.3.3 Implementing the Mapping as a Neural Network

This section addresses the question of how to implement the conceptual plant model as a neural network. Two aspects are discussed: (1) implementing the entire plant model as a single neural network or representing each plant model equation individually and (2) implementing the mappings directly or indirectly.

Comparison of Vector and Scalar Functions Whichever model type is taken, a design decision about the global structure of the neural network model has to be made. There are two basic alternatives:

1. Implement the entire equation system, viewed as a vector mapping. For example, implement the second-order input–output model (5.70) as a single neural network (Fig. 5.1).
2. Implement each equation, viewed as scalar function of multiple variables, as an individual network (Fig. 5.2).

The main advantage of the single-network alternative is the possibility for hidden units to learn representations that are common to all equations of the system. In pattern recognition tasks, hidden units can be expected to represent implicit invariances within the pattern space. Dynamic models, partial nonlinearities, or entire hidden states may be identified.

On the other hand, the multiple network scheme is more flexible. In particular, it provides an important means for expressing causality and independence.

Let us first briefly discuss what causality really is. An extensive discussion of causality concepts as well as an axiomatic theory thereof has been presented by Pearl [114]. Some of Pearl's observations are relevant for determining the structure of a plant model. The dependence view, usually implicitly adopted in control theory, is based on the dichotomy between coupled and decoupled

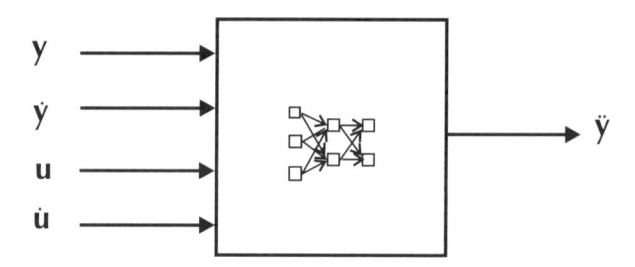

Figure 5.1 Single neural network representation of a plant.

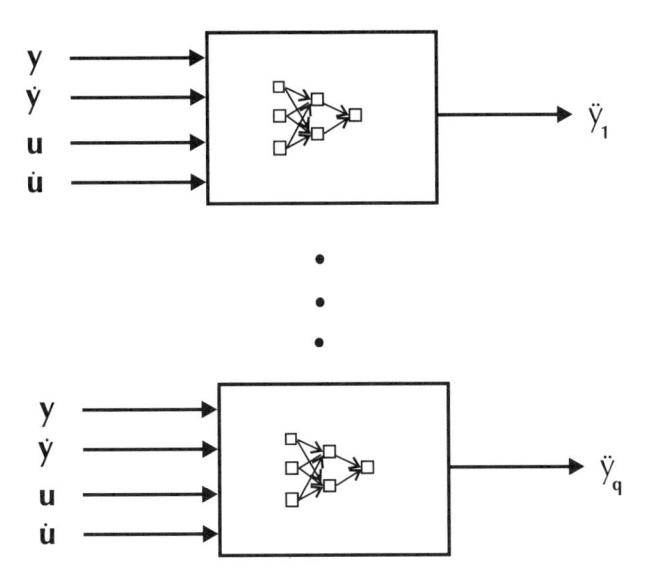

Figure 5.2 Multiple neural network representation of a plant.

variables. From this viewpoint, a variable depends on another if the behavior of the former is influenced by that of the latter. This property is *transitive*; the dependence of X on Z results from X depending on Y and Y depending on Z. By contrast, in Pearl's framework, it is important to investigate, for coupled variables, whether there are other variables that may mediate the dependence. For example, for a triple of variables X, Y, and Z, the dependence of X on Y is investigated in the following way: If the trajectory of X depends on the trajectory of Z and if this dependence, for a given trajectory of Z, is the same with various trajectories of Y, the dependence of X on Y is mediated by Z. In Pearl's terminology, X is then independent from Y, even if Z itself depends on Y. This concept of dependence is *nontransitive*. Obviously, it is the nontransitive dependence that should underlie the design of the structure of a model.

An important contribution of correct model structuring with the help of nontransitive dependences is the reduction of the total quantity of dependences without reducing the expressive power of the model. This can be illustrated by the drive train test bench (Fig. 5.3) of Chapter 13. A drive train consisting of an engine, a gear box, and an axle is fixed on a test bench. The test bench is designed to follow predefined trajectories of axle torque and angular velocity. To reach this, the test bench is capable of manipulating the accelerator lever of the engine and of applying load to the axle. The measurable variables are the load l, the accelerator position a, as well as the torques t and angular velocities w indexed by e for engine, g for gear box, and a for axle. The input to the plant is the reference torque t_r and reference angular velocity w_r. There are obviously no decoupled parts of the plant. With the

118 PLANT IDENTIFICATION

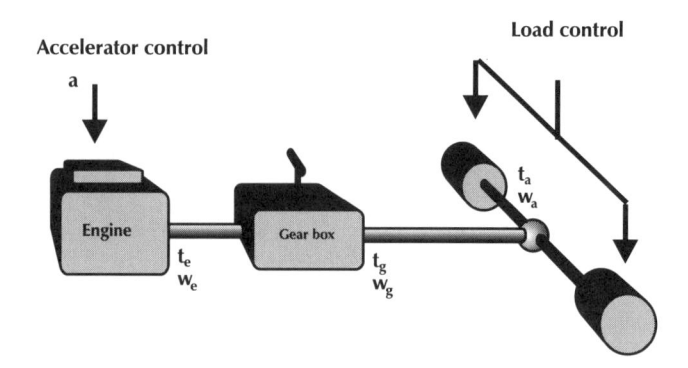

Figure 5.3 Scheme of the drive train test bench.

transitive definition of causality, all variables depend on all others (including themselves, since the models are dynamic) as well as on both inputs—a total of $8 \times 10 = 80$ transitive dependences. On the other hand, it is obvious that the axle is influenced by the engine, but only *via* the gear box. In other words, the axle state does not depend on the engine state if nontransitive concepts are used. With the nontransitive definition, there are six dependences for the engine (i.e., the torque depends on angular velocity, accelerator position, and itself, angular velocity depends on torque, accelerator position and itself), eight for the gear box, ten for the axle, and twelve for the test bench control unit (each of both action variables, accelerator position and load, depends generally on all six state variables of the drive train), a total of 36 nontransitive dependences.

It can be recommended that whenever knowledge of plant structure is available (the author has scarcely experienced the contrary, so far), it should be exploited. For complex problems, the boundary between feasible and infeasible computation may lie between the more expensive unstructured single network model and a more economical structured multiple network model.

Direct and Indirect Mapping Representation The most straightforward way to represent a mapping with the help of neural networks is to identify the input of the mapping with the input layer and the output from the mapping with the output layer.

In Sections 3.2 and 3.3.1, alternative representations have been proposed. Their outstanding property is their ability to represent the zero equilibrium point of the mapping. It has also been stated that mappings resulting from physical laws frequently have this property. One of the proposed mappings has been called gain vector network. It represents an arbitrary mapping

$$\mathbf{z} = f(\mathbf{x}) \tag{5.73}$$

MODEL STRUCTURE 119

with property

$$0 = f(0) \qquad (5.74)$$

in the equally general form

$$z = f'(x)x \qquad (5.75)$$

In addition to the equilibrium point property, there are two further features that make a plant model representation in this form interesting.

The first of them concerns a simplified representation of linear mapping. It occurs if the vector function f' is represented by an asymmetric multilayer perceptron with linear output units A, then multilayer perceptron can be written as

$$g = w_0 + Wh(x) \qquad (5.76)$$

where w_0 is the vector of output layer thresholds, W is the output layer weight matrix, and $h(x)$ is the vector mapping of the input layer on the output of the hidden layer.

Then, Eq. (5.75) can be written as

$$z = [w_0 + Wh(x)]x = w_0 x + Wh(x)x \qquad (5.77)$$

The first summand is linear so that with zero W, a linear mapping is represented. In this way, system representation can be trained successively by (1) training the parameters w_0 representing the linear term and keeping W zero, and (2) training the parameters W to represent the difference from linearity. This procedure is inspired by Taylor's expansion and is closely related to what has been discussed in Section 3.3.1. The first term, constituting the first-order approximation, is determined, followed by the simultaneous computation of an approximation of the remaining terms by multilayer perceptron. The separate computation is justified by the fact that the first term of the Taylor expansion is orthogonal to the remaining ones. This difference-from-linearity representation certainly makes the optimization task easier (see Section 4.3) and improves the chance for success.

The second reason for using representation by gain vector network is that it separates the nonlinear gain from linear dynamics. This provides a flexibility for representing special configurations by using different vectors for (1) input into the network to determine the gain and (2) multiplication by the gain vector:

$$z = f'(x_1)x_2 \qquad (5.78)$$

120 PLANT IDENTIFICATION

The following two examples illustrate what such special configurations may look like.

1. If the plant output depends on control action $\mathbf{u}_t$ only linearly, with nonlinear dependence from plant state or output, the plant inversion can be performed explicitly. This has been proposed by Narendra and Mukhopadhyay [104] for the discrete-time decoupling model of the form

$$\begin{aligned} \mathbf{y}^*_{t+1} = & f_1(\mathbf{y}_t, \mathbf{y}_{t-1}, \ldots, \mathbf{y}_{t-n_o+1}, \mathbf{u}_{t-1}, \ldots, \mathbf{u}_{t-n_o+1}) \\ & + f_2(\mathbf{y}_t, \mathbf{y}_{t-1}, \ldots, \mathbf{y}_{t-n_o+1}, \mathbf{u}_{t-1}, \ldots, \mathbf{u}_{t-n_o+1}) \mathbf{u}_t \end{aligned} \quad (5.79)$$

and Freund [37] for a continuous-item decoupling model of the form

$$\mathbf{y}^* = f_1(\mathbf{x}) + f_2(\mathbf{x})\mathbf{u} \quad (5.80)$$

Algebraic methods for controller design with plant models of this form are addressed in Section 6.7.

2. Plant behavior may be influenced for slowly changing parameters $\mathbf{p}$ such as temperature or age. These plant parameters could be included in the model, but then they would influence only the gains of the model rather than enter the state or output vector. This can be immediately expressed in the gain vector model of the form

$$\dot{\mathbf{x}} = f(\mathbf{x}, \mathbf{u}, \mathbf{p}) \begin{bmatrix} \mathbf{x} \\ \mathbf{u} \end{bmatrix} \quad (5.81)$$

5.4 MODEL QUALITY CRITERION

After the commitment to a certain model structure, the problem arises of formulating an appropriate measure of model quality. A standard answer might be "square the error." However, squaring the error is not the whole answer. The first open question is: Square the error of what? Furthermore, with the squared error standardized, what are the nonstandard measures, and when should they be used? These are the questions discussed in this section.

Let us denote the mapping representing the plant as

$$\mathbf{z}_{t+1} = F(\mathbf{z}_{e_t}, \mathbf{u}_t, \omega_f) \quad (5.82)$$

This mapping has the form of a discrete-time state model, but, for the goals of this section, it should be understood in a more general sense: It stands for any operational model. The vector $\mathbf{z}_{t+1}$ may be either a state vector or an output vector at the next time period $t+1$, that is, *the forecast*. The vector $\mathbf{z}_{e_t}$ may be the state vector or the history of the output vectors; it is only important that it contains information known at time t. The vector $\mathbf{u}_t$ is the input in the general sense because it can comprise the history of inputs. The vector ω_f is the vector of free parameters of mapping f. It is not directly relevant to the topic of this section and will be frequently omitted.

For example, the general framework Eq. (5.82) may be specialized if a second-order input–output model such as Eq. (5.71) is used,

$$\mathbf{y}_{t+1} = 2\mathbf{y}_t - \mathbf{y}_{t-1} + \Delta_t^2 f\left(\mathbf{y}_t, \frac{\mathbf{y}_t - \mathbf{y}_{t-1}}{\Delta_t}, \mathbf{u}_t, \frac{\mathbf{u}_t - \mathbf{u}_{t-1}}{\Delta_t}\right) \quad (5.83)$$

Say we further have a criterion of identification quality that is applied to time steps $1, \ldots, T$, where T is the *identification horizon*

$$C_f = \sum_{t=1}^{T} c_t(\mathbf{z}_t, \mathbf{d}_t) \quad (5.84)$$

Widespread forms of $c_t(\mathbf{z}_t, \mathbf{d}_t)$ are the mentioned squared error

$$\sum_i (z_{i,t} - d_{i,t})^2 \quad (5.85)$$

or absolute value of error

$$\sum_i |z_{i,t} - d_{i,t}| \quad (5.86)$$

that is, the sum of squares or absolute values of deviations between forecast states $\mathbf{z}_t$ and measurements $\mathbf{d}_t$.

In Section 5.4.1, two alternative ways of computing $\mathbf{z}_t$, recursive and non-recursive, are discussed. The topic of Section 5.4.2 is the actual form of the error term. Section 5.4.3 presents the special case where the appropriate error criterion has a completely different framework: the system with a chaotic attractor.

5.4.1 Comparison of Single-step and Multistep Criterion

There are two different ways to compute the forecast $\mathbf{z}_{t+1}$ in Eq. (5.84). The more conventional approach makes forecasts with the help of the last-measured state:

$$\mathbf{z}_{t+1} = F(\mathbf{d}_{e_t}, \mathbf{u}_t) \quad (5.87)$$

The forecast is then a function of $\mathbf{d}_{e_t}$ and $\mathbf{u}_t$. The extended measurement vector $\mathbf{d}_{e_t}$ may be the measured state if a state model is used or the measurement

122 PLANT IDENTIFICATION

history if an input–output model is used. This approach is used in classical explicit formulas for the least squares of autoregressive models [8, 91] or the usual neural network models [e.g., 109].

With the *multistep criterion*, the state forecast is computed with help of the last *model state*, that is, with help of the last forecast or forecast history,

$$\mathbf{z}_{t+1} = F(\mathbf{z}_{e_t}, \mathbf{u}_t) \qquad (5.88)$$

Now, the forecast is a function of $\mathbf{d}_0$ and $\mathbf{u}_s, s = 1, \ldots, t$.

In the former case, a single-step simulation of a neural network is performed (or a sequence of single-step simulations, each taking the measured state $\mathbf{d}_{e_t}$ as a basis). In the latter case, recursive, multistep simulation [e.g., 11, 51, 105] is done. Alternative terms for plant identification with the help of recursive models are output-error or parallel models, in contrast to the model-error or series-parallel approach being the alternative terms for static simulation-based identification [see 101, 108].

In other words, with the multistep criterion, the error is accumulated through the state sequence. In this way, errors that are small but lead to large errors in subsequent states have a particularly large effect on the cost function value. Consequently, the optimization method focuses on reduction of such particularly harmful errors.

To decide what criterion is preferable, it is necessary to take into account the neurocontroller training method. If a method is used that does not use a long-term dynamic forecast of the state, such as the open-loop methods related to plant inversion [104; see also Section 6.5.1 in this book], there are no particular arguments in favor of one or the other approach. Then, the single-step method may gain points by its simplicity.

On the other hand, neurocontroller training methods based on genuine closed-loop optimization (see Section 6.5.1) evaluate the controller training cost function based on a long-term, dynamic simulation of the closed loop made of the plant model and the neurocontroller. In other words, they optimize whole control trajectories. Then, the plant model quality criterion must ensure that the behavior of the controller is identical both in training simulations and in the real plant. So the plant should be able to

> *correctly simulate a whole trajectory of plant states given an initial state and a time series of inputs (e.g., controller actions).*

In other words, a good plant model should be capable of a multistep (typically of the order of several hundred steps) forecast of plant behavior. For criterion of identification quality to be adequate for controller training, we should evaluate such trajectory correspondence. This is exactly what the multistep criterion does.

The most important reason why classical least-squares theory prefers the single-step criterion is certainly that concise expressions of analytic optima can be found in a linear setting. Another reason is that linear plants have a constant gain (or a constant gain matrix). Therefore, the consequences of the same magnitude of deviation (e.g., caused by a measurement error of constant variance, uncorrelated with the states) are approximately equally serious in any state.

By contrast, in nonlinear plant modeling, an error in a low-gain working point leads to much more serious errors in future steps than the same error in a high-gain working point (Fig. 5.4). Explicit a priori weighting of errors is impossible because the model and its gains in particular points are themselves subject to identification. So assessment of the seriousness of the present error is possible only by considering the error of a multistep forecast. This is why the usual single-step least-squares criterion is frequently inadequate in the nonlinear case; plant models trained with a single-step criterion fail to make precise multistep forecasts.

Computational Experiment The hypothesized superiority of the multistep criterion has been confirmed by computational experiments done by this author. One of them has been the drive train identification for the test bench application of Chapter 13. Its identification with the multistep approach has been compared with a traditional neural network identification using a single-step mean-square error. A 500-step simulation (500-step forecasts for a given initial state of state variables and a given series of external inputs or accelerator positions) of the traditional method (Fig. 5.5a) and multistep method with optimization horizons of 30 (Fig. 5.5b) and 300 (Fig. 5.5c) 17-ms time steps shows the superiority of the temporal sequence method. Note that the traditional method did not capture the most important aspect of engine identification, that is, the gain variations of the plant (the slope in steep segments of the charts is poorly reproduced). The multistep method with a horizon of 30 steps provided substantially better results but started to diverge after about 300 steps. Very good correspondence has been reached with a horizon of 300.

The ultimate goal of plant model identification has been to provide a good basis for neurocontroller training. The plant models received by the multistep identification method have led to good neurocontrollers. By contrast, no acceptable results have been achieved with traditional mean-square-based neural identification. For more details on this neurocontrol application, see Chapter 13.

5.4.2 Choice of Error Term

The error term $c_t(\mathbf{z}_t, \mathbf{d}_t)$ will always, in some form, contain the deviation vector $\|\mathbf{z}_t - \mathbf{d}_t\|$. Its concrete form has the following features: weighting deviations of individual variables $z_{i,t}$; power of an error; and measurement correlations.

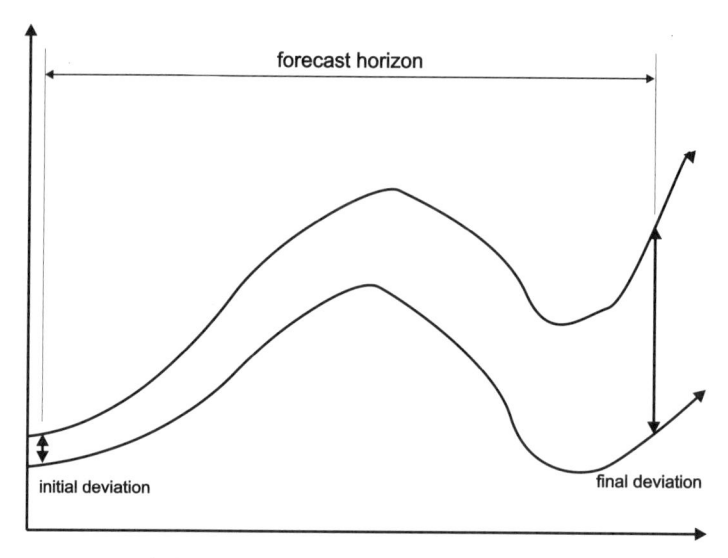

Figure 5.4 Consequences of the forecast error for low- and high-gain working points.

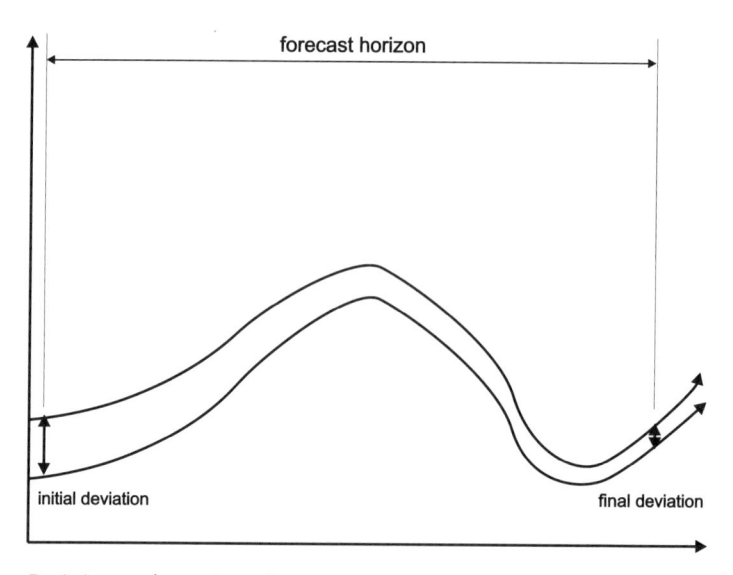

Weighting Deviations of Individual Variables $z_{i,t}$. In theory, a deterministic model can be approximated by a multilayer perceptron or an RBF network. Then, the global minimum of every error measure is zero. In practice, (1) the approximation is imperfect, (2) real-world global optimization algorithms do not guarantee a complete coverage of the parameter space, and

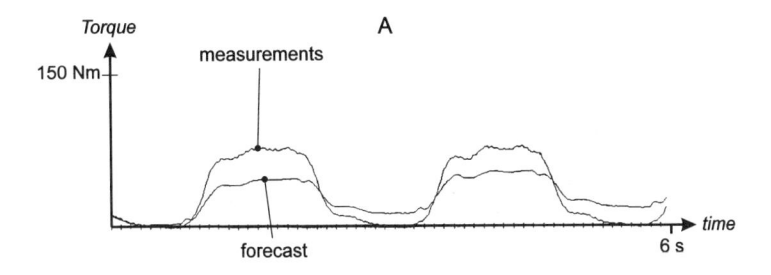

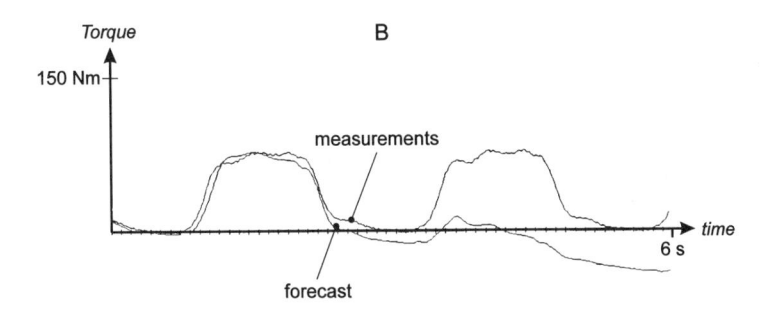

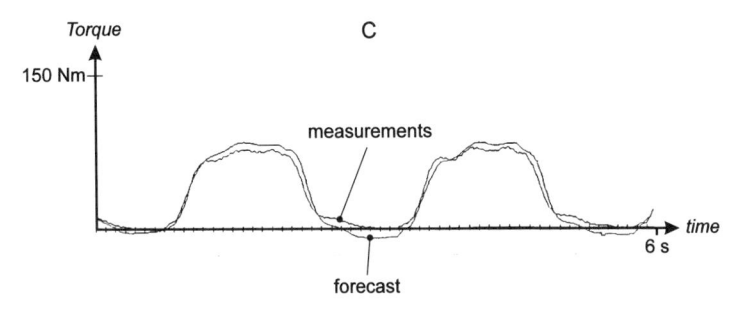

Figure 5.5 Measurement series and a simulation of an engine torque.

(3) measurements are subject to random errors. Then, the minimum is always a trade-off between the forecast errors of individual variables. The optimization error tends to find a solution in which absolute forecast errors are at least of a similar order of magnitude. This may lead to very peculiar solutions if individual variables have very different ranges. A remedy for this is to weight individual deviations, for example, by standard deviations:

$$\frac{\|z_{i,t} - d_{i,t}\|}{\sqrt{\mathrm{Var}(d_i)}} \tag{5.89}$$

Power of an Error The most frequent error type used is the squared error. With second-order numerical optimization methods, this is a good choice, since the cost function is then well approximated by a quadratic form, at least in some neighborhood of the minimum. On the other hand, the squared error is sensitive to measurements outliers. From statistical point of view, using the squared error is justified for maximum-likelihood estimates with a Gaussian distribution of measurement errors. Among the various causes for measurement outliers, there are also real-time software errors producing arbitrary, uniformly distributed, signals. The resulting distribution of all outliers will then exhibit larger tails than Gaussian [117; Section 15.7]. This is an argument in favor of taking the power of an error less than two, for example, one.

Measurement Correlations Knowledge of the correlation between individual measurement errors can be exploited by the error criterion as

$$(\mathbf{z}_t - \mathbf{d}_t)^T \mathbf{C}^{-1} (\mathbf{z}_t - \mathbf{d}_t) \tag{5.90}$$

To assess the practical importance of these three features, the correct scaling is the key factor. Whether the robustness of absolute error trades off the potentially reduced efficiency of optimization remains an open question, and this author's computational experience has not provided an appropriate relevant conclusion. Assumptions about correlations between measurement errors rarely make important contributions to model quality.

5.4.3 Chaotic Systems

To show that the error criterion can take on forms completely different from the squared error, the special case of *chaotic systems* will be considered in this section. The topic is certainly far from being ready for routine industrial applications, but it will illustrate the importance of deliberate construction of error criteria for nonstandard cases.

Chaotic systems are a relatively broad class of nonlinear systems. This class can roughly be characterized by two properties: the boundedness of the state variables and instability. The formal characteristics of the latter property is one or more eigenvalues with absolute value exceeding one (for discrete systems) or one or more eigenvalues with positive real part (for continuous systems). Instability and boundedness are a contradiction for linear systems. For nonlinear systems, coexistence of both properties leads to a folding of the state space: Since an unstable state variable cannot grow arbitrarily because of boundedness, it must change the direction across the state space. Then, it will necessarily come into the neighborhood of past segments of its trajectory.

Chaotic systems are notorious (and feared) for their nonpredictability, which is a direct consequence of state space folding. Their behavior cannot be exactly forecast by identified models, since a deviation in a model's parameters leads to an exponential forecast error [e.g., see 110]. This results from the fact that a small deviation in the initial state x_0 of a chaotic system leads to an exponentially growing deviation in the state x_t for $t \to \infty$. A small deviation in model parameters leads, for an identical initial state, to a small deviation in the state x_1 and has thus the same consequences as a small deviation in the initial state.

However, dissipative systems are not completely unpredictable. If a chaotic system is *dissipative*, its energy (or some abstract equivalent of energy) must decrease. This leads to a particular behavior: The volume enclosing a family of trajectories in the state space shrinks with time [e.g., 130]. As a consequence, even a chaotic dissipative system is attracted to a region of smaller dimension than the complete state space, the so-called *strange attractor*. This attractor is typically a set that is thin (in fact, infinitely thin) in some directions and thick in others. An example of this topology is the Lorenz attractor of Figure 5.6. It consists of two regions that can be viewed as approximately planar: a horizontal cycle and a vertical cycle.

This property of dissipative systems makes a certain forecast of the system behavior possible. We cannot know where the system is exactly, but we can try to:

- Identify the attractor; that is, look for a model whose attractor has similar topology to that of the real system
- Determine the position along thin, predictable dimensions

Because practically all technical, and many natural, systems are dissipative, solving this problem would bring a substantial impetus to our ability to forecast the behavior of nonlinear systems.

The problem with all *additive* mean square criteria is that the inherently necessary error (given by the unpredictability of the exact position within the strange attractor) in dimensions where the attractor is thick may prevent minimization of the error in other thin dimensions. The reason for this is that even the smallest casual improvement along the former dimension (e.g., caused by a minor local adaptation to the actual measurements) may be numerically larger than a substantial improvement along the latter. This suggests the use of nonadditive minimizing criteria.

Error Volume Criterion Dissipative systems are characterized by the contraction of an initially n-dimensional phase space to a (possibly chaotic) attractor of dimension lower than n. The intuition behind the following developments is that

> although the forecast error cannot be reduced down to zero, it can be reduced to fill a volume of dimension lower than n.

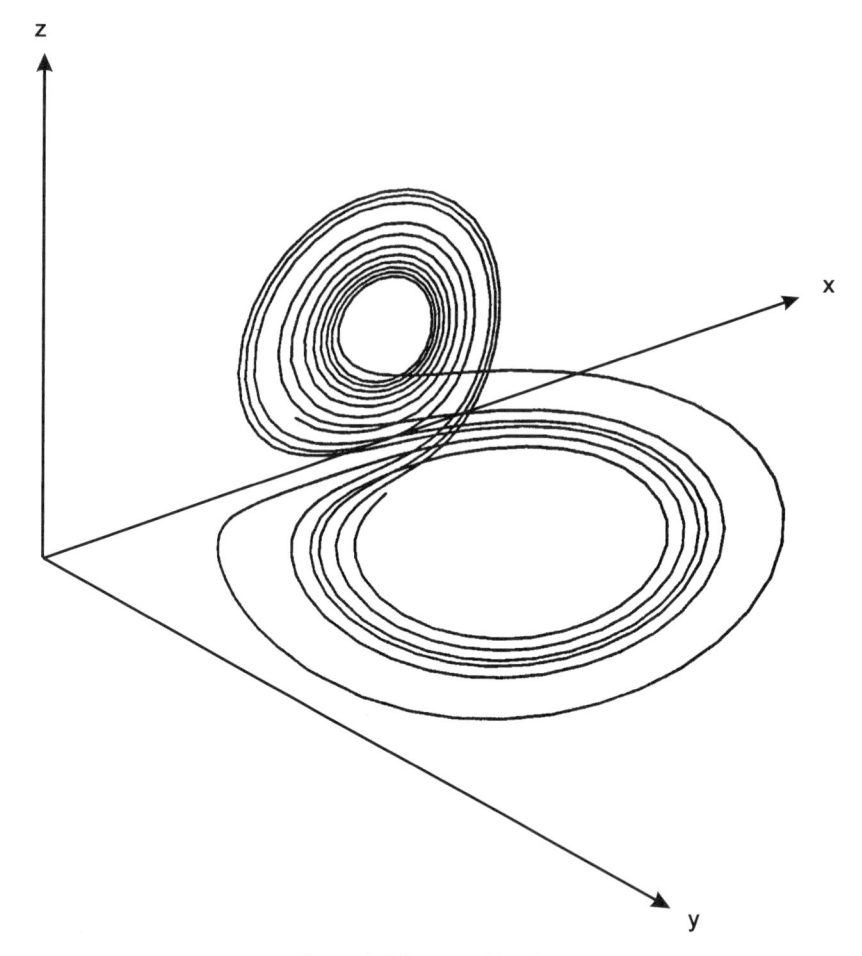

Figure 5.6 Lorenz attractor.

This is why an optimization criterion appropriate for identifying chaotic dissipative systems must have the following, volumetric properties:

1. A large deviation in one direction can be offset by a small deviation in another direction, in a multiplicative way
2. These directions need not coincide with individual state variables

The second criterion is more important because the flat dimension of a strange attractor may be oriented arbitrarily. This excludes the use of a simple multiplicative criterion of the form

$$\prod_i (z_i - d_i)^2 \qquad (5.91)$$

To determine whether the forecast errors fall into a set with a certain n-dimensional volume, n independent forecasts are necessary. Such forecasts can be received by slightly perturbing the initial state in n directions, for example, in the directions of the n unit vectors. This amounts to perturbing each of n state variables z_i by a small δ_i.

The appropriate volume measure is the determinant of the matrix of positions $z_{i,j}$, $i,j = 1,\ldots,n$ of n multistep forecasts with slightly perturbed initial conditions, with i being the state variable index and j the perturbation index

$$V = \det\left[z_{i,j} - d_i\right] \tag{5.92}$$

If the perturbations are orthogonal and of the same norm d, they span, together with their origin, the nonperturbed initial state, a hypercube of dimension n and side d. The determinant is equal to the volume of the n-dimensional parallelogram defined by the coordinate origin (the measurement with zero error) and the n errors of perturbed forecasts.

This criterion has the following properties, resulting from the properties of determinants:

1. It is proportional to the error of every state variable (assuming that this error is of similar size for all initial condition perturbations).
2. It is zero if all errors are in a hyperplane, that is, if the model and the real system are in the same flat attractor. This can also be seen by considering the equation (for the three-dimensional case)

$$\begin{bmatrix} x_1 & y_1 & z_1 & 1 \\ x_2 & y_2 & z_2 & 1 \\ x_3 & y_3 & z_3 & 1 \\ 0 & 0 & 0 & 1 \end{bmatrix} \begin{bmatrix} a \\ b \\ c \\ d \end{bmatrix} = \begin{bmatrix} a \\ b \\ c \\ d \end{bmatrix} \tag{5.93}$$

which describes the condition that the points $[x_i, y_i, z_i], i = 1, 2, 3$ (the error coordinates for three initial state perturbations) and the point $[0, 0, 0]$ (the origin for errors) are all in a plane described by the equation

$$ax + by + cz + d = 0 \tag{5.94}$$

This homogeneous linear equation has a solution for variables a, b, c, d iff the determinant of this matrix is zero. By expanding along the last row, this determinant can be reduced to

$$\begin{vmatrix} x_1 & y_1 & z_1 \\ x_2 & y_2 & z_2 \\ x_3 & y_3 & z_3 \end{vmatrix} \tag{5.95}$$

which is identical with Eq. (5.92).

Criterion (5.92) is zero if the errors with n perturbated initial states are in a hyperplane of dimension lower than n. Unfortunately, this is also the case for a trivial forecast consisting of a fixed constant value. (This fixed value, forecast for all perturbations, and the measured value d_i enclose a set of dimension one.)

Let us define D as the radius of the smallest sphere within which are found all forecasts. For the trivial fixed-point forecast mentioned above D is zero (the sphere contains just the fixed point). For a strange attractor, D cannot be reduced below a certain lower bound given by the spatial extension of the attractor. For nonchaotic, genuine point attractors, D shrinks to zero for the optimum forecast.

To cancel the undesirable minimum connected with the trivial forecast, let us seek an error criterion of the form

$$E = \frac{|V|}{D^k} \quad (5.96)$$

For a strange attractor and a nontrivial forecast, D has a positive lower bound D_0, and the minimum of $|V|$ is also a minimum of E for arbitrary k.

The following constraints apply to k:

1. For a trivial forecast in the strange attractor case, both $|V|$ and D^k will converge to zero during optimization. For small D, $|V|$ will shrink in all but one dimension and thus develop with D^{n-1}. For E not to converge to zero, $|V|$ has to shrink slower than D^k. This imposes the constraint $k > n - 1$.

2. For point attractors, and also for forecasts of chaotic systems over the time horizon during which chaotic behavior has not developed, convergence is desirable. The error volume contracts to a zero-dimensional point, developing with D^n. To enable the (desirable) convergence of E to zero, $|V|$ has to decrease faster than D^k. This implies the constraint $k < n$.

Both constraints can be summarized to

$$n - 1 < k < n \quad (5.97)$$

Equations (5.96) and (5.97) constitute the ultimate form of the error volume criterion.

The identification algorithm minimizes criterion (5.96). Forecasts $z_{i,j}$ in Eq. (5.92) are multistep forecasts with a time horizon sufficiently long for nonlinear behavior to fully develop.

Note 5.6 The arguments of this section are based on the assumption that the attractor is locally approximated by a hyperplane. Of course, general strange attractors are no hyperplanes. As an example, let us, once more, observe the Lorenz attractor of Figure 5.6. The error volume criterion will work as long as both the measurement and the forecasts are either both in the horizontal cycle or both in the vertical cycle. Then, it is possible to forecast the flat dimension. However, the divergence can go so far that, for example, the measurement is in the horizontal planar region and the forecast in the vertical one. In this case, the chaos is complete, and even the error volume criterion cannot be expected to help. However, it is then still possible to model the attractor, that is, to look for a model whose strange attractor (as a state space subset) is a good approximation of that of the real system. A (frequently satisfied) precondition is that the system be structurally stable; that is, a small perturbation in model parameters does not change the qualitative type of the attractor. All that can be done is *to extend the horizon of the possible forecast* by the use of the error volume criterion.

Computational Experiments The concepts discussed in this section are particularly illustrative if applied to the Roessler model described by the equations

$$\dot{x} = -y - z$$
$$\dot{y} = x + ay \qquad (5.98)$$
$$\dot{z} = bx - cz + xz$$

with parameter values $a = 0.38, b = 0.3, c = 4.82$. For this parameter combination, the Roessler model has a strange attractor (Fig. 5.7). Moreover, with this parameter combination, the system is in the proximity of a homoclinic point [110], in which chaotic trajectories pass through a small bottleneck. In Figure 5.7, the homoclinic point is in the center of the circular attractor region in the xy plane. This configuration makes the forecast particularly difficult. The results are given in Figures 5.8 and 5.9. Figure 5.8 shows the forecast of variable Z for 1000 time periods that are 100 ms in length with an additive multistep criterion (the bold line). Figure 5.9 shows the same for the criterion described by Eq. (5.96). Obviously, some parts of the attractor are flat in the Z direction, whereas others consist of peaks of chaotic height. A successful forecast should identify the flat parts of the trajectory with great precision, while the height of the peaks remains after reaching a certain unpredictable time horizon.

It can be seen that although the additive criterion model loses contact with reality after approximately 350 steps (site I), the error volume criterion model loses the contact only temporarily, rejoining the measurement after approximately 380 steps (site J). In other words, the error volume criterion model fails to forecast only the peak between sites I and J. The forecast deteriorates only

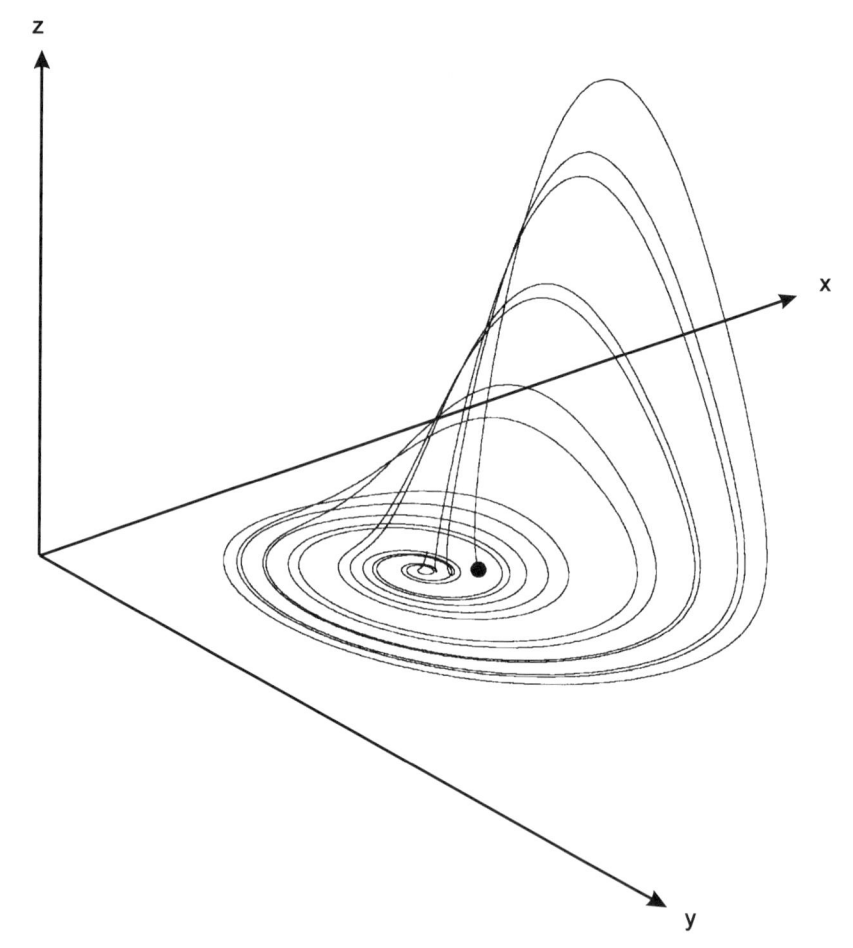

Figure 5.7 Roessler attractor.

after approximately 600 steps (site K), but once more correctly identifies the peak at the 800th step (site L).

To measure the quality of the identification of the attractor itself, viewed as a subset of state space, the state space has been divided into 18,018 cells, and the visits of the real system and of the model in individual cells have been counted. For each cell, the minimum of both counts has been determined. The sum of all minima, if related to the total count of iterations, is a simple measure of correspondence of both attractors: the sum is equal to the iteration count if both attractors are identical within the given observation grain and is zero if the attractors are completely disjoint. In other words, it is a measure of intersection of both attractors. For the additive squared error, the intersection is only 63.87%, whereas for the error volume criterion it is 83.20%.

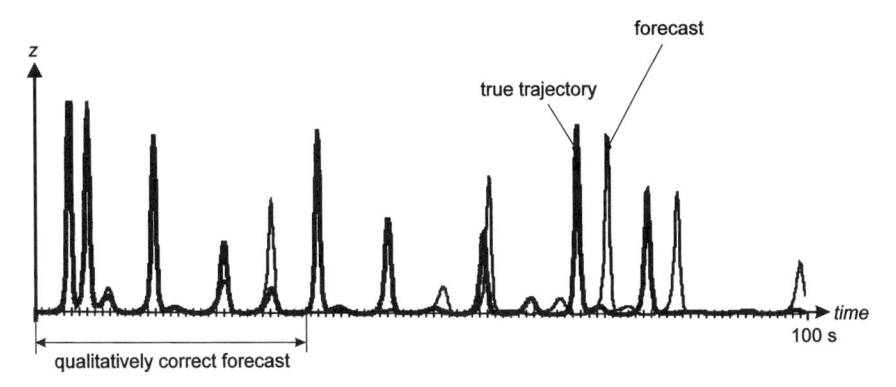

Figure 5.8 Roessler system identification—additive squared error.

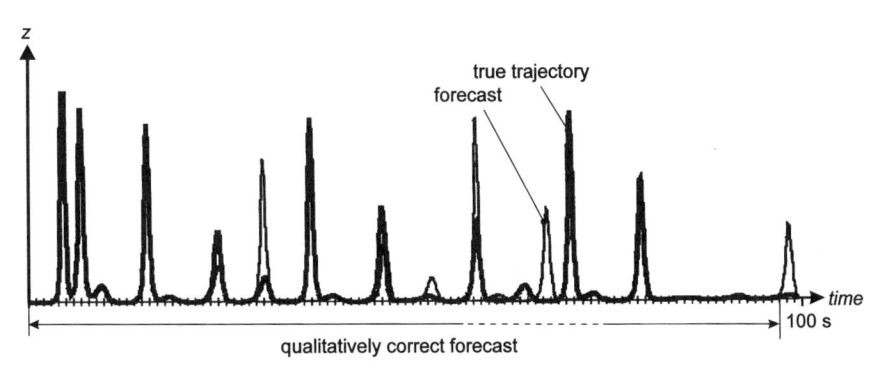

Figure 5.9 Roessler system identification—error volume criterion.

5.5 ALGORITHMS FOR IDENTIFICATION

The task of an identification algorithm is to minimize the cost function

$$C_f = \sum_{t=1}^{T} c_t(\mathbf{z}_t, \mathbf{d}_t) \tag{5.99}$$

with vectors $\mathbf{z}_t$ computed as

$$\mathbf{z}_{t+1} = F(\mathbf{z}_{e_t}, \mathbf{u}_t, \omega_f) \tag{5.100}$$

Terms $c_t(\mathbf{z}_t, \mathbf{d}_t)$ usually have the form

$$\sum_i (z_{i,t} - d_{i,t})^2 \tag{5.101}$$

Model (5.100) is an operational model. It represents the complete simulation from time t to time $t+1$.

134 PLANT IDENTIFICATION

The algorithms are different for incremental schemes, on the one hand, and batch schemes, on the other. The basic steps for incremental and batch schemes are presented in Sections 5.5.1 and 5.5.2, respectively. Analytic gradient computation is treated as a separate topic in Section 5.5.3.

Besides the stability of the plant model, the stability of the identification algorithm must be considered. An unstable identification algorithm may lead to a deterioration instead of improvement in model quality. Theoretical results for general nonlinear systems are still not available, but analogies to linear systems show the trends briefly discussed in Section 5.5.4.

A comparison of incremental and batch schemes is given in Section 5.5.5.

5.5.1 Incremental Schemes

A general incremental scheme for plant identification using neural networks is given in Figure 5.10. It is characteristic for incremental plant identification schemes that changes of plant parameters are done after each sample. (This can be modified to making changes after a small number of samples.) In this way, the information contained in the sample is incorporated into the weight change, and the sample is discarded.

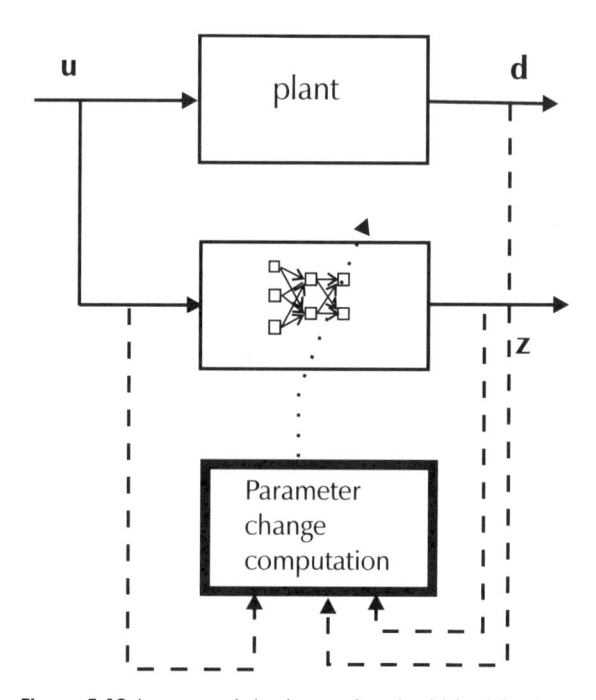

Figure 5.10 Incremental scheme for plant identification.

Note 5.7 An exception to the rule of changing plant parameters after each sample are incremental schemes working with a fixed training set. Arguments against such schemes are briefly discussed in Note 4.8.

More formally, the algorithm consists of the following steps, iteratively performed at each sampling period t:

1. Measurement vector **d** is sampled.
2. The current model, represented by a neural network, is simulated from time $t - 1$ to t, using as input the extended measurement vector at time $t - 1$, that is, $\mathbf{d}_{e_{t-1}}$ [with the static version shown by Eq. (5.87)] or the past value of the extended state vector $\mathbf{z}_{e_t}$ [with dynamic version shown by Eq. (5.88)].
3. The error term $c_t(\mathbf{z}_t, \mathbf{d}_t)$ is evaluated.
4. The change of model parameters is computed. This is usually done using a gradient of c_t with regard to the model parameter vector:

$$\Delta \omega_f = -\alpha \nabla_{\omega_f} c_t \qquad (5.102)$$

where α is a constant or variable learning rate. A more sophisticated way to use the gradient is Kalman training (Section 4.4).
5. New parameter vector is computed.

5.5.2 Batch Schemes

Batch schemes differ from incremental schemes in the way the information supplied by individual measurements is summarized to reach an optimum covering the whole state space.

In contrast to incremental schemes, batch procedures first collect data of sufficient quantity and then perform the minimization of criterion (5.23) summed over all measurements. The structure of a typical batch plant identification is given in Figure 5.11. Here simulation of the model and generation of model forecasts **z** are done within the optimization algorithm block, according to the needs of the particular optimization algorithm used.

The steps of a batch procedure are basically not iterative (although the optimization algorithm of Step 4 may, or may not, be iterative):

Step 1 One or more measurement time series of appropriate length are collected.

Step 2 A procedure for evaluating cost function (5.99) as a sum over the time series length is made available. The cost function definition can imply static (5.87) or dynamic (5.88) computation.

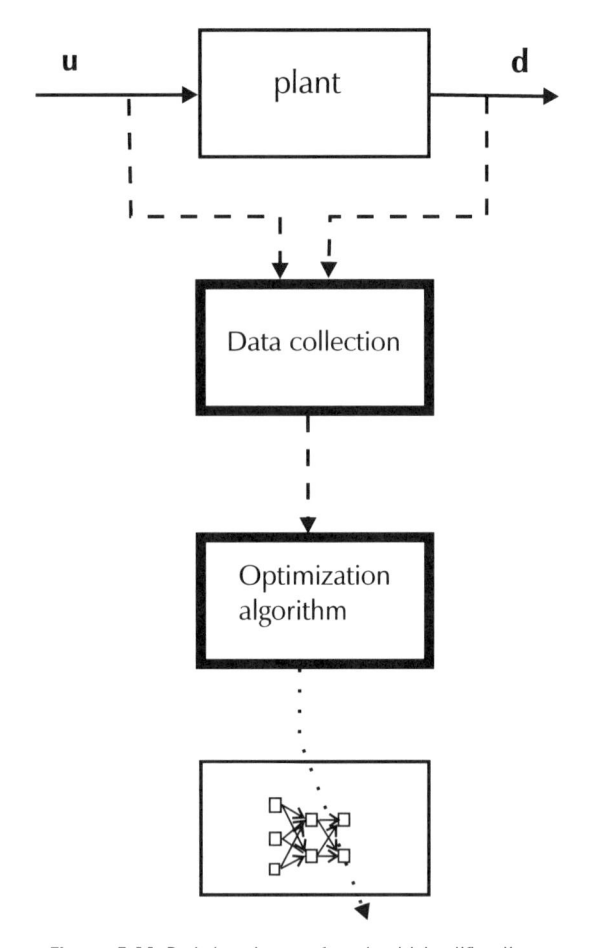

Figure 5.11 Batch scheme for plant identification.

Step 3 If necessary, a procedure for gradient computation is made available.

Step 4 An optimization algorithm using both procedures is started.

Usually, multiple measurement time series are given. Each of them defines matrix $\mathbf{D}_k, k = 1, \ldots, K$ with rows $\mathbf{d}_{k,t}$ for $t = 0, \ldots, T_k$. Let the corresponding inputs constitute matrices $\mathbf{U}_k$ with rows $\mathbf{u}_{k,t}, t = 0, \ldots, T_k - 1$. Then, Eq. (5.99) can attain the form

$$C_f = \sum_{k=1}^{K}\sum_{t=1}^{T_k} c_t(\mathbf{z}_{k,t}, \mathbf{d}_{k,t}) = C_f(\mathbf{D}_1, \ldots, \mathbf{D}_K, \mathbf{U}_1, \ldots, \mathbf{U}_K, \omega_f) \qquad (5.103)$$

ALGORITHMS FOR IDENTIFICATION   **137**

The latter term expresses that the cost function C_f depends only on the measurement series, the input series, and the parameter vector.

The computed states $z_{k,t}$ do not occur because they are a function of the initial state $z_{e_{k,0}}$, which is identical to the measurement in period 0, $\mathbf{d}_{e_{k,0}}$, and the input series $\mathbf{U}_k$. Recall that index e is used to denote states and measurements extended by the measurement history up to the value of the observability index if a higher-order input–output model is used.

5.5.3 Gradient Computation

Numeric Gradient With a batch scheme, a numeric gradient can be used. In the simplest case, it is computed as a vector

$$\nabla_{\omega_f} C_f = \left[\frac{C_{f_i} - C_{f_0}}{\Delta_i}\right] \tag{5.104}$$

with

$$C_{f_i} = C_f(\mathbf{D}_1, \ldots, \mathbf{D}_K, \mathbf{U}_1, \ldots, \mathbf{U}_K, [\omega_1, \ldots, \omega_i + \Delta_i, \ldots, \omega_o]) \tag{5.105}$$

and

$$C_{f_0} = C_f(\mathbf{D}_1, \ldots, \mathbf{D}_K, \mathbf{U}_1, \ldots, \mathbf{U}_K, [\omega_1, \ldots, \omega_i, \ldots, \omega_o]) \tag{5.106}$$

(for more, see Section 4.1.3). For recommended values of Δ_i, see Eqs. (4.20) and (4.21). The numeric gradient is computationally more expensive. However, it has two decisive advantages over the analytic gradient. The first is that it is perfectly consistent with the function value even if the (preferable) multistep model error criterion with dynamic simulation is used. For larger sample horizons T_k, numeric roundoff and truncation errors may lead to the analytic gradient pointing in a direction where the function C_f does not improve at all. This will probably upset all optimization methods.

Another advantage is its simplicity. Compare formula (5.104), which is universal for all model types, network types, and even analytic or partially analytic models, with the formulas the next subsection, which incompletely suggests how to compute the analytic gradient with a single network type!

Analytic Gradient The cost function in batch schemes is shown by Eq. (5.103). It is a sum of terms c_t. Therefore, the gradient can be computed as a sum of gradients of c_t, such that

$$\nabla C_f = \sum_{k=1}^{K} \sum_{t=1}^{T_k} \nabla c_t(\mathbf{z}_{k,t}, \mathbf{d}_{k,t}) \tag{5.107}$$

In incremental schemes, neither the cost function expressed by Eq. (5.99) nor its gradient are evaluated directly because the samples are not stored. Instead, the gradient is computed for every c_t and used to perform parameter change.

For both cases, the gradient formulas for partial error c_t are needed. In the case of a batch scheme, they are summed up correspondingly. The gradient of partial error c_t is a vector of derivatives

$$\frac{\partial c_t}{\partial \omega_{jk}} \quad (5.108)$$

Because c_t is a chain of nested functions of ω (the index f with ω will be omitted in this subsection), the derivatives are to be computed backward through this chain. Part of this computation is advantageously organized in a widespread backpropagation algorithm [126, 146].

With static computation of states z_t, the chain of nested function is the following:

1. The output of neural network is a function of network parameters ω.
2. The model output z_t is a function of the output of a neural network. If a discrete-time model is used and directly represented by a neural network, this function is trivial. If, for example, the first-order input–output model (5.69) with mapping f is directly implemented as a neural network, then the model output is proportional to the neural network output by factor Δ_t. A more complex relationship occurs if a gain vector network, as in Eq. (5.75), is used.
3. The partial cost function c_t is a (usually quadratic) function of model output z_t.

The computation of gradient is done in an inverse order:

1. The derivative of c_t with regard to model output is computed.
2. The derivative of c_t with regard to network output is computed.
3. The derivative of c_t with regard to network weights is computed.

For illustration, let us consider the simplest case of network output being identical with model output vector. The backpropagation scheme for a gradient computation consists in repeatedly applying the of chain rule to computing derivatives. The recursion proceeds backward (i.e., from output layer toward input layer) over network layers. For example, let us have a perceptron with input vector α, output vector γ, and a single hidden-layer vector $\mathbf{h}$. Let derivatives $\partial c_t / \partial \gamma_j$ of function c_t with regard to network output vector γ be as already computed. The input vector α of the perceptron will consist of past plant outputs and inputs. The output vector γ will be the model output or a numeric derivative thereof. (With the gain vector network, the network output would be the vector of gain coefficients.)

Derivatives $\partial c_t / \partial w_{jk}$ of function c_t with regard to output layer weights w_{jk} are

$$\frac{\partial c_t}{\partial w_{jk}} = \frac{\partial c_t}{\partial \gamma_j} h_k \tag{5.109}$$

The derivatives with regard to the hidden layer outputs are

$$\frac{\partial c_t}{\partial h_k} = \sum_j \frac{\partial c_t}{\partial \gamma_j} w_{jk} \tag{5.110}$$

If a sigmoid function hidden layer is considered, with activation function

$$h = s(y) = \frac{1}{1 + e^{-y}} \tag{5.111}$$

and derivative

$$\frac{dh}{dy} = h(1 - h) \tag{5.112}$$

then, the partial derivative with regard to the hidden layer inputs is

$$\frac{\partial c_t}{y_k} = \sum_j \frac{\partial c_t}{\partial \gamma_j} w_{jk} h_k (1 - h_k) \tag{5.113}$$

Finally, the derivatives with regard to the hidden layer weights are

$$\frac{\partial c_t}{\partial v_{ki}} = \sum_k \frac{\partial c_t}{\partial y_k} \alpha_i = \sum_k \sum_j \frac{\partial c_t}{\partial \gamma_j} w_{jk} h_k (1 - h_k) \alpha_i \tag{5.114}$$

Formulas (5.109) and (5.114) are used for gradient computation.

Derivatives $\partial c_t / \partial \gamma_j$ are computed in dependence of the specific error function. If the network output γ directly corresponds to a state vector $\mathbf{y}$ and $c_t = \sum_j (d_j - y_j)^2$, they are

$$\frac{\partial c_t}{\partial \gamma_j} = 2(y_j - d_j) \tag{5.115}$$

For $c_t = \sum_j |d_j - y_j|$, they are

$$\frac{\partial c_t}{\partial \gamma_j} = \text{sign}(y_j - d_j) \tag{5.116}$$

The dynamic scheme (see the series model of Narendra [101] and also multistep criterion of Section 5.4.1) differs from the static scheme by introducing a recursive dependence of model outputs on previous model outputs.

1. The output of a neural network is a function of the network parameters ω and of previous model outputs.
2. The model output z_t is a function of the output of a neural network and of previous model outputs.
3. The partial cost function c_t is a function of model output z_t.

The computation of gradient is done in an inverse order:

1. The derivative of c_t with regard to model output at time t is computed.
2. The derivatives of $c_s, s \geq t$ with regard to model output at time t are computed and summed up. For this, derivatives of network output $z_s, s > t$ with regard to network output z_t are used.
3. The summed derivative of $c_s, s \geq t$ with regard to network output is computed.
4. The summed derivative of $c_s, s \geq t$ with regard to network weights is computed.

First, let us consider a first-order model without delays. Let the Jacobian matrix of plant model in time t be

$$\mathbf{K}_t = \left[\frac{\partial z_{t+1,i}}{\partial z_{t,j}} \right] \tag{5.117}$$

The vector of partial derivatives of c_t with regard to z_t, $[\partial c_t / \partial z_{ti}]$ is denoted as $\mathbf{g}_t$. The derivatives of c_t with regard to $\mathbf{z}_t$ for all $t = 1, \ldots$ can be determined by the recursive formula

$$\frac{\partial e}{\partial \mathbf{z}_t} = \mathbf{g}_t + \mathbf{K}_t \frac{\partial e}{\partial \mathbf{z}_{t+1}} = \sum_{s=t}^{T} \prod_{r=t}^{s-1} \mathbf{K}_r \mathbf{g}_s \tag{5.118}$$

Note that the computation via a Jacobian matrix is not an efficient way to implement Eq. (5.118). Rather, the multiplication by the Jacobian matrix must be viewed as an operator acting upon a gradient vector. It can be implemented in an analogy to the static backpropagation formulas

$$\frac{\partial c_t}{\partial z_{ei}} = \sum_k \frac{\partial c_t}{\partial y_k} v_{ki} = \sum_k \sum_j \frac{\partial c_t}{\partial z_j} \omega_{jk} h_k (1 - h_k) v_{ki} \tag{5.119}$$

It has been, once more, assumed that the plant model is represented directly by a multilayer perceptron. This scheme has been called *backpropagation through time* [145]. An alternative scheme is *dynamic backpropagation* [105].

For higher-order input–output models, multiple Jacobian matrices

$$\mathbf{K}_{t,k} = \left[\frac{\partial z_{t+1,i}}{\partial z_{t-k,j}}\right] \qquad (5.120)$$

for $k = 0, \ldots, p-1$ with p being the order of the model, arise. Then, the recursive formula (5.118) becomes

$$\frac{\partial c_t}{\partial \mathbf{z}_t} = \mathbf{g}_t + \sum_{k=0}^{\min(p-1, T-1-t)} \mathbf{K}_{t,k} \frac{\partial c_t}{\partial \mathbf{z}_{t+k+1}} \qquad (5.121)$$

With

$$\sum_{s \leq t} \frac{\partial c_s}{\partial z_{t,i}} \qquad (5.122)$$

substituted for $\partial c_t/\partial z_{t,i}$, formulas (5.114) and (5.109) can be applied in an analogous way.

5.5.4 Stability of Adaptive Algorithms

Stability of Incremental Schemes Stability is a topic of crucial importance for incremental schemes. A globally stable incremental scheme is such that whatever a sequence of measurements the plant produces (provided the plant is sufficiently excited in a sense that can be made mathematically precise, at least for linear plants), the prediction error converges to zero. Lack of global stability may lead to situations in which some measurement sequences, for example, those along particularly turbulent trajectories, lead to such large changes of network weights that the model forecasts become worse for all working points except for the present one, or even for all working points including the present one.

So far, besides globally stable adaptive rules for linear systems [103], few rules have been found for nonlinear neural-network-based plant identification. One of them is the adaptive rule of Narendra [102] working with RBF networks. In the case of a single input and a single output, of a relative degree equal to one and of an observability index n_o, the plant model is assumed to have the following invertible form:

$$y_{t+1} = \sum_i \alpha_{i=0}^R r_i(y_t, \ldots, y_{t-n_o-1}) + \sum_{j=0}^{n_0-1} \beta_j u_{t-j} = \omega \mathbf{x} \qquad (5.123)$$

where ω is the concatenated vector of both α_i and β_j and $\mathbf{x}$ is the concatenated vector of r_i and u_j. This model is linear in coefficients weighting the outputs of

RBF units r_i, and in plant inputs u_t. Adaptation is limited to the coefficients α_i and β_j, which are linear with regard to the output. The adaptive rule

$$\delta \omega_{t+1} = -\epsilon \frac{(\mathbf{y}_{t+1} - \mathbf{d}_{t-1})\mathbf{x}}{1 + \|\mathbf{x}\|^2} \qquad (5.124)$$

assumes that the parameters of RBF units are fixed. The output $\mathbf{y}_{t+1}$ is computed with the help of past measurements $\mathbf{d}_{t-j}$ substituted into Eq. (5.123), that is, it is a single-step rule.

The adaptive rule expressed by Eq. (5.124) is, in fact, the gradient rule with learning rate $\epsilon/(1 + \|\mathbf{x}\|^2)$. The formulation is sufficiently general to be applied to other configurations with linear output layers, such as a multilayer perceptron with a single hidden layer and linear output layer. For adaptation of the RBF layer itself, no stable rule has been proposed. For low-dimensional network input, the RBF layer can be fixed, for example, with RBF unit centers set to nodes of a multidimensional grid. In practical terms, the dimension of input would have to be as low as two or three. This might be, for example, an input–output plant model with observability index one, one input variable, and one or two output variables. For higher dimensions of network input, the RBF layer would have to be determined by other principles, such as self-organization.

The fact that the rule (5.124) is closely related to the usual gradient rule can justify an intuitive expectation that adaptive gradient rules are stable within a particular basin of attraction if the rules have

1. A single-step criterion, that is, a series-parallel model
2. A bounded norm of the hidden layer output
3. A linear output layer
4. A learning rate of hidden layer parameters substantially lower than that of the output layer

However, global stability would require convergence throughout the parameter space. This can be reached using the gradient method only if the cost function is convex, which is rather the exception with nonlinear neurocontrol problems. This is no contradiction to the generality of rule (5.124); the assumption of a fixed hidden hidden layer or its very slow learning may be the obstacle in reaching the global error minimum.

In addition to global stability of the idealized continuous adaptation, further sources of instability are in the discrete implementation of the algorithm. Gradient search uses discrete steps in the direction of the steepest gradient. For very small steps, the convergence of the algorithm is slow. For very large steps, an overshoot of the error minimum is possible; the procedure is then unstable even if its continuous counterpart is stable. If the error is a reasonably smooth function of network parameters, it can frequently be viewed, at least in the neighborhood of the error minimum, as a quadratic. Then, both convergence speed and stability can be substantially improved by using second-order information, that is, a Hessian matrix in the deterministic case or a correlation matrix in the stochastic case. Incremental schemes for second-order gradient descent have been proposed by Singhal and Wu [133] and used by Puskorius and Feldkamp [120].

The third source of instability of incremental schemes is the stochastic character of measurements. Measurement errors arise for many reasons; one omnipresent source is the discretization error caused by analog–digital converters. So incremental schemes for error minimization try to fix the minimum of a function with information received by single, randomly distorted samples. As stated in Section 4.4, the goal is to reach stable weight, which is the case for stochastic approximation. Stochastic approximation theory [25, 73, 124] states, in commonsense terms, that if the network parameters are to converge to stable fixed values, the step size for the gradient method must be a decreasing sequence such that

The steps decrease slowly enough to be able to pass the distance to the error minimum

The steps decrease rapidly enough to become zero, that is, to no longer react to the stochastic perturbations in the measurement, once the minimum is reached

One possible set of mathematical conditions imposed on the step size Δ_t is

$$\sum_{t}^{\infty} \Delta_t = \infty$$
$$\sum_{t}^{\infty} \Delta_t^2 < \infty$$
(5.125)

Stochastic approximation theory says nothing about the speed of convergence and thus is inappropriate for literal implementation by mortal users. Some results from pattern recognition theory [139] concern learning algorithms with maximum speed in approaching the error minimum. It is interesting to point out that these algorithms frequently do not satisfy the stochastic approximation conditions and are thus unstable in the stochastic approximation sense. This makes it clear that stability and convergence speed are contradictory objectives in incremental error minimization working with stochastic measurements.

Note 5.8 The maximum convergence speed addressed above concerns the maximum *within* incremental algorithms. The factors influencing convergence speed of incremental and batch algorithms are discussed in Section 4.4.

Stability of Batch Schemes For incremental schemes, stability has been a key point of interest. Let us recall the three sources of instability:

1. Lack of global stability of an idealized continuous model
2. Changing network parameters in discrete steps
3. Random disturbances in measurements

What about the stability of batch schemes? In an incremental scheme, the genuine value of cost function (5.23) cannot be computed because the past measurements are forgotten. An exception would be incremental schemes that store the measurements, which is possible, but then, the incremental scheme provides no advantage over truly batch schemes. By contrast, a batch optimization algorithm can evaluate the cost function whenever necessary. This has the immediate consequence that the last sources of instability, the random disturbances, are no threat for batch schemes. The only difference from the completely deterministic case is that the error minimum is equal to the noise variance (for the quadratic error criterion) rather than to zero.

A further consequence of the availability of the cost function value is that the optimization algorithm can always figure out whether a new solution is *globally better* than the the last one and always has the option not to make a step. A more sophisticated variant used by sophisticated local optimization algorithms is the search for a optimum step size by line search in a given direction (e.g., the direction of gradient). In this sense, also the first two sources of instability are harmless for batch schemes.

This encouraging statement is true provided one observes two caveats. In repeated applications of a batch scheme, and particularly in the process of updating data used in individual batch computing sessions, it is important for the stability that the collection of individual data batches be increasingly representative for the whole state space. Otherwise, a sequence of singular working-point-specific models would be received. Such models arbitrarily vary in dependence on the data batch. Consequently, the data-updating algorithm must take care of improving the coverage of state space by the consecutive data batches. (This requirement is something of a tautology: It makes no sense to apply a batch scheme repeatedly if we expect later iterations not to bring about better models than the earlier ones.) This improved coverage can be trivially reached if past measurements are never discarded. However, this will clearly lead to a training set growing without bounds. This question is briefly discussed in Chapter 8.

The second caveat concerns the optimization method. Most popular optimization methods are local. They start from a prespecified point and converge to a particular minimum, in whose attractor is found the initial point. Obviously, there is an element of arbitrariness in this approach. Although the optimum from the last batch is probably a good starting point for the next batch, the model can be drawn away from the global optimum by similar (although more subtle and less analytically tractable) mechanisms as in globally unstable incremental schemes. It is thus recommended that global

ALGORITHMS FOR IDENTIFICATION   145

optimization algorithms be used (see Chapter 4) or at least past solutions discarded as starting points.

An Example   The relationship between batch and incremental as well as static and dynamic schemes can be illustrated on a simple nonneural example. Let us consider an unstable nonlinear plant containing an exponential term. It is described by the differential equation

$$\dot{z} = cfe^{z}(u - z) \qquad (5.126)$$

with c and f being constants. This plant is excited by a sinusoidal signal of variable amplitude a and constant frequency f so that

$$u = a\sin(2\pi ft) \qquad (5.127)$$

This plant is approximated by a quadratic model of the form

$$\dot{z} = cf(1 + pz)(u - z) \qquad (5.128)$$

with a single free parameter p.

The plant model has been identified by two batch-oriented and two incremental algorithms:

1. Batch minimization of cost function (5.85) with the static model in the discrete time form

$$z_{t+1} = d_t + \Delta_t cf(1 + pd_t)(u_t - d_t) \qquad (5.129)$$

   with d_t being measurements from the plant (5.126).

2. Batch optimization of cost function (5.85) with a dynamic model

$$z_{t+1} = z_t + \Delta_t cf(1 + pz_t)(u_t - z_t) \qquad (5.130)$$

3. Incremental adaptation of static scheme (5.129) with the adaptive rule

$$p_{t+1} = p_t + \alpha cf(d_{t+1} - z_{t+1})d_t(u_t - d_t) \qquad (5.131)$$

   with positive learning rate α

4. Incremental adaptation of dynamic scheme (5.130) with the adaptive rule

$$p_{t+1} = p_t + \alpha cf(d_{t+1} - z_{t+1})z_t(u_t - z_t) \qquad (5.132)$$

Adaptive schemes (5.131) and (5.132) correspond to the gradient method. The derivative of the squared error $e^2 = (d_{t+1} - z_{t+1})^2$ is

$$\frac{de^2}{dp} = -cf(d_{t+1} - z_{t+1})d_t(u_t - d_t) \tag{5.133}$$

for Eq. (5.131) and

$$\frac{de^2}{dp} = -cf(d_{t+1} - z_{t+1})z_t(u_t - z_t) \tag{5.134}$$

for Eq. (5.132). The constant cf is subsumed in learning rate α. Batch schemes have been optimized with a simple golden search optimization routine given by Press et al. [117].

The continuous-time version of the static scheme is globally stable if the real plant model has a structure consistent with model (5.129), that is,

$$d_{t+1} = d_t + \Delta_t cf(1 + p_0 d_t)(u_t - d_t) \tag{5.135}$$

This can be shown by considering the system of Eqs. (5.129), (5.131), and (5.135):

$$\begin{aligned} d_{t+1} &= d_t + \Delta_t cf(1 + p_0 d_t)(u_t - d_t) \\ z_{t+1} &= d_t + \Delta_t cf(1 + p d_t)(u_t - d_t) \\ p_{t+1} &= p_t + \alpha cf(d_{t+1} - z_{t+1})d_t(u_t - d_t) \end{aligned} \tag{5.136}$$

The first two equations can be substituted into the last, yielding

$$\begin{aligned} p_{t+1} &= p_t + \alpha \Delta_t cf(p_0 - p)d_t(u_t - d_t)d_t(u_t - d_t) \\ &= p_t + \Delta_t \alpha [cfd_t(u_t - d_t)]^2(p_0 - p) \end{aligned} \tag{5.137}$$

or, in continuous time,

$$\dot{p} = \alpha [cfd(u - d)]^2(p_0 - p) \tag{5.138}$$

The time derivative of the Lyapunov function $V = (p_0 - p)^2$ is

$$\dot{V} = -(p_0 - p)\alpha[cfd(u-d)]^2(p_0 - p) = -\alpha[cfd(u-d)(p_0 - p)]^2 \tag{5.139}$$

is obviously semidefinite. However, this global stability proof depends on the assumption that the real model is expressed by Eq. (5.135). With the real model being Eq. (5.126), stability cannot be proved. By contrast, the dynamic adaptive scheme with (5.130) and (5.132) cannot be proved to be globally stable even for identifying a plant described by (5.135).

A computational experiment has been performed in which (1) the quality of the identified model and (2) the stability of convergence of the four identification methods has been evaluated. The criterion for quality of identification has been the average squared error of the model trajectory over ten periods of the sinusoid input. Also stability was indicated by an average squared error: with a small squared error, the procedure converged to a reasonable model, whereas a large squared error indicated convergence to a spurious solution, such as a model with constant output. The initial value for parameter p was one, which is the value received by the first-order Taylor expansion of the exponential gain term of Eq. (5.126). Then, lack of convergence cannot be justified by an inappropriate initial value.

To produce cases of various degrees of difficulty, the amplitude of the input signal has been varied. For small amplitudes, the model trajectory is also approximately a sinusoid. For high amplitudes, the gain greatly varies between small values with negative z and high values with positive z. Exponential gain variations make the task difficult. For a given sampling rate of $\Delta_t = 10^{-5}$ and Euler method of numeric integration, even the simulation of the plant itself shown by Eq. (5.126) has been numerically stable up to an amplitude of approximately nine. So the values of amplitude a have been varied between 0 and 9. The constants have been assigned values $c = 5$ and $f = 10$.

The results are summarized in Table 5.1 and illustrated in Figures 5.12–5.15. The following observations can be made:

1. The dynamic models have better trajectory correspondence in both incremental and batch schemes. (With incremental schemes, this is true only as long as the dynamic model convergence remains stable, that is, for $a = 5$.) It can be observed that dynamic schemes (Figs. 5.13 and 5.15) generally converge to different models than do static ones (Figs. 5.12 and 5.14). Within the dynamic forecast group, as long as learning converges (Figs. 5.15A and 5.13A, the cases B and C do not converge for the incremental scheme), the forecasts are very similar, and so they are within the static forecast group (Figs 5.14A and 5.12A, the cases B and C do not converge for one or both of the static schemes). This

Table 5.1 Plant identification example: Mean squared error for various input amplitude

Amplitude	Static Batch		Dynamic Batch		Static Incr.		Dynamic Incr.	
	p	MSE	p	MSE	p	MSE	p	MSE
1.0	1.01	0.0003	0.91	0.0003	1.00	0.0003	1.13	0.0005
5.0	0.84	0.7547	0.50	0.3967	0.81	0.6959	1.89	2.2520
7.0	0.75	1.8571	0.42	0.9201	0.74	1.7754	1.11	3.1676
8.0	0.72	2.5393	0.40	1.2282	0.73	2.5581	0.87	3.3250
8.5	9.84	4.3988	3.91	1.3898	0.98	4.3986	0.78	3.3569
9.0	4.40	8.1352	3.81	1.5236	4.45	8.0685	0.71	3.3160

MSE, mean squared error.

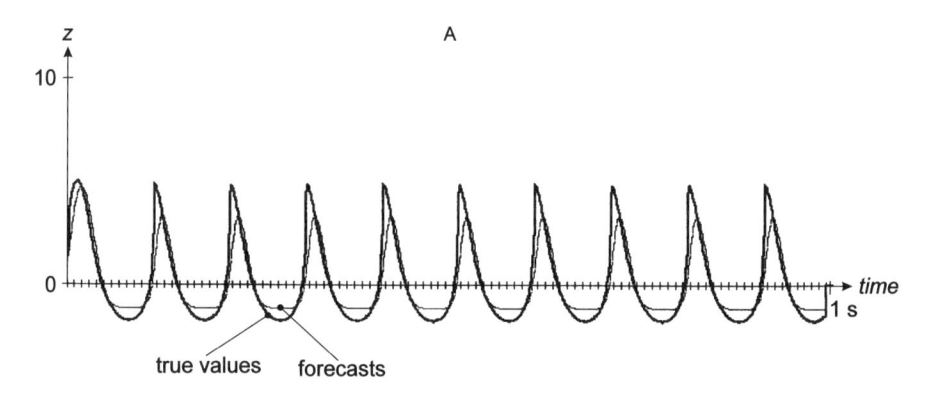

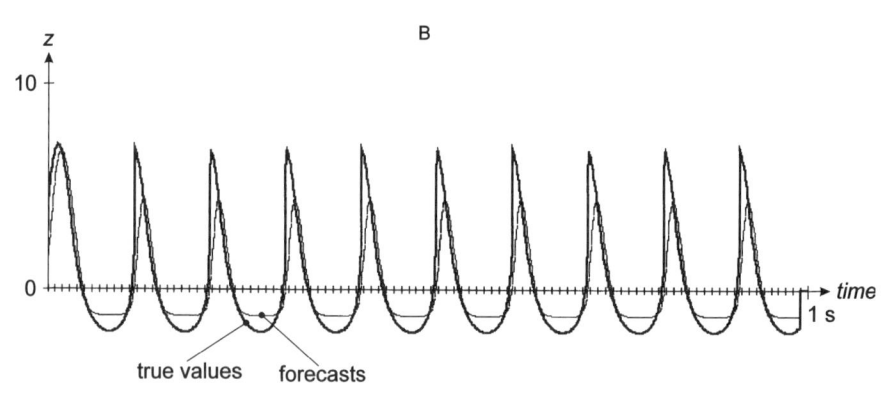

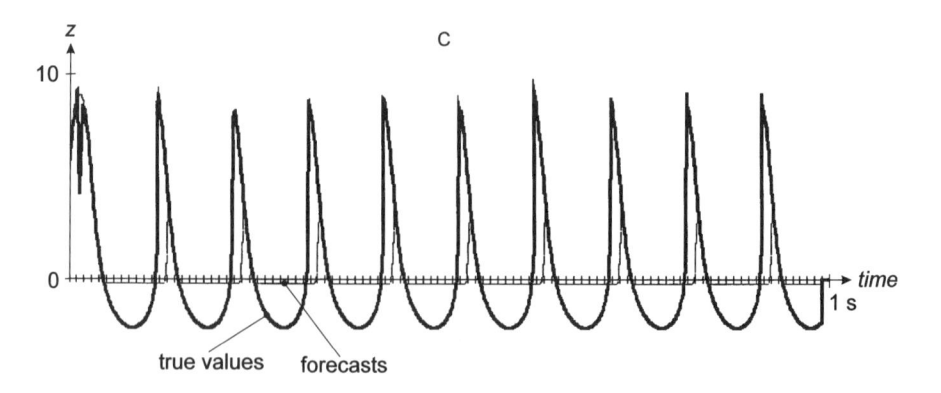

Figure 5.12 Identification: batch scheme, static model. (A) $a = 5$. (B) $a = 7$. (C) $a = 9$.

reflects the general forecast superiority of the multistep dynamic error criterion—the parameter values found by single-step, static schemes lead to poorer correspondence between the trajectories of the model and the real plant whether an incremental or a batch scheme is used.

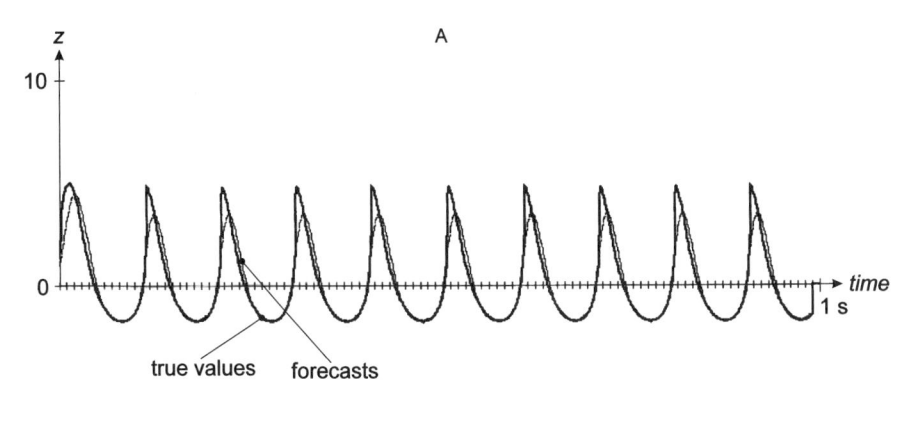

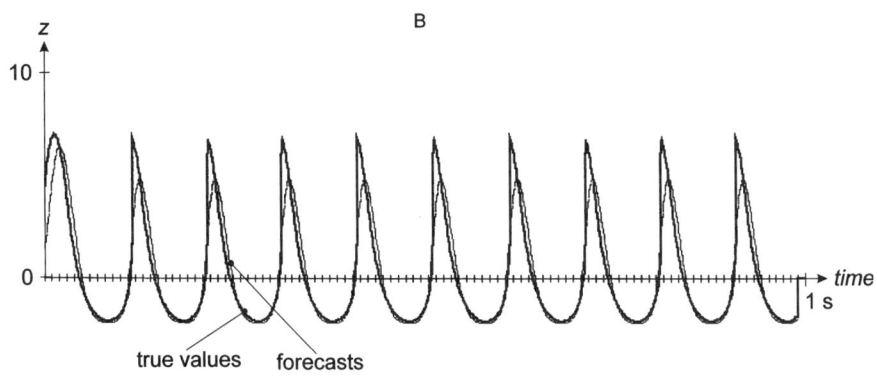

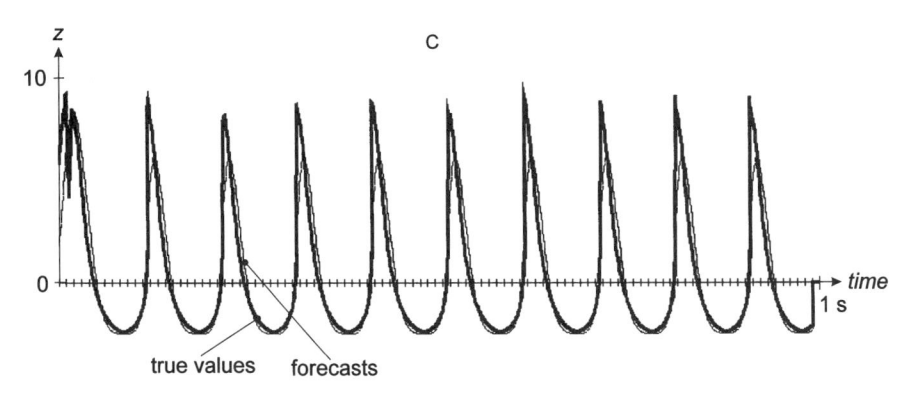

Figure 5.13 Identification: batch scheme, dynamic model. (A) $a = 5$, (B) $a = 7$, (C) $a = 9$.

2. In batch schemes, superiority of the dynamic model over the static one increases with the task difficulty. Although the dynamic model is acceptable, in particular, in low-gain parts of the trajectory, even for $a = 9$, the static model deteriorates in low- as well as high-gain phases for $a \geq 7$.

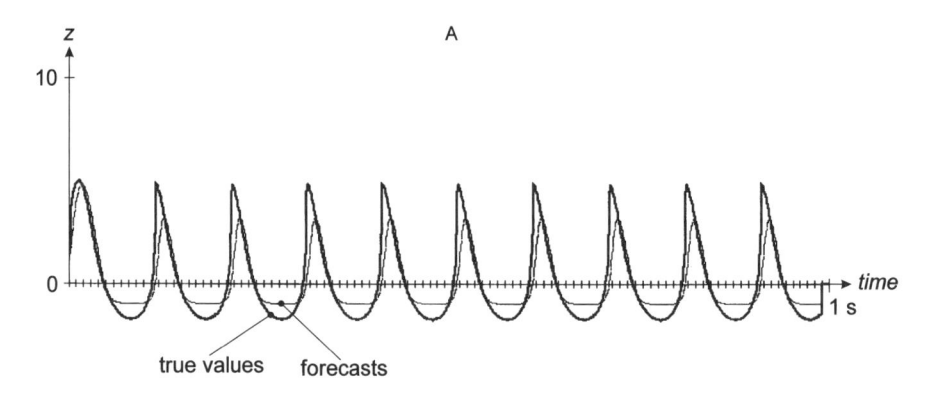

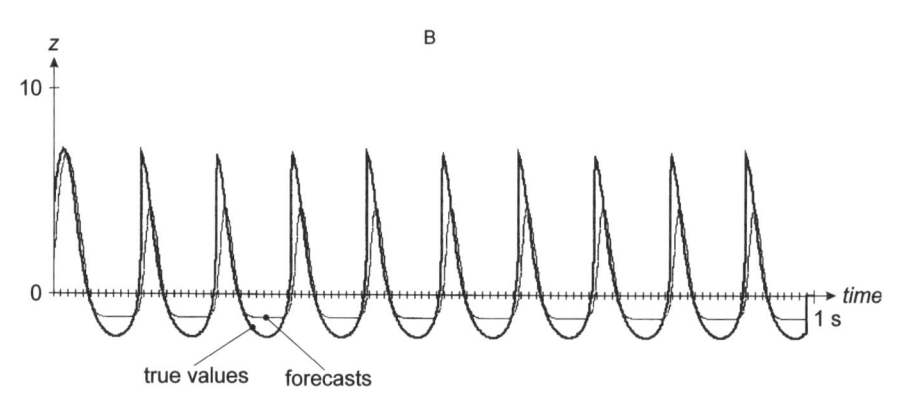

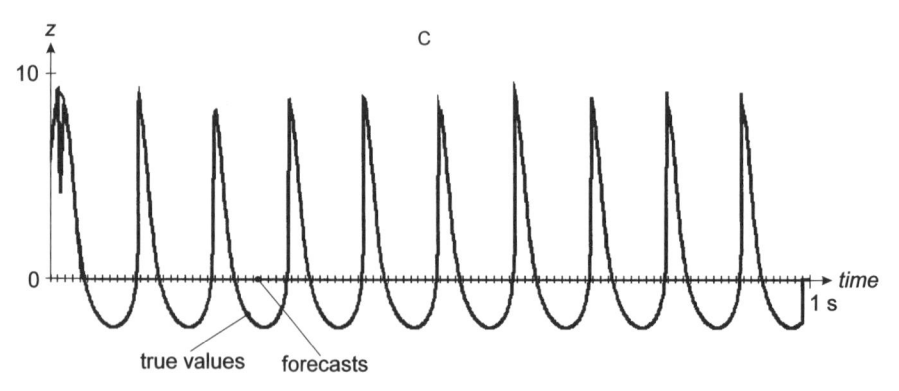

Figure 5.14 Identification: incremental scheme, static model. (A) $a = 5$, (B) $a = 7$, (C) $a = 9$.

3. In incremental schemes, the better quality of the dynamic model exists at the expense of stability. It becomes unstable for $a \geq 7$ (the forecasts are uniformly zero, see Figs. 5.15B and 5.15C), whereas the static model converges acceptably until $a = 7$.

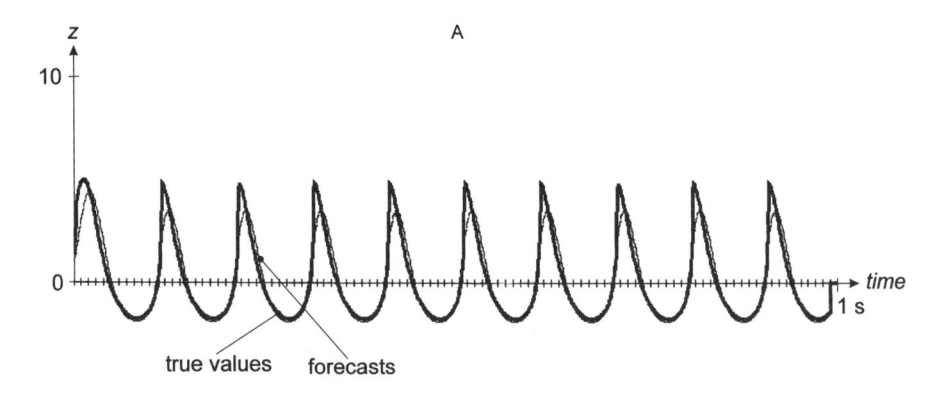

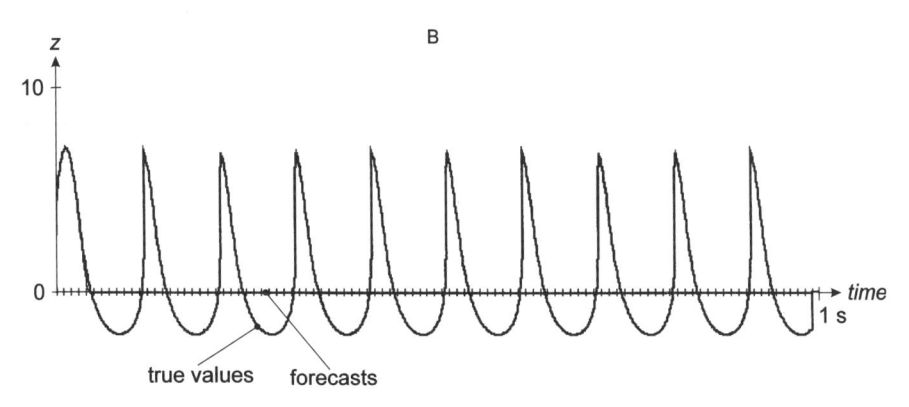

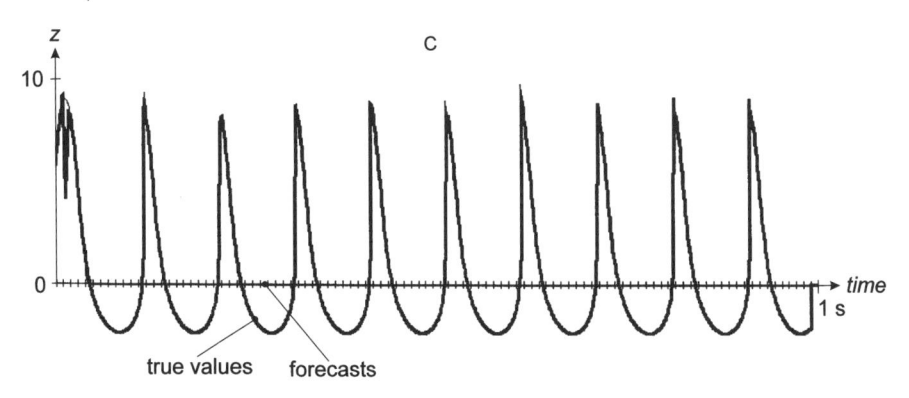

Figure 5.15 Identification: incremental scheme, dynamic model. (A) $a = 5$, (B) $a = 7$, (C) $a = 9$.

As far as computing efficiency is concerned, batch schemes are clearly superior. Computational expense for incremental schemes has been several hundred times larger than that for batch schemes.

5.5.5 Advantages and Disadvantages of Incremental and Batch Schemes

Choosing between incremental and batch schemes depends on the size and difficulty of the target problem and on the target computational platform. Incremental solutions are the light ones. They require little storage and can usually be computed online in the sampling rate of the control. Software administration of tasks of different priorities, such as frequent high-priority control tasks and less frequent, low-priority identification update tasks, can be avoided.

There are theoretical uncertainties concerning stability that can easily cause painful numeric convergence problems. The overhead for supervising such instability may negatively trade off the simplicity of the basic algorithm. More complex control problems can easily be beyond the boundaries of tractability with incremental schemes.

In incremental schemes, a static identification approach is preferable for two reasons. First are the stability aspects: static algorithms are more likely to be stable than dynamic ones. The second reason concerns gradient computation. For dynamic approaches with a longer time horizon for which an analytic gradient is recursively computed, the numeric precision of the analytic gradient deteriorates rapidly. Because there is no feasible way to compute the numeric gradient instead of the analytic one in incremental schemes, the best way is to avoid dynamic identification despite the fact that this can reduce the quality of the model with regard to neurocontroller training (see Section 5.4).

Note 5.9 The recommendation to avoid analytic gradient computations with dynamic approaches is based on arguments of Section 4.1.3. This recommendation may be controversial in the neurocontrol community— some authors report good results using this combination. A direct comparison of computational experience is difficult because benchmark applications exhibit varying complexity and the quality of neurocontrol solutions has only limited comparability. The reader, therefore, is invited to make his or her own judgment.

Batch solutions require care and implementational overhead with data sampling if used offline and with data updates if used iteratively. To exploit their advantageous properties, sophisticated optimization must be implemented. This makes either an offline operation or real time multitasking advisable.

Stability problems are substantially less serious or can even be completely avoided with batch schemes. Plants of considerable complexity can be reliably identified if sophisticated numeric methods are used.

For identification of plants with strongly nonlinear behavior, dynamic identification usually leads to better results than static one. With batch schemes, numeric inaccuracy of recursive computations of the analytic gradient is no obstacle: Optimization methods with no gradient or with a numeric gradient can be used.

From personal experience, the author prefers batch schemes whenever problems of industrial size are addressed.

5.6 SAMPLED DATA

Within a framework defined by the model structure, the only information that plant identification has to work with is gained from sampled measurements. It is clear that these data cannot be arbitrary: Their quantity and quality substantially influence the quality of the identified plant model.

The question of what data are necessary for identification of an arbitrary nonlinear plant has not so far been theoretically addressed. However, there are three viewpoints from which useful hints can be obtained:

1. The data should be sufficient in quantity to determine all model parameters.
2. The data should be sufficiently variable to provide information about the dynamics of the plant.
3. The data should represent the whole state space.

To determine a model with p parameters, it is clear that at least p samples are necessary. Otherwise, the identification task must be reduced to a set of q equations for p variables with $q < p$. This equation system might have an infinite number of solutions (although this cannot be guaranteed for nonlinear equations). These solutions would not be equivalent in the sense of representing the same dynamic model; they would merely be fitted just to these p samples. In practice, it is recommended that the number of samples used be a large multiple of the number of parameters. For plant models with typically tens to hundreds of neural network weights, several thousand samples are the minimum.

Another viewpoint concerns the variability of data sufficient to show the dynamic behavior. Obviously, a constant measurement series with constant input can be matched by an infinite number of models of various orders. In terms of minimization of error, many models can correspond to the same minimum, a zero forecast error. Clearly, these models, except for the single correct one, cannot behave consistently with the plant in more excited states. It is not only the constant measurement series that is insufficient to instantiate a dynamic model. The necessary conditions for data to be sufficient for *linear* plant identification have been formalized in the concept of persistent nth-order excitation. The single variable discrete-time definition by Åström and Wittenmark [8] is the following: The *input signal* u_t is persistently exciting of nth order if the matrix of input autocovariances $\text{cov}_s = \text{cov}(u_t, u_{t-s})$ with delays $i, i = 0, \ldots, n-1$,

154 PLANT IDENTIFICATION

$$\mathbf{C}_n = [c_{ij}] = [\text{cov}_{i+j-2}] \tag{5.140}$$

is positive definite. Another variant is the definition using nth-order differential operator L. If for all nth-order differential operators L,

$$\lim_{t \to \infty} \frac{1}{t} \left(\sum_{k=1}^{t} L u_k \right)^2 > 0 \tag{5.141}$$

then the input signal is persistently exciting of order n. This topic is discussed in depth by Narendra and Annaswamy [103].

A persistently exciting input signal of order n will generate data that sufficient to identify a linear plant model of order n. For some simple signal types, this order [8] is:

One for a step function
n for a mixture of n sinusoids (with different frequencies)
Infinity for a random input signal

It may be difficult to excite a plant by a completely random signal, so the most practical advice is to use n sinusoids. In batch schemes, this can be materialized simply by taking n time series with n different sinusoid inputs. In incremental schemes, either a mixture of n sinusoids or cyclically repeated single-sinusoid series should be used.

Note 5.10 The concept of persistent excitation was developed in the context of incremental plant identification. A persistently exciting input signal of sufficient order is a necessary condition for the convergence of model parameters to genuine plant parameters. That is, the guarantee of convergence can be given only together with the guarantee of stability of the adaptive scheme (see Section 5.5.4). In batch schemes, a persistently exciting input signal guarantees that the squared error minimum (i.e., zero) is reached for (and only for) the model with correct parameters, assuming the time series is sufficiently long to determine all parameters.

The last judgment about the number of samples necessary for successful plant model identification has to do with nonlinear functions and their approximation by neural networks. If a plant could be identified for which the model mapping (i.e., the mapping on the right side of the system equation) is *known* to belong to a certain parametrized class of mappings, it would be sufficient, with some reservation, to provide a certain number of (persistently excited) samples. For example, if the state equation were known to be quadratic of the form

$$z_{t+1} = az_t - bz_t^2 + cu_t \qquad (5.142)$$

The three free model parameters a, b, c could be determined by five samples $z_s, s = t, \ldots, t+4$ and $u_s, s = t, \ldots, t+3$, that is, four samples to build three equations with three variables and the fifth sample to disambiguate multiple solutions, if necessary. Except for the requirement of persistent excitation, it makes no difference from what region of state space the samples come. For coefficients $a = b = c = 1$ and $u = 1, 0, 2, 0$, it is unimportant whether the state sequence is $1, 0, 2, -2$ (for initial state 0) or $-5, -30, -928, -862, 112$ (for initial state 3) ; both are sufficient to identify the model.

The situation is completely different if the model mapping is approximated. Even theoretically well-understood approximations such as a Taylor series have bounded errors only on a certain interval. The error bounds typically depend on the bounds of higher derivatives on this interval. The statement that, for example, a multilayer perceptron can approximate an arbitrary non-linear function on an interval allows no inferences about the quality of approximation outside this interval.

Also, approximation refers to the function represented with a certain density by input–output pairs. The interpolation between (and extrapolation beyond) specified input–output pairs depends on the smoothness of real mapping, smoothness of neural network mapping, and distances from specified points. Interpolation-related concepts, not immediately applicable to control, for RBF networks have been investigated, for example, by Poggio and Girosi [115], but it can be stated that there are no reliable practical means of checking the quality of interpolation in plant identification other than by testing. The only way to obtain a good approximation of a nonlinear state equation is by a dense coverage of the state space. So the data should be consistent with persistent excitation conditions at *all working points* of the plant. To satisfy this condition literally is, of course, impossible—the continuum of working points can, at best, be substituted by a set of selected, obviously distinct, working points. The upper bound (which is advisable to approach closely) is set by computing resources. In the author's experience, several thousands of samples do the work in most cases.

Note 5.11 Narendra [101] points out that if a model is used only to propagate *instantaneous* derivatives at the current working point for controller training in an incremental scheme, persistent excitation conditions can be relaxed because convergence to true parameters of a global nonlinear model is not required. Then, the model varies from one working point to another; it is sufficient that it be correct at the current working point. Strictly speaking, this will not lead to *global* optimization of the controller; the sequence of *locally* optimal actions is not identical with a globally optimal trajectory. This is the case even if dynamic backpropagation is used: The optimization can refer only to the environment of the current working point (for which the model is correct) and not to the future consequences of the control action, where the closed-loop trajectory passes other working points (for which the model is, now, potentially incorrect). This approach may deliver good practical results in online adaptations.

5.7 USING EXPLICIT MODELS WITH FREE PARAMETERS

For many industrial control problems, sophisticated nonlinear models have been developed mostly for simulation. Such simulation models can be used directly for neurocontrol training. The correspondence between such models and the real plant is usually checked by visual inspection of charts. This is also sufficient if the simulation is intended to capture only qualitative aspects of plant behavior.

For neurocontrol training, quantitative fitting of the model to the plant is required if the neurocontroller is to exploit, for example, the margins between speed and stability. Then, analytic simulation models can be fitted to sampled time series.

Formally, the algorithm is identical with that for neural network based identification. The plant model is analogous to Eq. (5.82):

$$\mathbf{z}_{t+1} = F_a(\mathbf{z}_{e_t}, \mathbf{u}_t, \omega_f) \qquad (5.143)$$

Vector mapping F_a is now an expression of arbitrary complexity, for example, a Runge-Kutta simulation of an equation system. The vector ω_f is the vector of arbitrary free parameters, such as:

Time constants of first- or higher-order dynamic subsystems
Delays
Parameters of nonlinear mappings

An example of such a model is the engine model of Section 13. This model contains a time constant of low pass, a delay, and a nonlinear function expressing the stationary dependence of engine torque on accelerator position and angular velocity, with 16 free parameters such as lower and higher bounds of outputs, and slopes of individual segments.

For simple models, there is a possibility of deriving the analytic cost function gradient with help of model Jacobian matrix. However, in most cases, this would make the approach too expensive in terms of manual labor—to set up the simulation is enough work. Therefore there is a strong bias in favor of numeric derivatives.

The most advantageous aspect of this approach is the prestructuring of the model. The prestructuring is equivalent to reducing the space in which the functional approximation F_a is searched. Whereas with a neural-network-based identification, the search goes over all dynamic systems with given inputs and outputs, an explicit model with free parameters is sought only on the space of all dynamic models of a given structure. A frequent consequence is a reduction in the number of free parameters.

Another advantage is that free parameters can mostly be assigned meaningful initial values, so a purely local search has a good chance for success (or, at least, is not so hopeless as for neural-network-based identification of a complex plant).

Prestructuring can also become a disadvantage. It is clear that in a reduced search space, nothing can be found that is outside of this space. It is important to be very careful in capturing all possible features of the plant behavior (e.g., all possible delays).

Both approaches, neural and analytic, can be combined easily. A good starting point is identifying a neural-network-based plant with explicit delays. A delay in the real plant introduces a degenerate order and may increase the number of network parameters drastically if the attempt is to identify it implicitly. By contrast, in explicit notation, it is a single free parameter.

6

CONTROLLER TRAINING

In Chapter 2, several fundamental approaches to neurocontrol were presented. Each approach consisted of a framework into which the controller, as a function approximator, was embedded and subsequently optimized. The basic approaches differed in the way a general cost or utility function of control was operationalized by transforming into one or more fundamental cost functions that can be minimized with the help of formalized information available to the control engineer.

However, these abstract frameworks lacked the details necessary to solve any particular control task. Several abstract concepts have to be given a concrete form before the solution can go on. One of them, specific for the closed-loop optimization approach, is the plant model identification discussed in Chapter 5. Another is the optimization method itself addressed in Chapter 4. The discussion of the most important remaining basic solution steps is the topic of this chapter. Special topics concerning robust design, adaptive frameworks, and stability proofs are treated in Chapters 7, 8, and 9, respectively.

What design decisions are necessary to fix a control problem?

1. First, there is the question of structural feasibility. For example, without effective action variables, no control is possible. Also, it is obvious that not every system can be manipulated to follow every trajectory.
2. The problem has to be formulated in such a way that obvious relationships need not to be discovered by expensive training. For example, known equilibrium points of a closed loop can be incorporated a priori into the controller structure.

3. The designer's preferences have to be formalized in a way that reflects these preferences completely and in correct relationships.
4. The continua, such as the whole state space, should be represented by countable (frequently on the fingers of one's hands, for efficiency reasons) sets of representative points.
5. Finally, optimization methods are not omnipotent, and some formulations are more optimization-friendly than others.

As in the case of plant identification, a brief survey of the most relevant results of classical linear control theory is the starting point. For space reasons, only continuous-time results are presented. Discrete-time results are almost direct analogies, with delays substituted for time derivatives.

6.1 CLASSICAL VIEW OF CONTROL

Classical control theory provides several frameworks for controller design that are instructive for making sound neurocontrol formulations. There are two basic types of control tasks that can be investigated separately.

1. Stabilization of the plant around a stationary value
2. Trajectory following, also called *tracking*

The following two subsections briefly survey how these tasks can be solved if plants and controllers are linear.

6.1.1 Plant Stabilization

The stabilization task is formulated in the following way: A plant with an n-dimensional state vector $\mathbf{z}$ and a p-dimensional action vector $\mathbf{u}$ of the form

$$\dot{\mathbf{z}} = \mathbf{Az} + \mathbf{Bu} \qquad (6.1)$$

with a q-dimensional vector of measurable variables

$$\mathbf{y} = \mathbf{Cz} \qquad (6.2)$$

is to be stabilized at the point $\mathbf{y} = \mathbf{w}$.

A plant is called *controllable* if it can be brought from an arbitrary initial state to an arbitrary final state by an appropriate trajectory of action vector $\mathbf{u}$. This is the case if the plant satisfies the condition that the matrix

$$\left[\mathbf{B}, \mathbf{AB}, \ldots, \mathbf{A}^{n-1}\right] \qquad (6.3)$$

has rank n. A controllable plant (even if unstable) can be stabilized at the point characterized by **w** by linear state feedback, that is, by a controller performing actions in linear dependence from the current state **z** and the desired measured state **w**, or formally

$$\mathbf{u} = -\mathbf{Rz} + \mathbf{Mw} \tag{6.4}$$

The matrices **R** and **M** play different roles in control. The matrix **R** represents the feedback part of the controller and its task is the stabilization of the behavior of the closed loop. The stability behavior of every linear system is determined by the eigenvalues of its system matrix. If all eigenvalues have negative real parts, the system is stable. If some of them have a positive real part, the system is unstable. The eigenvalues of the plant, that is, those of matrix **A**, are fixed. However, the closed loop consisting of plant (6.1) and controller (6.4) has different eigenvalues. One of the fundamental tasks of controller design is to make the eigenvalues of the closed loop more advantageous than those of the plant—making an unstable plant stable, or a plant with slow reactions to one with fast reactions, and so on.

The way that closed-loop eigenvalues can be manipulated is particularly transparent for plants with a single input variable. Such plants, if controllable, can always be transformed by a linear transformation of state and input variables into the controllable normal form with system matrices

$$\mathbf{A} = \begin{bmatrix} 0 & 1 & 0 & \cdots & 0 \\ 0 & 0 & 1 & \cdots & 0 \\ \vdots & & & & \vdots \\ -a_0 & -a_1 & -a_2 & \cdots & -a_{n-1} \end{bmatrix} \tag{6.5}$$

and

$$\mathbf{B} = \begin{bmatrix} 0 \\ \vdots \\ 0 \\ 1 \end{bmatrix} \tag{6.6}$$

The eigenvalues s of the plant on the open loop are solutions of the characteristic equation, which is in a transparent relationship to the matrix (6.5):

$$s^n + \sum_{i=0}^{n-1} a_i s^i = 0 \tag{6.7}$$

With a linear feedback matrix

$$[r_1 \ldots r_n] \qquad (6.8)$$

the closed-loop system matrix is

$$\mathbf{A} - \mathbf{BR} = \begin{bmatrix} 0 & 1 & 0 & \ldots & 0 \\ 0 & 0 & 1 & \ldots & 0 \\ \vdots & & & & \vdots \\ -(a_0 + r_1) & -(a_1 + r_2) & -(a_2 + r_3) & \ldots & -(a_{n-1} + r_n) \end{bmatrix} \qquad (6.9)$$

resulting in a characteristic equation

$$s^n + \sum_{i=0}^{n-1} (a_i + r_{i+1}) s^i = 0 \qquad (6.10)$$

So the coefficient of the characteristic polynomial can be set to arbitrary values. Consequently, arbitrary eigenvalues for the closed loop can be specified. They define a particular characteristic polynomial, which, in turn, defines the elements r_i of the feedback controller matrix. Therefore, from the stability viewpoint, we can reach the goal of stabilizing the closed loop, that is, making the real parts of all closed-loop eigenvalues negative in infinitely many ways. For plants with p inputs, the redundancy is even larger. The np elements of matrix $\mathbf{R}$ are used to determine n eigenvalues. This redundancy gives the designer the possibility to make closed loops not only stable but also fast and nonoscillating in the transient phase, and so on.

At Which Points Can the Plant Be Stabilized? So far, it has been only investigated whether the closed loop with a certain controller is stable. A complementary question is *at which point* the closed loop will be stabilized.

The closed loop with a controller made of the matrix $\mathbf{R}$ alone may be stable, but, in the most cases, it does not settle down to point $\mathbf{w}$ but to point $\mathbf{0}$. The matrix

$$\mathbf{M} = \left[\mathbf{C}(\mathbf{BR} - \mathbf{A})^{-1} \mathbf{B} \right]^{-1} \qquad (6.11)$$

the so-called *prefilter matrix*, shifts the controller action $\mathbf{u}$ so that the closed-loop equilibrium point is at $\mathbf{w}$. The construction of the prefilter matrix is possible if the matrix $\mathbf{C}(\mathbf{BR} - \mathbf{A})^{-1}\mathbf{B}$ is nonsingular. Otherwise, formula (6.11) cannot be applied. There may be two reasons for this matrix not to be nonsingular. The first is that the number of measured variables is not equal to the number of action variables—the matrix is then not a square matrix. The

CLASSICAL VIEW OF CONTROL   163

second reason may be that some state variables *cannot* have arbitrary values in the stationary state.

This can be elucidated if the stationary state is described as a system of linear equations

$$0 = \mathbf{Az} + \mathbf{Bu}$$
$$\mathbf{w} = \mathbf{Cz} \tag{6.12}$$

If a controller of the form

$$\mathbf{u} = -\mathbf{Rz} + \mathbf{v} \tag{6.13}$$

is assumed, with $\mathbf{v}$ being the part defining the equilibrium point, then Eq. (6.14) can be written as

$$0 = (\mathbf{A} - \mathbf{BR})\mathbf{z} + \mathbf{Bv}$$
$$\mathbf{w} = \mathbf{Cz} \tag{6.14}$$

This is a system of $n + q$ equations with $n + p$ variables $\mathbf{z}$ and $\mathbf{v}$. The following three cases can occur:

1. For $q > p$, $\mathbf{w}$ cannot be chosen arbitrarily for the solution of Eq. (6.14) to exist. In other words, $\mathbf{w}$ has to be consistent with some stationary state.
2. If $q = p$, $\mathbf{w}$ can be chosen arbitrarily, assuming that the matrix

$$\begin{bmatrix} \mathbf{A} - \mathbf{BR} & \mathbf{B} \\ \mathbf{C} & 0 \end{bmatrix} \tag{6.15}$$

is nonsingular. Trivial reasons for the singularity of Eq. (6.15) might be (a) linear dependences in $\mathbf{B}$ caused by redundant action variables or (b) linear dependences in $\mathbf{C}$ resulting from redundantly defined measured variables. These are formal flaws of the model that can be corrected by the elimination of redundant variables. However, nontrivial reasons for the singularity may occur, too. They may have the form of linear dependence between the rows of the $[\mathbf{A} - \mathbf{BR}, \mathbf{B}]$ submatrix and the $[\mathbf{C}, 0]$ submatrix. This can be illustrated on a simple double integrator plant described by the state equation

$$\begin{bmatrix} \dot{x}_1 \\ \dot{x}_2 \end{bmatrix} = \begin{bmatrix} 0 & 0 \\ 1 & 0 \end{bmatrix} + \begin{bmatrix} 1 \\ 0 \end{bmatrix} u \tag{6.16}$$

and a feedback control matrix, say, $[-1 - 1]$. If the variable x_1 can be assigned a goal value in the stationary state, the matrix (6.15) becomes

$$\begin{bmatrix} -1 & -1 & 1 \\ 1 & 0 & 0 \\ 1 & 0 & 0 \end{bmatrix} \tag{6.17}$$

The second and the third rows are identical, and the equation system has no solution except for a goal value $w = x_1 = 0$. This is also common sense: x_1 is, in fact, a derivative of x_2 and cannot thus assume values other than zero in a stationary state. It is generally true that variables that are nothing but derivatives of other state variables cannot appear in the vector **w** unless the corresponding element of **w** is zero.

3. The case of $q < p$ implies that **w** can be chosen arbitrarily as long as matrix (6.14) is of rank $n + q$.

This can be summarized from a different viewpoint. The goal state defined by vector **w** has to be consistent with a possible state of the plant. Even if this is the case, the prefilter matrix can be computed according to Eq. (6.11) only as long as $p = q$.

Note 6.1 Here it may be useful to see what the term *measured variables* really means. In the present context, it is the set of linear combinations of state variables for which a goal stationary state is specified. This is independent of whether these variables can be measured or not—it can be a subset of genuinely measurable variables.

Incomplete Feedback To be able to feed the complete state to the controller, the state must be completely measurable. If it is not the case, and the complete state is observable with the help of a given vector of measured variables **y**, the use of a state observer is indicated. A *state observer* is a linear dynamic filter (with its own dynamic behavior). There are design methods for state observers that guarantee that its own eigenvalues will not shift the eigenvalues of the closed loop. Nevertheless, the pair observer + controller (which can, altogether, be viewed as a controller with dynamic behavior) increases the order of the closed loop, which may negatively influence the dynamic behavior of the closed loop, for example, by making its reactions slower. Another way to observe the state is by considering the history of inputs and outputs (or higher-order derivatives) up to the observability index, as discussed in Sections 5.1.1 and 5.3. This approach is limited by the numeric accuracy of measurements (5.3.2).

This, and the fact that observer design is an additional task whose solution requires effort and brings about the risk of inaccuracies arising from an imprecise plant model, have lead to the idea of *incomplete feedback*, or output feedback. It uses only directly measurable variables:

$$\mathbf{u} = -\mathbf{Ry} + \mathbf{Mw} \tag{6.18}$$

This commitment has consequences for the design freedom of the controller. The matrix **R** has now only *qp* (instead of *np*) elements. These elements are to be determined so that (at least) the stability of the closed loop is guaranteed, that is, so that all closed-loop eigenvalues have negative real parts. This is a system of *n* nonlinear constraints for *qp* variables. Because of nonlinearity, this cannot be guaranteed even if $qp \geq n$. On the other hand, the probability of nonexistence of a solution is rather low and is frequently neglected.

This controller structure can be written in the following form

$$\mathbf{u} = \mathbf{R}(\mathbf{w} - \mathbf{y}) + (\mathbf{M} - \mathbf{R})\mathbf{w} = \mathbf{R}(\mathbf{w} - \mathbf{y}) + \mathbf{Sw} \quad (6.19)$$

In this representation, the action is separated into (1) the equilibrium point action **Sw**, which persists after the stable state has been reached, and (2) the state error term $\mathbf{R}(\mathbf{w} - \mathbf{y})$. The advantages of this representation for neurocontrol are discussed below.

Systematic Disturbances and Error Integration Complete state feedback and incomplete output feedback share the property that they are not robust against systematic disturbances, which are equivalent to a constant measurement offset.

Systematic disturbances can be viewed as additional state variables; they can be identified by a state observer, or, equivalently, by making the controller dynamic. This approach is popular in the form of frequently used PI controllers. PI controllers for a single state variable $(w - z)$ have the form

$$u = a(w - z) + b \int_t (w - z) dt \quad (6.20)$$

They can be transformed to the *velocity form*

$$\dot{u} = a(\dot{w} - \dot{z}) + b(w - z) \quad (6.21)$$

It is the error integration (either before the controller or after it) that leads to the convergence to the state $w = z$.

These ideas have inspired an extension of the state feedback controller [31, 68] by an additional error integrator

$$\dot{\mathbf{e}} = \mathbf{w} - \mathbf{y} = \mathbf{w} - \mathbf{Cz} \quad (6.22)$$

This constitutes, together with Eq. (6.1) an extended plant of the form

$$\begin{bmatrix} \dot{\mathbf{z}} \\ \dot{\mathbf{e}} \end{bmatrix} = \begin{bmatrix} \mathbf{A} & \mathbf{0} \\ -\mathbf{C} & \mathbf{0} \end{bmatrix} \begin{bmatrix} \mathbf{z} \\ \mathbf{e} \end{bmatrix} + \begin{bmatrix} \mathbf{B} \\ \mathbf{0} \end{bmatrix} \mathbf{u} + \begin{bmatrix} \mathbf{0} \\ \mathbf{I} \end{bmatrix} \mathbf{w} \quad (6.23)$$

This dynamic system can be stabilized by the controller

$$\mathbf{u} = -\mathbf{Rz} + \mathbf{Se} + \mathbf{Mw} \tag{6.24}$$

An outstanding property of this controller is that the closed loop stabilizes at $\mathbf{y} = \mathbf{w}$ even if the model is not exact (see Section 7.1). In particular, this is the case even if the term $\mathbf{Mw}$ is omitted.

Controllers with Integrated Control Action Analogous to the velocity form of a pure I-controller with a complete state feedback is the controller

$$\dot{\mathbf{u}} = -\mathbf{Pz} + \mathbf{Tu} + \mathbf{M}(\mathbf{w} - \mathbf{y}) = -(\mathbf{P} + \mathbf{MC})\mathbf{z} + \mathbf{Tu} + \mathbf{Mw} \tag{6.25}$$

Setting $\mathbf{R} = \mathbf{P} + \mathbf{MC}$ and omitting the constant term $\mathbf{Mw}$, which has no influence on the dynamics, we receive a closed-loop description

$$\begin{aligned} \dot{\mathbf{z}} &= \mathbf{Az} + \mathbf{Bu} \\ \dot{\mathbf{u}} &= -\mathbf{Rz} - \mathbf{Tu} \end{aligned} \tag{6.26}$$

The eigenvalue shift, once more, can be illustrated on the single-input case. The system matrix of the closed loop shown in Eq. (6.26) is

$$\mathbf{A} = \begin{bmatrix} 0 & 1 & 0 & \cdots & 0 & 0 \\ 0 & 0 & 1 & \cdots & 0 & 0 \\ \cdot & & & & \cdot & \cdot \\ \cdot & & & & \cdot & \cdot \\ \cdot & & & & \cdot & \cdot \\ -a_0 & -a_1 & -a_2 & \cdots & -a_{n-1} & 1 \\ -r_1 & -r_2 & -r_3 & \cdots & -r_n & -t \end{bmatrix} \tag{6.27}$$

The characteristic equation is

$$\det|\mathbf{A}| = \begin{vmatrix} s & -1 & 0 & \cdots & 0 & 0 \\ 0 & s & -1 & \cdots & 0 & 0 \\ \cdot & & & & \cdot & \cdot \\ \cdot & & & & \cdot & \cdot \\ \cdot & & & & \cdot & \cdot \\ a_0 & a_1 & a_2 & \cdots & s + a_{n-1} & 1 \\ r_1 & r_2 & r_3 & \cdots & r_n & s+t \end{vmatrix} \tag{6.28}$$

Expanding the determinant by the last column, we get the characteristic equation

CLASSICAL VIEW OF CONTROL 167

$$\left(s^n + \sum_{i=0}^{n-1} s^i a_i\right)(s+t) - \sum_{i=0}^{n-1} s^i r_{i+1} = 0 \qquad (6.29)$$

The characteristic polynomial can be transformed into the form

$$\left(s^n + \sum_{i=0}^{n-1} a_i s^i\right)s + \left(s^n + \sum_{i=0}^{n-1} a_i s^i\right)t - \left(\sum_{i=0}^{n-1} r_{i+1} s^i\right)$$

$$= \left(s^{n+1} + \sum_{i=1}^{n} a_{i-1} s^i\right) + \left(ts^n + \sum_{i=0}^{n-1} ta_i s^i\right) - \left(\sum_{i=0}^{n-1} r_{i+1} s^i\right)$$

$$= \left(s^{n+1} + a_{n-1} s^n + \sum_{i=1}^{n-1} a_{i-1} s^i\right) + \left(ts^n + \sum_{i=1}^{n-1} ta_i s^i + ta_0\right) - \sum_{i=1}^{n-1} r_{i+1} s^i + r_1$$

$$= s^{n+1} + (a_{n-1} + t)s^n + \sum_{i=1}^{n-1} (a_{i-1} + ta_i - r_{i+1})s^i + (ta_0 + r_1)$$

$$(6.30)$$

resulting in the characteristic equation

$$s^{n+1} + (a_{n-1} + t)s^n + \sum_{i=1}^{n-1} (a_{i-1} + ta_i - r_{i+1})s^i + (ta_0 + r_1) = 0 \qquad (6.31)$$

All eigenvalues of a closed loop can be set arbitrarily by making appropriate choices of r_i and t, just as for the state feedback with static controller.

The stationary state of the closed loop with state equation (6.26) is defined by the conditions $\dot{z} = 0$, $\dot{u} = 0$. If there are no redundant states and actions, the system matrix is nonsingular. Without additional terms defining the control action offset for a given goal state, the state equation becomes a homogeneous equation for state z and action u whose only solution is $z = 0$, $u = 0$.

In addition to the output error feedback, this controller contains the state feedback. This is why its convergence to $y - w$ is not guaranteed. This would be the case if the complete state feedback were omitted:

$$\dot{u} = M(w - y) \qquad (6.32)$$

This form, in turn, does not guarantee stability for an arbitrary linear plant. However, stability may be guaranteed for certain plants, in the same sense as for incomplete feedback. The advantage, unlike the usual output feedback, is that no prefilter matrix is required.

6.1.2 Trajectory Following

Trajectory-following task can be formulated as a requirement for the output **y** of the plant described by Eqs. (6.1) and (6.2) to be identical or close to the time-varying reference vector **w**. This formulation implies the requirement of stability to cope with small deviations from the reference trajectory but goes beyond this. A controller stabilizing the closed loop may be too slow for rapidly changing parts of the reference trajectory. In other words, a controller for trajectory following must satisfy requirements involving time derivatives of the reference vector.

A very general approach for such controller design is the decoupling controller design proposed by Falb and Wolowich [27]. It is based on the decoupling model mentioned in Section 5.1.1. In contrast to the input–output form (5.20), the continuous-time model usually taken as a basis is in the state form

$$\mathbf{y}^* = \mathbf{C}^*\mathbf{x} + \mathbf{D}^*\mathbf{u} \qquad (6.33)$$

where $\mathbf{y}^*$ is, analogically to Eq. (5.20), the vector of output variable derivatives of order equal to the relative degree of the output variable

$$\mathbf{y}^* = \begin{bmatrix} y_1^{(r_1)} \\ \vdots \\ y_n^{(r_n)} \end{bmatrix} \qquad (6.34)$$

If the reference trajectory is given in the relative degree form

$$\mathbf{w}^* = \begin{bmatrix} w_1^{(r_1)} \\ \vdots \\ w_n^{(r_n)} \end{bmatrix} \qquad (6.35)$$

and matrix $\mathbf{D}^{-1}$ is invertible, a controller

$$\mathbf{u} = \mathbf{R}^*\mathbf{x} + \mathbf{D}^{*-1}\mathbf{w}^* \qquad (6.36)$$

can be, under some additional conditions, designed so that it is both stable and follows the trajectory given by $\mathbf{w}^*$ exactly. The matrix $\mathbf{R}^*$ can always be determined so that (1) the closed loop is stable and (2) the dynamic behavior of y_i^* depends only on w_i^*. (This is the reason the controller is called the *decoupling controller*.)

The reasons for which matrix $\mathbf{D}^*$ may not be invertible, or more generally, why some reference trajectories $\mathbf{w}^*$ cannot be followed, are similar to those for inaccessible states in stabilization tasks. The number of output variables q being equal to the number of input variables p is a precondition, which, if satisfied, reduces the set of such infeasible trajectories to singular cases.

With the help of observability of the state vector $\mathbf{z}$ via $\mathbf{y}_{(i)}, i = 0, \ldots, n_o - 1$ and $\mathbf{u}_{(i)}, i = 0, \ldots, n_o - 2$ (see Section 5.1.1), the controller (6.36) can be written in the form

$$\mathbf{u} = \sum_{i=0}^{n_o-1} \mathbf{P}_i \mathbf{y}_{(i)} + \sum_{j=0}^{n_o-2} \mathbf{Q}_j \mathbf{u}_{(j)} + \mathbf{D}^{*-1} \mathbf{w}^* \qquad (6.37)$$

or

$$\mathbf{u}^{(n_o)} = \sum_{i=0}^{n_o-1} \mathbf{P}_i \mathbf{y}_{(i)} + \sum_{j=0}^{n_o-1} \mathbf{Q}_j \mathbf{u}_{(j)} + \mathbf{D}^{*-1} \mathbf{w}^* \qquad (6.38)$$

Note that the stabilization controller shown in Eq. (6.4) and the trajectory-following controller described by Eq. (6.36) are structurally identical. Both can also be transformed to the input–output form of type (6.38) under the assumption of observability.

The effects of incomplete feedback concern stability similarly to that in stabilization tasks.

Error integration can also be included. The fact that error integration usually makes the response of the closed loop slower has more serious consequences for trajectory following than has been the case for stabilization. Because the attainable precision of trajectory following is lower than that of stabilization in a stationary point, the importance of error integration as a precision enhancer is relatively low.

6.1.3 Summary on Classical Control

The results of Sections 6.1.1 and 6.1.2 can be verbally summarized in the following way:

1. With complete state feedback, controllers for both stabilization and trajectory following can be designed with guaranteed stability properties.
2. If the plant is observable, state feedback can be substituted by the history, or higher-order derivatives, of plant inputs and outputs.
3. Not every stationary state or trajectory is consistent—inconsistent states cannot be stabilized, and inconsistent trajectories cannot be followed. That a state or trajectory is consistent is more probable (but not guaranteed) if it is characterized by a number of outputs equal to the number of inputs. If the number of specified reference outputs is larger

than that of inputs, the consistency becomes less probable (but not impossible).
4. Incomplete feedback, or output feedback, is not guaranteed to be sufficient for stability of *any* plant, though it works in most cases. The accessibility of states or trajectories specified with the help of output is not reduced with output feedback.
5. Error integration is a means of introducing robustness against omnipresent modeling inaccuracies and systematic disturbances—exact stabilization with an inexact model in an arbitrary consistent state is possible *only* with error integration. This is why error integration (i.e., possibly in velocity form, through controller action integration) should be included in every stabilization controller. The less importance of high accuracy in trajectory following, and the negative effects of error integration on the speed of a closed-loop response make error integration less recommendable for trajectory-following tasks.

6.2 NEUROCONTROL FORMULATION OF THE CONTROL PROBLEM

Neurocontrol is a subset of nonlinear control. Basic laws for nonlinear control are equally valid for neurocontrol. In the absence of theoretical results for nonlinear control, analogies to linear control can be exploited.

As in plant identification, the first aspect we turn to, in Section 6.3, is the controller's representation. The controller to be represented by a neural network can work only if the information it receives as input is appropriate for a particular task. The topic encompasses (1) determining a mapping (i.e., the mapping of what to what) that represents a theoretically founded controller, and (2) representation of this mapping by a neural network.

The view of neurocontroller training as an optimization process makes clear the importance of a cost function. The cost function for neurocontroller training, discussed in Section 6.4, has to express all explicit and implicit goals of control. An analogy to plant identification is limited to standard control tasks such as stabilization and tracking, where the closed loop follows a given trajectory (a constant trajectory in the stabilization case). Then, the cost function contains terms evaluating some type of deviation. However, for many practical control tasks, no goal trajectory is given. In other cases, its following is only one, sometimes not the dominant one, of several criteria of control quality. Unlike plant identification, controller training in the closed-loop optimization approach (e.g., in contrast to critic-based approaches) does not use sampled measurements in the form of time series. Rather, the measurement time series are incorporated into the model. Controller training takes place in hypothetical situations constructed in order to cover the relevant cases that the controller will be expected to operate on. These training cases, or training examples, and their distribution are, in fact, a part of the cost function.

Note 6.2 Although measured time series do not enter neurocontroller training in a closed-loop optimization approach as sequences, individual measured samples may be used as initial states for the neurocontroller. Equally well, they can be substituted by an arbitrary, representatively selected set of initial states.

With controller representation determined, the plant model fixed by plant identification, and the cost function formulated, the last component for controller training is the training algorithm, discussed in Section 6.5. It is a direct application of one of the optimization methods discussed in Chapter 4. For some of them, in particular the incremental ones, an analytic gradient is useful. The formulas for simple cases are given in Section 6.5.

To be complete, the possibilities for using an analytic model for neurocontroller training and a neural network plant model for analytic controller design are mentioned in Sections 6.6 and 6.7, respectively.

6.3 CONTROLLER STRUCTURE

6.3.1 Basic Theoretical Schemes

State Models and Input–Output Models Once more, classical control theory provides guidelines for determining the controller structure (i.e., the set of controller inputs and outputs) appropriate for standard control tasks. Narendra and Mukhopadhyay [104] have proposed two theoretically fundamental schemes for a nonlinear discrete-time plant model of the form

$$\mathbf{z}_{t+1} = f(\mathbf{z}_t, \mathbf{u}_t) \tag{6.39}$$

with measurement vector

$$\mathbf{y}_t = h(\mathbf{z}_t) \tag{6.40}$$

Both of them assume controllability of the state by the vector $\mathbf{u}$ as well as observability by vector $\mathbf{y}$. Nonlinear controllability and observability are defined around a certain point in state space in analogy to their linear counterparts by using the Jacobian matrices in a state space point instead of state matrices. For theorems and proofs, see Levin and Narendra [86].

The first scheme incorporates the complete state feedback:

$$\mathbf{u}_t = g(\mathbf{z}_t, \mathbf{w}_t) \tag{6.41}$$

The second scheme uses the fact that, in an observable plant, the current state can be computed from the history of actions and measurements, that is, from the n_o last values of $\mathbf{y}$ and of $\mathbf{u}$. The integer $n_o \leq n$ is the *observability index*,

172 CONTROLLER TRAINING

saying just how many past measurements and actions are necessary to determine the state. The controller now has the form of a dynamic system

$$\mathbf{u}_t = g(\mathbf{y}_t, \ldots, \mathbf{y}_{t-n_o+1}, \mathbf{u}_{t-1}, \ldots, \mathbf{u}_{t-n_o+1}, \mathbf{w}_t) \qquad (6.42)$$

These two forms have guaranteed properties, at least in a certain neighborhood of a fixed state space point: With appropriately designed or trained controller, the closed loop is stabilized at point $\mathbf{w}$.

The controllers for trajectory following are structurally analogous. The state controller has the form

$$\mathbf{u}_t = g(\mathbf{z}_t, \mathbf{w}_t^*) \qquad (6.43)$$

and its input–output counterpart is

$$\mathbf{u}_t = g(\mathbf{y}_t, \ldots, \mathbf{y}_{t-n_o+1}, \mathbf{u}_{t-1}, \ldots, \mathbf{u}_{t-n_o+1}, \mathbf{w}_t^*) \qquad (6.44)$$

The vector $\mathbf{w}_t^*$ is now the vector of future reference states $w_{i,t+r_i}$. This controller can be trained so that the closed loop exactly follows the trajectory given by the time series of $\mathbf{w}_t^*$; that is, it follows a time series $\mathbf{w}_t$ with the minimum accessible delay resulting from the relative degrees of individual elements of measurement vector.

If a neurocontroller is to be designed for a plant modeled by a neural network, it will benefit from the fact that the neural network model has been gained from measurements. Except that the user *wishes* to design a controller that does not require all of the measurements used for plant identification, the complete state characterized by the measurement history is available to the controller. Then, the difference between the state controller (6.41) and input–output controller (6.42) is only formal. In the following, only form (6.42) will be considered.

Comparison of Controllers Using Delays and Those Using Time Derivatives
Discrete-time formulas such as Eq. (6.42) contain delayed values of state or action variables. As discussed in Section 5.3.1, the values of the same variable with various delays by small multiples of the sampling period are highly autocorrelated. This causes convergence problems in optimization. This is why the use of numeric derivatives

$$\mathbf{x}_t^{(k)} = \frac{1}{\Delta_t}(\mathbf{x}_t^{(k-1)} - \mathbf{x}_{t-1}^{(k-1)}) \qquad (6.45)$$

where $\mathbf{x}_t^{(0)} = \mathbf{x}_t$ is preferable. Controllers using numeric derivatives are (exactly like the plant models with numeric derivatives) equivalent in theoretical properties to those using delays. The formulas can be received from those using the complete state by substituting the input–output state

$$\mathbf{z}_t = z_{\text{func}}(\mathbf{y}_t, \ldots, \mathbf{y}_t^{(n_o-1)}, \mathbf{u}_{t-1}, \ldots, \mathbf{u}_{t-1}^{(n_o-2)}) \qquad (6.46)$$

For stabilization, it is the controller

$$\mathbf{u}_t = g(\mathbf{y}_t, \ldots, \mathbf{y}_t^{(n_o-1)}, \mathbf{u}_{t-1}, \ldots, \mathbf{u}_{t-1}^{(n_o-2)}, \mathbf{w}_t) \qquad (6.47)$$

or its equivalent form with output orthogonal to inputs

$$\mathbf{u}_t^{(n_o-1)} = g(\mathbf{y}_t, \ldots, \mathbf{y}_t^{(n_o-1)}, \mathbf{u}_{t-1}, \ldots, \mathbf{u}_{t-1}^{(n_o-2)}, \mathbf{w}_t) \qquad (6.48)$$

For trajectory following, it is the controller

$$\mathbf{u}_t = g(\mathbf{y}_t, \ldots, \mathbf{y}_t^{(n_o-1)}, \mathbf{u}_{t-1}, \ldots, \mathbf{u}_{t-1}^{(n_o-2)}, \mathbf{w}_t^*) \qquad (6.49)$$

or its equivalent orthogonal form

$$\mathbf{u}_t^{(n_o-1)} = g(\mathbf{y}_t, \ldots, \mathbf{y}_t^{(n_o-1)}, \mathbf{u}_{t-1}, \ldots, \mathbf{u}_{t-1}^{(n_o-2)}, \mathbf{w}_t^*) \qquad (6.50)$$

with $\mathbf{w}_t^*$ now being the vector of the r_ith derivatives of reference states $w_{i,t}$. If the relative degrees are not known (and they are usually not), the derivatives of $\mathbf{w}$ are to be included up to the supposed maximum relative degree $r \leq n$, resulting in the controller form

$$\mathbf{u}_t = g(\mathbf{y}_t, \ldots, \mathbf{y}_t^{(n_o-1)}, \mathbf{u}_{t-1}, \ldots, \mathbf{u}_{t-1}^{(n_o-2)}, \mathbf{w}_t, \ldots, \mathbf{w}_t^{(r)}) \qquad (6.51)$$

or

$$\mathbf{u}_t^{(n_o-1)} = g(\mathbf{y}_t, \ldots, \mathbf{y}_t^{(n_o-1)}, \mathbf{u}_{t-1}, \ldots, \mathbf{u}_{t-1}^{(n_o-2)}, \mathbf{w}_t, \ldots, \mathbf{w}_t^{(r)}) \qquad (6.52)$$

An additional reason for using neurocontroller structures based on temporal derivatives is that the continuous-time version of relative degree is more grounded in the properties of real plants as physical systems (see Section 5.3.1).

Note 6.3 Like for plant models (see Note 5.2), the *conceptual structure* of the controller should be distinguished from the *operational structure*. A call of the operational controller consists of three steps:

1. Computing the required numerical derivatives from sampled measurements
2. Evaluating the conceptual controller containing derivatives
3. If necessary, integrating the controller output given in the form of a derivative to receive the next action vector

Robustness by Error Integration The identified plant model is always of limited precision. This leads to a requirement of robustness. The state feedback is sufficient for stabilization but does not imply robustness. Besides plant model inaccuracy, systematic disturbances of unknown form are often present. One could try explicit or implicit observers that increase the order of the dynamic controller, but they may fail because of the inaccuracy of higher-order measurements. Also, using third-order derivatives of measurements as additional controller inputs might theoretically help to train the controller to observe the disturbance but, in fact, will fail because of numeric imprecision of measured third derivatives.

A simple and universal means to cope with systematic disturbances and plant model error equivalent to them is to use the integral-of-error inputs to the controller, or equivalently, control error as input together with integration of controller action.

The nonlinear counterpart of the linear controller using the integral of error described by Eq. (6.24) is

$$\mathbf{u} = g_1(\mathbf{e}, \mathbf{z}, \mathbf{w}) \tag{6.53}$$

where

$$\dot{\mathbf{e}} = \mathbf{w} - \mathbf{y} \tag{6.54}$$

The only stationary state of the plant with the additional state equation (6.54) and controller (6.53) is at $\mathbf{e} = \mathbf{0}$, that is, at $\mathbf{y} = \mathbf{w}$.

Another possibility is controller action integration. The analogy to Eq. (6.25) is

$$\dot{\mathbf{u}} = g_2(\mathbf{w} - \mathbf{y}, \mathbf{z}, \mathbf{u}) \tag{6.55}$$

This form is guaranteed to stabilize the controller but does not guarantee the convergence to $\mathbf{y} = \mathbf{w}$. However, its incomplete feedback form

$$\dot{\mathbf{u}} = g_2(\mathbf{w} - \mathbf{y}) \tag{6.56}$$

has the only stationary state at $\mathbf{y} = \mathbf{w}$.

Note 6.4 Integration of controller action is preferable to integration of error for the following reason: The error integral can grow without bounds and remain large even after the goal state is reached. This problem is well-known in the control design practice and is usually solved by resetting the integral to a zero value. This problem may be a serious obstacle in using error integration in neurocontrol. Integration of controller action does not suffer from this.

CONTROLLER STRUCTURE 175

Incomplete Feedback The idea of incomplete feedback can be used also for nonlinear systems. In fact, with models identified from data, the feedback is always incomplete because some part of the plant dynamics has certainly been neglected through imperfect measurements of a plant order that is too low. For stabilization tasks, the nonlinear analogy to Eq. (6.19) is

$$\mathbf{u}_t = g(\mathbf{y}_t, \mathbf{w}_t) = g'(\mathbf{w}_t - \mathbf{y}_t, \mathbf{w}_t) \tag{6.57}$$

This controller can be viewed as the input–output controller (6.47) with input–output history reduced to the last output vector.

For the decoupling controller (6.51), a less incomplete feedback is received by leaving off the history of **u**

$$\begin{aligned}\mathbf{u}_t &= g(\mathbf{y}_t, \ldots, \mathbf{y}_t^{(r_o-1)}, \mathbf{w}_t, \ldots, \mathbf{w}_t^{(r_o)}) \\ &= g'(\mathbf{y}_t - \mathbf{w}_t, \ldots, \mathbf{y}_t^{(r_o-1)} - \mathbf{y}_t^{(r_o-1)}, \mathbf{w}_t, \ldots, \mathbf{w}_t^{(r_o)})\end{aligned} \tag{6.58}$$

with r_o being the maximum of d_o and r. Both controllers are interesting in that they decompose the input into control errors, that is, differences between the reference values and the measured output, and reference values themselves. In the latter variant, the trajectory is compared up to the maximum relative degree.

Note 6.5 Throughout this chapter, the term *measured output* is used in contrast to the previous model output, to differentiate between static models making a single-step forecast, on the one hand, and dynamic models making multistep forecasts, using its own past outputs in a simulation loop over a certain time horizon. This measured output is used only for the initial state of the model—it is not compared with the model output during controller training.

As mentioned in Note 6.2, it is not necessary that measured output be even measured (although this will frequently be the case, particularly in online schemes). It can equally well be an artificial initial state.

As mentioned above controllers (6.53) and (6.55) can be collapsed into the output feedback forms, respectively,

$$\mathbf{u} = g_1(\mathbf{e}, \mathbf{w} - \mathbf{y}) \tag{6.59}$$

and

$$\dot{\mathbf{u}} = g_2(\mathbf{w} - \mathbf{y}) \tag{6.60}$$

As stated for linear controllers, incomplete feedback cannot guarantee stability for all plants. This does not mean that it does not perfectly stabilize particular plants. On the contrary, the arguments of Section 6.1 suggest that this is probable.

Equilibrium Point View of the Control Problem A stable nonlinear system has an equilibrium point at which the derivatives of the system equation are zero. A closed loop made of the plant model

$$\dot{\mathbf{z}} = f(\mathbf{z}, \mathbf{u}) \tag{6.61}$$

and a controller

$$\mathbf{u} = g(\mathbf{z}, \mathbf{w}) \tag{6.62}$$

described by

$$\dot{\mathbf{z}} = f[\mathbf{z}, g(\mathbf{z}, \mathbf{w})] \tag{6.63}$$

can, like any other system, be viewed in this way [e.g., 32]. The stabilization task can be formulated so that the system (6.63) has an equilibrium point at point $\mathbf{z}_w$, for which

$$\mathbf{0} = f[\mathbf{z}_w, g(\mathbf{z}_w, \mathbf{w})] \tag{6.64}$$

and

$$\mathbf{w} = h(\mathbf{z}_w) \tag{6.65}$$

For a given $\mathbf{w}$, the closed loop (6.63) can be transformed into the form

$$\dot{\mathbf{e}} = f'[\mathbf{e}, g'(\mathbf{e})] \tag{6.66}$$

with $\mathbf{e} = \mathbf{z} - \mathbf{z}_w$ with equilibrium point at the origin. The reference vector is, in this form, always zero and so has been omitted.

There is a good reason for considering the closed-loop equilibrium points. Limited accuracy is a property of every neurocontroller, as well as, of course, any controller designed by numeric methods, which is every industrial controller in the world. To assess the consequences of limited accuracy, it is useful to realize that controller design can be conceptually decomposed into two parts: designing the dynamic behavior, such as the stability and speed of response, and designing the static behavior characterized by the existence of desired equilibrium points at $\mathbf{y} = \mathbf{w}$. The nature of both subtasks results in different requirements on the accuracy. The dynamic behavior design is a constraint-satisfaction and optimization task: The controller should be such that the stability constraints on closed-loop eigenvalues are satisfied with some reserve and the response is as good as possible. By contrast, errors concerning the existence of appropriate equilibrium points are more disastrous: The controller can either reach the goal state but remain unstable there or reach another state and remain stable there. This suggests that the equilibrium point definition must be paid particular attention. The best principle to follow is to shift the desired state to a fixed point whose existence is guaranteed by the construction of the controller, for example, the coordinate origin.

If the complete state were known and if the reference vector specified the complete goal state, this problem could be easily solved in the following way. The controller g' would be transformed into the equilibrium point form

$$\mathbf{u} = g_1(\mathbf{z} - \mathbf{w}) + g_2(\mathbf{w}) \tag{6.67}$$

with two parts:

1. Mapping $g_1(\mathbf{z} - \mathbf{w})$ would be a function of the deviation of the measured state from the reference state. This g_1 would have the property of mapping a zero vector on a zero vector.
2. Mapping $g_2(\mathbf{w})$ would be a function of the reference vector alone, and would map $\mathbf{w}$ on action vector φ that is consistent with the reference state $\mathbf{w}$, or formally, to φ for which

$$\mathbf{0} = f[\mathbf{w}, g_2(\mathbf{w})] \tag{6.68}$$

This would guarantee the existence of a closed-loop equilibrium point at which $\mathbf{z} = \mathbf{w}$:

$$f[\mathbf{w}, g_1(\mathbf{w} - \mathbf{w}) + g_2(\mathbf{w})] = f[\mathbf{w}, g_2(\mathbf{w})] = \mathbf{0} \tag{6.69}$$

A generalized equilibrium point also can be defined for the case of trajectory following. The term *equilibrium point* is to be understood now relative to the trajectory. Once more, the controller must be decomposed into the state feedback part maintaining stability and the feedforward part computing the action for ideal trajectory. Then, the mapping g_2 requires $\dot{\mathbf{w}}$ as an additional argument and is defined by the equation

$$\dot{\mathbf{w}} = f[\mathbf{w}, g_2(\mathbf{w}, \dot{\mathbf{w}})] \tag{6.70}$$

The resulting controller is

$$\mathbf{u} = g_1(\mathbf{e}) + g_2(\mathbf{w}, \dot{\mathbf{w}}) \tag{6.71}$$

Both of these controllers can be designed by appropriate choice of mapping g_1, so that the closed loop is stable—the argument of g_1, the error $\mathbf{z} - \mathbf{w}$, represents a complete state feedback. However, because the feedback is relative to $\mathbf{w}$, the concrete form of mapping g_1 would have to depend on $\mathbf{w}$.

Note that the existence of mapping g_2 defined by Eq. (6.68) or (6.70) is not guaranteed for arbitrary $\mathbf{w}$. This is the analogy to the linear case where, at best, the same number of outputs can be independently specified as there are control actions.

Similar concepts can also be applied to input–output models (6.48) and (6.52). The equilibrium point is applied to **y** and its derivatives as well as to **u** and its derivatives. This means that instead of **y**, $\mathbf{e} = \mathbf{y} - \mathbf{w}$ is substituted, and instead of **u**, $\mathbf{v} = \mathbf{u} - \boldsymbol{\varphi}$ is taken, such that

$$\mathbf{v}_t^{(n_o-1)} = g_1(\mathbf{e}_t, \ldots, \mathbf{e}_t^{(n_o-1)}, \mathbf{v}_{t-1}, \ldots, \mathbf{v}_{t-1}^{(n_o-2)}) \tag{6.72}$$

For stabilization tasks, $\boldsymbol{\varphi} = g_2(\mathbf{w})$ is defined by setting the input–output model to the stationary state at $\mathbf{y} = \mathbf{w}$:

$$\mathbf{0} = f[\mathbf{w}, \mathbf{0}, \ldots, \mathbf{0}, g_2(\mathbf{w}), \mathbf{0}, \ldots, \mathbf{0}] \tag{6.73}$$

Because all derivatives of both **w** and consequently also of $\boldsymbol{\varphi}$ are zero, Eq. (6.72) becomes

$$\mathbf{u}_t^{(n_o-1)} = g_1(\mathbf{y}_t - \mathbf{w}_t, \ldots, \mathbf{y}_t^{(n_o-1)}, \mathbf{u}_{t-1} - \boldsymbol{\varphi}_{t-1}, \ldots, \mathbf{u}_{t-1}^{(n_o-2)}) \tag{6.74}$$

For trajectory following, the controller action is computed by

$$\mathbf{u}_t^{(n_o-1)} = \boldsymbol{\varphi}_t^{(n_o-1)} + g_1(\mathbf{e}_t, \ldots, \mathbf{e}_t^{(n_o-1)}, \mathbf{v}_{t-1}, \ldots, \mathbf{v}_{t-1}^{(n_o-2)}) \tag{6.75}$$

The feedforward component $\boldsymbol{\varphi}$ is now a dynamic system

$$\boldsymbol{\varphi}_t^{(n_o-1)} = g_2(\mathbf{w}_t, \ldots, \mathbf{w}_t^{(n_o-1)}, \boldsymbol{\varphi}_{t-1}, \ldots, \boldsymbol{\varphi}_{t-1}^{(n_o-2)}) \tag{6.76}$$

Note 6.6 Strictly speaking, n_o is to be chosen so that it is not lower than the highest relative degree expected.

So far, the controller has been consisting of two parts: a feedforward part defining the equilibrium point, and a feedback part responsible for stability around this equilibrium point. The value of this approach is just in this separation. The feedforward part can be trained in advance. For stabilization tasks, data about stationary states of the open loop are sufficient for training. By such separation, the training of the feedback part is facilitated. This is a procedure that was working well in practice with trajectory-following tasks, for example, with the elastomer test bench of Chapter 12. However, some stabilization tasks may require higher precision of the equilibrium point. Then, it is desirable to make this precision independent from any training—the controller should possess this equilibrium point by definition.

With the knowledge of a complete state and a complete reference vector for this state, the closed loop made of model (6.61) and controller

$$\dot{\mathbf{u}} = g(\mathbf{z} - \mathbf{w}) \qquad (6.77)$$

has an equilibrium point at $\mathbf{z} - \mathbf{w}$ if the mapping f is such that $\mathbf{0} = g(\mathbf{0})$. Note that the feedback is incomplete—the complete one would have to contain the state $\mathbf{u}$ in the form

$$\dot{\mathbf{u}} = g(\mathbf{z} - \mathbf{w}, \mathbf{u}) \qquad (6.78)$$

or

$$\dot{\mathbf{u}} = g(\mathbf{z} - \mathbf{w}, \mathbf{u} - \boldsymbol{\varphi}) \qquad (6.79)$$

The reason for the omission is that form (6.78) has no equilibrium point at $\mathbf{z} - \mathbf{w}$ (for nonzero $\mathbf{u}$, $\dot{\mathbf{u}}$ may be nonzero), whereas the form (6.79) requires the knowledge of $\boldsymbol{\varphi}$ (which is to be avoided). With an input–output model, the controller (6.77) would be

$$\mathbf{u}_t^{(n_o-1)} = g_1(\mathbf{y}_t - \mathbf{w}_t, \ldots, \mathbf{y}_t^{(n_o-1)}, \mathbf{u}_{t-1}^1, \ldots, \mathbf{u}_{t-1}^{(n_o-2)}) \qquad (6.80)$$

The only difference to Eq. (6.74) consists in omitting the unknown term $\mathbf{u}_{t-1} - \boldsymbol{\varphi}_{t-1}$. A reduced, but frequently sufficient, form of Eq. (6.80) is

$$\mathbf{u}_t^{(1)} = g_1(\mathbf{y}_t - \mathbf{w}_t, \ldots, \mathbf{y}_t^{(n_o-1)}) \qquad (6.81)$$

omitting also the feedback by the derivatives of $\mathbf{u}$. This form is static and thus substantially easier to train.

It is important to point out that also for input–output models, the mapping g_1 is specific for a given equilibrium point. That is, it is additionally dependent on the reference state or trajectory. The complete dependence, in state representation, is then

$$g_1(\mathbf{e}_e, \mathbf{w}_e) \qquad (6.82)$$

In input–output representation, the extended error vector $\mathbf{e}_e$ stands for output errors and their derivatives as well as for the action offsets $\mathbf{v}$. With trajectory following, the extended reference vector $\mathbf{w}_e$ may include the derivatives. The difference between both arguments is in their relationship to the equilibrium point—only the first argument is required to satisfy

$$g_1(\mathbf{0}, \mathbf{w}) = \mathbf{0} \qquad (6.83)$$

Note 6.7 An important implicit assumption underlying the equilibrium point-based controller structures has been that reference values or trajectories are specified for *all output variables*. The importance of guaranteeing the consistency of the reference vector has been pointed out already.

Except for singular cases, the set of stationary states of a system with n state variables and p action variables is a p-dimensional subspace of the n-dimensional state space. This means that, with p action variables, p output variables can be assigned independent reference states. This is true for stabilization tasks. In trajectory-following tasks, exact trajectory following of p outputs cannot be guaranteed. The reason for this is that the input affects the derivative of the output given by the relative degree. (Additionally, the output depends on the state.)

If the number of output variables is less that p, the reference state is mostly consistent within the range given by plant nonlinearities. However, then the mapping f_2 is not unique; the whole subspace of possible actions leads to the same stationary state (a problem known from robotics, where redundant servos for individual joints are available). With the number of output variables exceeding p, the consistency of reference state or trajectory has to be checked.

Additional Requirements on Control Plant stabilization and trajectory following are the most frequent tasks for neurocontrol. However, they rarely appear in their pure theoretical form. Frequently constraints are added, or replace, the reference values. For example, the steering control task might be to keep the vehicle's longitudinal axis within an offset of 50 cm from the middle lane of the road while minimizing the lateral acceleration.

This introduces the question of what should the role of reference values be in such settings. One approach is by equivalence classes. In the steering problem, the equivalence class is constituted by all offsets that are less than or equal to 50 cm. However, such an equivalent class cannot be passed as an argument to a function directly. The reference value must be in a defined relationship to the measurements for the controller to be able to determine whether the goal has been reached or not. It is possible to define an equivalence-class-based representation of the difference between the reference value w_{crit} that represents, for example, an upper bound of admissible values, and the output value y, of the form $\max(y - w_{\text{crit}}, 0)$. For $y > w_{\text{crit}}$, the computed value represents a measure of constraint violation; otherwise, it is zero, representing the equivalence class of admissible control. Of course, the violation measure can be arbitrarily complex in practice.

Using such controller inputs may be practical in solving such nonstandard control tasks. However, it must be kept in mind that such representations do not possess guaranteed properties of standard stabilization or trajectory-following tasks. For example, the stabilization in the sense of nonoscillating trajectories cannot be guaranteed. Also, information about a hidden state via so manipulated output variables may be incomplete and lead to further complications.

Another extension of standard control tasks is making the controller dependent on some exogenous inputs that cannot be actively affected but can be measured. These inputs can be one of the following:

Slowly changing inputs that affect the gains of the plant, as represented by the vector **p** of the model (5.81). Such a parameter is, for example, temperature or device age.

Inputs that have the character of disturbances, such as lateral wind.

Both types of plant inputs can be viewed as extended states that are included in the controller input vector. The controller can then be optimized for performance under such variable conditions. (This roughly corresponds to the idea of *gain scheduling* in classical control [8].)

The first type of input, the slowly changing parameters, are sufficiently accounted for by including the actual (zeroth-derivative) values. The second type may require including the derivatives.

Note 6.8 With gain vector networks, the representation of external inputs is addressed in Note 6.10.

Summary on Controller Structures In the discussion of the preceding paragraphs, many diverse structures have been proposed. The following criteria helped determine which structures to consider for applications:

- Because most neurocontrollers are based on neural network models gained from measurements, that is, on input–output models, only structures using plant inputs and outputs are presented.
- The structures based on numerical derivatives are presented instead of those based on delays. This is justified by their advantageous properties for numeric optimization.
- Wherever relative degrees justify the controller structure, they are supposed not to be known exactly. Only their upper bound, or the maximum from the upper bound and the observability index, is to be specified.
- Only general structures are presented with arbitrary nonlinear plant models and complete feedback. The only exception is the incomplete feedback controller with integration of control action. It has been included because, in the author's experience, it has frequently been superior in performance to all other structures for stabilization tasks.

Now, the controllers chosen for consideration are as follows:

1. *Basic Controller—Stabilization Version*

$$\mathbf{u}_t^{(n_o-1)} = g(\mathbf{y}_t, \ldots, \mathbf{y}_t^{(n_o-1)}, \mathbf{u}_{t-1}, \ldots, \mathbf{u}_{t-1}^{(n_o-2)}, \mathbf{w}_t) \qquad (6.84)$$

2. *Basic Controller—Trajectory-Following Version*

$$\mathbf{u}_t^{(n_o-1)} = g(\mathbf{y}_t, \ldots, \mathbf{y}_t^{(n_o-1)}, \mathbf{u}_{t-1}, \ldots, \mathbf{u}_{t-1}^{(n_o-2)}, \mathbf{w}_t, \ldots, \mathbf{w}_t^{(r)}) \quad (6.85)$$

3. *Controller with Separate Feedforward Mapping g_2—Stabilization Version*

$$\mathbf{u}_t^{(n_o-1)} = g_1(\mathbf{y} - \mathbf{w}_t, \ldots, \mathbf{y}_t^{(n_o-1)}, \mathbf{u}_{t-1} - \boldsymbol{\varphi}_{t-1}, \ldots, \mathbf{u}_{t-1}^{(n_o-2)}; \mathbf{w}_t) \quad (6.86)$$

with $\boldsymbol{\varphi} = g_2(\mathbf{w})$. The semicolon separates the arguments that, if zero, imply that the mapping output be zero, from those that do not. The closed-loop equilibrium point is, by definition, at $g_1 = \mathbf{0}$. The accuracy of its consistency with the reference state depends only on the feedforward part g_2.

4. *Controller with Separate Feedforward Mapping g_2—Trajectory-Following Version*

$$\mathbf{u}_t^{(n_o-1)} = \boldsymbol{\varphi}_t^{(n_o-1)}$$
$$+ g_1(\mathbf{e}_t, \ldots, \mathbf{e}_t^{(n_o-1)}, \mathbf{v}_{t-1}, \ldots, \mathbf{v}_{t-1}^{(n_o-2)}; \mathbf{w}, \ldots, \mathbf{w}_t^{(n_o-1)}) \quad (6.87)$$

with

$$\boldsymbol{\varphi}_t^{(n_o-1)} = g_2(\mathbf{w}_t, \ldots, \mathbf{w}_t^{(n_o-1)}, \boldsymbol{\varphi}_{t-1}, \ldots, \boldsymbol{\varphi}_{t-1}^{(n_o-2)}) \quad (6.88)$$

and $\mathbf{v}_t = \mathbf{u}_t - \boldsymbol{\varphi}_t$. The index n_o denotes the maximum from the observability index and relative degree. The closed-loop equilibrium point is, by definition, at $g_1 = \mathbf{0}$. The accuracy of its consistency with the reference trajectory depends only on the feedforward part g_2.

5. *Controller with Error Integration*

$$\mathbf{u}_t^{(n_o-1)} = g(\mathbf{e}, \mathbf{y}_t, \ldots, \mathbf{y}_t^{(n_o-1)}, \mathbf{u}_{t-1}, \ldots, \mathbf{u}_{t-1}^{(n_o-2)}, \mathbf{w}_t) \quad (6.89)$$

with **e** being

$$\dot{\mathbf{e}} = \mathbf{w} - \mathbf{y} \quad (6.90)$$

This controller is robust against plant model inaccuracies as well as against constant systematic disturbances.

6. *Controller with Integration of Action*

$$\mathbf{u}_t^{(1)} = g_1(\mathbf{y} - \mathbf{w}_t, \ldots, \mathbf{y}_t^{(n_o-1)}; \mathbf{w}) \quad (6.91)$$

This controller is robust against plant model inaccuracies as well as against constant systematic disturbances. The closed-loop equilibrium point is, by definition, at $g_1 = \mathbf{0}$. This equilibrium point is completely consistent with an arbitrary reference state.

Important properties of these controllers are summarized in Table 6.1.

Whenever plant model inaccuracies are expected (e.g., because of noisy data used for plant identification) and high precision of stabilization is important, robust versions should be used. Expected presence of systematic disturbances is another reason for doing this.

Structures with defined equilibrium points provide computational advantages for reasons mentioned earlier in this section. There is usually more chance for getting good results with them than for structures with arbitrary equilibrium points. A comparison is given in the case study of Chapter 12.

The long argument lists defining these controller structures, particularly those with defined equilibrium points, does not make them appealing. However, their instantiations to a concrete plant model order seem much better. They will be listed in Section 6.3.2.

All of these controller schemes can be adapted to a particular situation:

1. The complete feedback requirement can be relaxed; incomplete feedback, if justified by practical considerations, is mostly sufficient.
2. Additional inputs such as distances from hard limits or constraints can be used.
3. The controller can be broken down to partial controllers if the structure of the problem allows it (see Section 6.3.2).

6.3.2 Representation of Controller Mapping by a Neural Network

Controller mapping in its general form can be represented, in the simplest case, by a neural network such as the multilayer perceptron. However, this is often not the best approach. For some applications, the controller can be structured to discard irrelevant dependences and thus make the optimization task easier than with the mechanical application of the general form. Another aspect

Table 6.1 Properties of controller structures

Controller Type	Equation for		Specified Fixed Point	Robust	Complete Feedback
	Stabilization	Tracking			
Basic schemes	(6.84)	(6.85)	No	No	Yes
Equilibrium point schemes	(6.86)	(6.87)	Yes	No	Yes
Error integration scheme	(6.89)	—	No	Yes	Yes
Action integration scheme	(6.91)	—	Yes	Yes	No

concerns the embedding of the neural network into the mapping. Additional requirements on the representation depend on the equilibrium-point forms of controllers. These are the questions addressed in this section.

Structuring the Controller In certain cases, general controller structures of Section 6.3.1 can be simplified by breaking them down into several partial controllers. This is analogous to representing partial plant models, as discussed in Section 5.3.3. However, there is an important difference. The plant model is open loop. In practice, each link of its chain of building blocks can frequently be identified separately. By contrast, a feedback controller acts, by its definition, as a closed loop. This is why it cannot simply be decomposed, for example, in accordance with the localization of control actions in particular plant components; the state vector of the closed loop may not be separable. While decomposition may be necessary for complexity reasons, it must be kept in mind that the controllability of the plant can be affected negatively.

Nevertheless, there are some situations in which controller structuring can legitimately reduce computational complexity without deteriorating the quality of the solution:

1. The plant structure may indeed be separable. This is the case if some part of the plant is physically independent from the rest, or if this dependence is negligible. For example, the steering control and the brake control of a vehicle can be separated because they act along different spatial coordinates and are thus somewhat separate.

2. The mutual dependences may be mediated by a limited number of state variables that can be viewed as measurable or unmeasurable disturbances. For example, the behavior of each wheel of a vehicle is influenced by the movement of the vehicle's body. This movement depends on all the wheels working together and thus each wheel depends on the other. However, the separate control of each wheel can consider the state of this wheel and the state of the body.

3. The system can be viewed as nested closed loops with different time scales relevant for the behavior. Then, outer loops can view inner loop controllers as parts of the plant. Inner loops, in turn, view the actions of outer loop controllers as external input, disturbance, or reference values. For example, in the biological wastewater treatment control of Chapter 15, the action components are aerators influencing the oxygen concentration. This, in turn, affects the biochemical processes in the plant. Because the oxygen concentration changes by aeration are substantially faster than those caused by biochemical processes, the aeration process can be viewed as an inner-loop process with reference values for oxygen concentration received from the outer-loop control.

Representation of Controller Structures with Specified Equilibrium Points

Basic schemes like Eqs. (6.84) and (6.85) can be implemented, say, as multilayer perceptrons in a straightforward way. However, it has been mentioned that using controller representations with defined equilibrium points as a part of the formulation may be a substantial help in attaining good results. As stated in Section 6.3.1, schemes such as Eqs. (6.86), (6.87), and (6.91) can define a closed-loop equilibrium point if the controller mapping g_1

$$g_1(\mathbf{x}_a, \mathbf{x}_b) \tag{6.92}$$

has the property

$$g_1(\mathbf{0}, \mathbf{x}_b) = \mathbf{0} \tag{6.92}$$

for arbitrary $\mathbf{x}_b$.

To reach this goal, networks should be used that allow the specification of which subset of input variables defines the equilibrium point and which does not. Alternative network configurations have been discussed in Section 3.2.5. One particularly flexible and transparent form is the gain vector network. The concrete representation of function g_1 is by

$$g_1(\mathbf{x}_a; \mathbf{x}_b) = g'_1(\mathbf{x}_a, \mathbf{x}_b)\mathbf{x}_a \tag{6.94}$$

This representation suggests the interpretation of variable linearization of the function in the working point specified by $\mathbf{x}_a, \mathbf{x}_b$. This view is particularly appropriate for control where, as is often the case, theorems about nonlinear systems (1) make statements about the behavior *around a certain working point* (rather than global statements) and (2) can be proved by linearization of this working point. In this view, the vector $[\mathbf{x}_a, \mathbf{x}_b]$ in the argument of vector function g'_1 would take the place of a working point definition, whereas the same vector multiplying this function would take the place of a linear analogy.

Note 6.9 In Eq. (6.94), the equilibrium-point-relative vector $\mathbf{x}_a$ is also an argument of function g'_1. This is necessary for the function also to be nonlinear in $\mathbf{x}_a$. If, instead, the form

$$g_1(\mathbf{x}_a, \mathbf{x}_b) = g'_1(\mathbf{x}_b)\mathbf{x}_a \tag{6.95}$$

were used, the function would represent a continuum of variable linear functions depending on $\mathbf{x}_b$. In terms of controllers (6.86), (6.87), and (6.91), this would amount to a *continuum of working-point-specific linear controllers*, that is, linear controllers that are specific for each reference state or each point on the reference trajectory. This restricts, to a certain extent, general nonlinear controllers, and it prevents optimal nonlinear (e.g., time-optimal) control strategies from being found by training.

On the other hand, this arrangement eliminates the multiplicative influence of state or output variables via argument of g'_1 and via vector multiplication. Since the reference trajectory is usually smooth, and measurements are usually noisy, the smoothness of control action and, consequently, the small-scale stability of the controller are frequently improved.

With gain vector networks, the controller structures with defined equilibrium points can be rewritten in the following way:

3a. *Controller with Separate Feedforward Mapping g_2—Stabilization Version (6.86)*

$$\mathbf{u}_t^{(n_o-1)} = g'_1(\mathbf{y}-\mathbf{w}_t,\ldots,\mathbf{y}_t^{(n_o-1)},\mathbf{u}_{t-1}-\boldsymbol{\varphi}_{t-1},\ldots,\mathbf{u}_{t-1}^{(n_o-2)},\mathbf{w}_t) \begin{bmatrix} \mathbf{y}-\mathbf{w}_t \\ \cdot \\ \cdot \\ \mathbf{y}_t^{(n_o-1)} \\ \mathbf{u}_{t-1}-\boldsymbol{\varphi}_{t-1} \\ \cdot \\ \cdot \\ \mathbf{u}_{t-1}^{(n_o-2)} \end{bmatrix}$$

(6.96)

with mapping g_2 represented directly, for example, by a multilayer perceptron.

4a. *Controller with Separate Feedforward Mapping g_2—Trajectory-Following Version (6.87)*

$$\mathbf{u}_t^{(n_o-1)} = \boldsymbol{\varphi}_t^{(n_o-1)} + g'_1(\mathbf{e}_t,\ldots,\mathbf{e}_t^{(n_o-1)},\mathbf{v}_{t-1},\ldots,\mathbf{v}_{t-1}^{(n_o-2)},\mathbf{w},\ldots,\mathbf{w}_t^{(n_o-1)}) \begin{bmatrix} \mathbf{e}_t \\ \cdot \\ \cdot \\ \mathbf{e}_t^{(n_o-1)} \\ \mathbf{v}_{t-1} \\ \cdot \\ \cdot \\ \mathbf{v}_{t-1}^{(n_o-2)} \end{bmatrix}$$

(6.97)

with mapping g_2 represented directly.

CONTROLLER STRUCTURE 187

6a. *Controller with Integration of Action (6.91)*

$$\mathbf{u}_t^{(1)} = g'_1(\mathbf{y} - \mathbf{w}_t, \ldots, \mathbf{y}_t^{(n_o-1)}, \mathbf{w}) \begin{bmatrix} \mathbf{y} - \mathbf{w}_t \\ \vdots \\ \mathbf{y}_t^{(n_o-1)} \end{bmatrix} \quad (6.98)$$

Note 6.10 If external inputs to the plant are to be included in the controller input vector (see Note 6.8), they are to be taken as the arguments of the mappings represented by neural networks, that is, of mappings g'_1 and g_2.

Their role in mapping g_2 is that action necessary to reach a reference state may be shifted. In mapping g'_1, additional inputs can modify the controller gain if the plant gain is changed. This is frequently the case with slowly changing parameters such as temperature or age, which modify the plant behavior, sometimes in a nonlinear way. By contrast, some disturbances, such as lateral wind in the vehicle-steering problem, do not affect the dynamic behavior except by introducing a control offset. Then, their inclusion into the input vector of feedforward mapping g_2 is sufficient.

Formulas for Plants with Observability Index 1 and 2 Finally, as promised above, let us turn to the actual formulas for low-order models. As argued in Section 5.3.2, these models cover much of what is feasible in practice.

Maximum Observability Index 2, Relative Degree 1

1. Controllers with a defined equilibrium point:
   a. Controller with separate feedforward mapping g_2—stabilization version (6.96)

$$\mathbf{u}_t^{(1)} = g'_1(\mathbf{y}_t - \mathbf{w}_t, \mathbf{y}_t^{(1)}, \mathbf{u}_{t-1} - \varphi_{t-1}, \mathbf{w}_t) \begin{bmatrix} \mathbf{y}_t - \mathbf{w}_t \\ \mathbf{y}_t^{(1)} \\ \mathbf{u}_{t-1} - \varphi_{t-1} \end{bmatrix} \quad (6.99)$$

   and

$$\varphi_{t-1} = g_2(\mathbf{w}_{t-1}) \quad (6.100)$$

   b. Controller with separate feedforward mapping g_2—trajectory following version (6.97)

$$\mathbf{u}_t^{(1)} = \varphi_t^{(1)} + g'_1(\mathbf{e}_t, \mathbf{e}_t^{(1)}, \mathbf{u}_{t-1} - \varphi_{t-1}, \mathbf{w}, \mathbf{w}_t^{(1)}) \begin{bmatrix} \mathbf{e}_t \\ \mathbf{e}_t^{(1)} \\ \mathbf{u}_{t-1} - \varphi_{t-1} \end{bmatrix} \quad (6.101)$$

and

$$\varphi_t^{(1)} = g_2(\mathbf{w}_t, \mathbf{w}_t^{(1)}, \varphi_{t-1}) \tag{6.102}$$

c. Controller with integration of action (6.98)

$$\mathbf{u}_t^{(1)} = g_1'(\mathbf{y}_t - \mathbf{w}_t, \mathbf{y}_t^{(1)}, \mathbf{w}) \begin{bmatrix} \mathbf{y}_t - \mathbf{w}_t \\ \mathbf{y}_t^{(1)} \end{bmatrix} \tag{6.103}$$

2. Controllers without a defined equilibrium point:
   a. Basic controller—stabilization version

$$\mathbf{u}_t^{(1)} = g(\mathbf{y}_t, \mathbf{y}_t^{(1)}, \mathbf{u}_{t-1}, \mathbf{w}_t) \tag{6.104}$$

b. Basic controller—trajectory following version

$$\mathbf{u}_t^{(1)} = g(\mathbf{y}_t, \mathbf{y}_t^{(1)}, \mathbf{u}_{t-1}, \mathbf{w}_t, \mathbf{w}_t^{(1)}) \tag{6.105}$$

c. Controller with error integration

$$\mathbf{u}_t^{(1)} = g(\mathbf{e}, \mathbf{y}_t, \mathbf{y}_t^{(1)}, \mathbf{u}_{t-1}, \mathbf{w}_t) \tag{6.106}$$

with **e** being

$$\mathbf{e}_t^{(1)} = \mathbf{w}_t - \mathbf{y}_t \tag{6.107}$$

Maximum Observability Index 1, Relative Degree 1

1. Controllers with a defined equilibrium point
   a. Controller with separate feedforward mapping g_2—stabilization version (6.96)

$$\mathbf{u}_t = g_2(\mathbf{w}_t) + g_1'(\mathbf{y}_t - \mathbf{w}_t, \mathbf{w}_t)[\mathbf{y}_t - \mathbf{w}_t] \tag{6.108}$$

b. Controller with separate feedforward mapping g_2—trajectory following version (6.97)

$$\mathbf{u}_t = g_2(\mathbf{w}_t, \mathbf{w}_t^{(1)})$$
$$+ g_1'(\mathbf{y}_t - \mathbf{w}_t, \mathbf{y}_t^{(1)} - \mathbf{w}_t^{(1)}, \mathbf{w}_t, \mathbf{w}_t^{(1)}) \begin{bmatrix} \mathbf{y}_t - \mathbf{w}_t \\ \mathbf{y}_t^{(1)} - \mathbf{w}_t^{(1)} \end{bmatrix} \tag{6.109}$$

c. Controller with integration of action (6.98)

$$\mathbf{u}_t^{(1)} = g'_1(\mathbf{y}_t - \mathbf{w}_t, \mathbf{w})[\mathbf{y}_t - \mathbf{w}_t] \qquad (6.110)$$

2. Controllers without a defined equilibrium point:
   a. Basic controller—stabilization version

   $$\mathbf{u}_t = g(\mathbf{y}_t, \mathbf{w}_t) \qquad (6.111)$$

   b. Basic controller—trajectory following version

   $$\mathbf{u}_t = g(\mathbf{y}_t, \mathbf{w}_t, \mathbf{w}_t^{(1)}) \qquad (6.112)$$

   c. Controller with error integration

   $$\mathbf{u}_t = g(\mathbf{e}, \mathbf{y}_t, \mathbf{w}_t) \qquad (6.113)$$

   with **e** being

   $$\mathbf{e}_t^{(1)} = \mathbf{w}_t - \mathbf{y}_t \qquad (6.114)$$

Like plant models using numeric derivatives, the controllers should be transformed to the operational form based on delays (see Note 6.3).

6.4 COST FUNCTION FOR NEUROCONTROLLER TRAINING

6.4.1 Basic Concepts

The cost function for closed-loop optimization is the function of variables characterizing the behavior of the closed loop. This function is minimized by an optimization or learning algorithm.

To represent an adequate cost function for this goal, the plant model has to be reformulated. Because the internal structure of the model is not relevant here, we proceed from the encapsulated, operational form of the plant model (5.82):

$$\mathbf{y}_{t+1} = F(\mathbf{y}_{e_t}, \mathbf{u}_t, \omega_f) \qquad (6.115)$$

It has been mentioned in Section 5.3.3 that a plant model can be extended by a modifier vector **p** of Eq. (5.81) that makes the model dependent on slowly changing, measurable parameters. It is also possible that not all plant inputs can be manipulated, that is, that not all elements of vector **u** in the identified plant model (6.115) are also control actions. For example, measurable external disturbances such as lateral wind in steering control (Notes 6.8 and 6.10) are plant inputs that cannot be affected by the controller.

The plant model parameters ω_f have been assigned fixed values. However, closed-loop optimization can simultaneously be performed for multiple models, identified under different conditions, to reach robustness (see

Chapter 7; a multiple-model neurocontroller approach has also be used by Feldkamp and Puskorius [29]). Then, the parameters ω_f, or their compressed representation, such as the model identifier, can enter controller training. These parameters may be viewed as measurable (if it is known in the real-time control environment which model variant is to be taken) or nonmeasurable (if it is not known).

To embed the plant model (6.115) in the closed loop, we extend Eq. (6.115) to

$$\mathbf{y}_{t+1} = F(\mathbf{y}_{e_t}, \mathbf{u}_{e_t}, \boldsymbol{\theta}, \boldsymbol{\zeta}) \qquad (6.116)$$

The definitions valid for (6.116) are the following: (1) $\mathbf{y}_{e_t}$ is the output feedback to the model, including all delayed outputs. (2) $\mathbf{y}_{t+1}$ is the next model output. (3) $\mathbf{u}_{e_t}$ is the vector of genuine action variables (i.e., outputs of the neurocontroller controller), including their delayed values. (4) $\boldsymbol{\theta}$ is the vector of all *measurable* plant inputs or parameters that are not control actions. This vector may include parts of parameter vector $\mathbf{p}$ of Eq. (5.81), parts of plant input $\mathbf{u}_t$ of Eq. (6.115), as well as measurable identifiers of model versions mentioned above. (5) $\boldsymbol{\zeta}$ is the vector of all *nonmeasurable* plant inputs or parameters. This vector usually consists of nonmeasurable identifiers of model versions mentioned above or other parameters and modifiers that express assumed plant variations. It may also contain parts of parameter vector $\mathbf{p}$ of Eq. (5.81) or parts of plant input $\mathbf{u}_t$ of Eq. (6.115), if the user decides that these variables, although measurable at the time of data sampling for identification, cannot be measured in real-time control.

Now, a similarly general formulation for the controller is given:

$$\mathbf{u}_t = G(\mathbf{y}_{e_t}, \mathbf{u}_{e_t}, \mathbf{w}_t, \boldsymbol{\theta}, \boldsymbol{\omega}_g) \qquad (6.117)$$

where $\mathbf{u}_t$ is the actual action vector; $\mathbf{w}_t$ is the generalized reference vector; and $\boldsymbol{\omega}_g$ is the vector or free parameters of the neurocontroller network; in accordance to the control problem type, $\mathbf{w}_t$ contains, for example, information about a reference state, trajectory, constraints or distance from them.

The optimization problem consists of looking for the value of parameter vector $\boldsymbol{\omega}_g$ that minimizes the control objective function

$$C_g = \int_\theta \int_\zeta \int_\mathbf{w} \int_{\mathbf{y}_{e_0}} \sum_{t=0}^{T-1} c_t(\mathbf{y}_{t+1}, \mathbf{u}_t, \mathbf{w}_t, \boldsymbol{\theta}, \boldsymbol{\zeta}) \rho(\mathbf{y}_{e_0}, \mathbf{w}, \boldsymbol{\zeta}, \boldsymbol{\theta}) d\mathbf{y}_{e_0} d\mathbf{w} d\boldsymbol{\zeta} d\boldsymbol{\theta} \qquad (6.118)$$

The integration goes over the following domains:

1. All values of measurable parameters $\boldsymbol{\theta}$. (More generally, all *sequences* of $\boldsymbol{\theta}_t$ might be considered, under a slight notation modification.)
2. All values of nonmeasurable parameters $\boldsymbol{\zeta}$.
3. All sequences $\mathbf{w}$ of reference vectors $\mathbf{w}_t$.
4. All initial states $\mathbf{y}_{e_0}$.

The distribution of parameters, reference trajectories, and initial states is given by the function p. In practice, this distribution is represented by a limited number of training examples so that the quadruple integration becomes a sum over all training examples. How to construct training examples is briefly discussed in Section 6.4.4.

Cost terms c_t can express the square deviation from goal value or goal trajectory, costs, penalties for violating constraints, and so on. This is the topic of Section 6.4.3.

The length of optimization horizon T has to do with the question of single-step and multistep cost criteria. It is addressed in Section 6.4.2.

Note 6.11 Besides this type of cost function, based on closed-loop simulation, different approaches are conceivable. Some examples concerning stability criteria in the cost function are given in Section 9.2. Even frequency-domain-based formulations are possible, as is briefly addressed in Section 7.4.4.

6.4.2 Single-Step and Multistep Criteria

The criterion for determining optimization horizon T is that credit assignment can be done. That is, during the time period up to the horizon, the complete control sequence with all its consequences of numerically relevant size can be terminated. There are different opinions about how long this sequence should be.

One extreme is to take a horizon equal to the relative degree of the plant. The training algorithm for this case has been proposed by Narendra and Mukhopadhyay [104], and it is presented in Section 6.5.1. This approach is based on the assumption of control action without practical limits, which would prevent the closed loop from reaching an arbitrary state within the time given by the relative degree. In this case, a model based on relative degree must be used: The model output $\mathbf{y}^*_{t+1}$ is the prediction by r_i steps, or the ith derivative of output (r_i being the relative degree of the ith output variable). In addition, $\mathbf{w}^*_{t+1}$ is the reference value r_i steps ahead, or the ith derivative of reference trajectory.

The model (6.116) is then reformulated as

$$\mathbf{y}^*_{t+1} = F(\mathbf{d}_{e_t}, \mathbf{u}_{e_t}, \boldsymbol{\theta}, \zeta) \tag{6.119}$$

that is, with the extended measurement vector $\mathbf{d}_{e_t}$ containing the measured output (in the sense of Note 6.5) and its history, instead of vector $\mathbf{y}_{e_t}$ containing the model output and its history. The same principle is applied to the computation of the controller (6.117); that is,

$$\mathbf{u}_t = G(\mathbf{d}_{e_t}, \mathbf{u}_{e_t}, \mathbf{w}^*_t, \boldsymbol{\theta}, \omega_g) \tag{6.120}$$

The cost function is computed as simply a static sum (i.e., forecasts containing no feedback). The time horizon T has nothing to do with credit assignment. Rather, it represents T separate training examples.

An alternative principle is based on a multistep evaluation of the credit assignment. The definitions of $\mathbf{y}_{t+1}$ and $\mathbf{w}_t$ now do not need to consider relative degree. The model and the controller are evaluated according to Eqs. (6.116) and (6.117), with model feedback.

For the closed-loop optimization algorithms of Section 6.5.1, this author has had good experience with horizons of tens to hundreds of sampling periods. Within this time frame, stability-relevant properties of the controller usually become explicit. This is true if the sampling period is at an order of magnitude one-tenth of that required by the sampling theorem, and the plant is not very stiff, that is, the time constants have comparable orders of magnitude.

For stiff plants with very fast and very slow processes, the sampling period is determined by the fastest and horizon by the slowest process.

6.4.3 How to Formulate the Cost Terms

The cost function for neurocontroller training is the platform for formulating the user's priorities. One of the big attractions of neurocontrol is that this cost function indeed can be arbitrary. Control engineers who have been working with specifications via closed-loop poles or, at best, quadratic functions, first have to discover the freedom to wish for anything. Nevertheless, care in formulating the cost function is necessary. The main caveats are the following:

> Only feasible goals should be formulated. Searching after the infeasible ones could prevent finding those that are feasible.
>
> All goals should be made explicit as well as those that are frequently implicit in classical design methods, such as stability.

c_t represents the user-defined part of the cost function. The fundamental arguments of c_t are the plant output $\mathbf{y}_t$ and the generalized reference vector $\mathbf{w}_t$. For standard stabilization and trajectory-following tasks, the cost terms simply may be the squared error

$$c_t = \|\mathbf{y}_{t+1} - \mathbf{w}_t\|^2 \tag{6.121}$$

or, more generally, the quadratic form

$$c_t = (\mathbf{y}_{t+1} - \mathbf{w}_t)\mathbf{C}_y(\mathbf{y}_{t+1} - \mathbf{w}_t)^T \tag{6.122}$$

The discussion of this case is analogous to that of Section 5.4.2. Correct scaling by appropriate weights corresponding to the diagonal of matrix $\mathbf{C}_y$ is crucial.

COST FUNCTION FOR NEUROCONTROLLER TRAINING 193

The argument $\mathbf{u}_t$ of cost terms may be used, as in classical linear-quadratic control, for minimization of expenses connected with the action. In contrast to linear-quadratic control, the dependence is not constrained to be quadratic. On the contrary, the dependence of financial or energetic costs is often linear; that is, the corresponding part of c_t may be

$$\sum_i \rho_i |u_{i,t}| \qquad (6.123)$$

with weights ρ_i.

Constraints of individual states of action variables can be implemented by terms of the type

$$\sum \pi_i \max(y_{i,t} - w_{i,\text{crit}}, 0)^k \qquad (6.124)$$

where $w_{i,\text{crit}}$ are the bounds, which should not be violated, and k is an appropriate exponent ensuring a progressive penalty. Cost function weights π_i assigned to these constraint terms should be relatively high so that they cannot be traded for good values of other cost terms, but they should not be so high as to overpower the optimization of each remaining term.

As stated above, stability requirements have to be explicitly specified in the cost function. There are several means to formulate such stability requirements, including these two possibilities:

1. Deviation from the reference vector can be weighted progressively, for example, exponentially over time. This can be used for step-response training in stabilization tasks where the deviation can be expected to decrease exponentially.
2. Higher-order derivatives (frequently the second derivatives) of deviation can be penalized. Higher-order derivatives act as high-pass filters; high-frequency oscillations are avoided, while the penalty for possibly unavoidable deviation itself remains low.

The cost terms c_t generally depend on plant parameters and external inputs θ and ζ so that they can represent different priorities for plant variations.

6.4.4 Training Examples

Training examples provide the means to represent the distribution $p(\mathbf{y}_{e_0}, \mathbf{w}, \zeta, \theta)$ of Eq. (6.118). In real computations, the integrals are substituted by the sum over all training examples indexed by $j = 1, \ldots, J$:

$$C_g = \sum_{j=1}^{J} \sum_{t=0}^{T-1} c_t(\mathbf{y}_{t+1}, \mathbf{u}_t, \mathbf{w}_t, \theta, \zeta) \rho_i(\mathbf{y}_{e_0}, \mathbf{w}, \zeta, \theta) \qquad (6.125)$$

where ρ_i is weight of individual examples. Each training example specifies the initial state $\mathbf{y}_{e_0}$, the reference trajectory $\mathbf{w}$, measurable parameters θ, and non-measurable parameters ζ.

Reference trajectories can be represented by a limited number of parameters. For example, if there were a single output y ranging from -1 to 1 and reference trajectories characterized by frequency θ ranging from 10 to 20, both uniformly distributed, a good choice would be nine equally weighted training examples [i.e., pairs (z,p)]: $(-1, 10)$, $(0, 10)$, $(1, 10)$, $(-1, 15)$, $(0, 15)$, $(1, 15)$, $(-1, 20)$, $(0, 20)$, $(1, 20)$. These examples represent a Cartesian product of 3×3 values. For high-dimensional tasks, Cartesian products become infeasible because of their exponentially growing size. Then, an appropriate subset must be determined by experience or knowledge of correlations between initial states and parameters. A minimum training example set is received by taking a mean value of all but one state variable or parameter, combined with several values of this variable or parameter. In our example, this would amount to $(-1, 15)$, $(1, 15)$, $(0, 10)$, $(0, 20)$. An appropriate compromise is to be found between this and the Cartesian-product training set. Most real-world problems can be solved within an order of magnitude of tens of training examples.

Note 6.12 The distribution of initial states is of importance for generating trajectories that visit a representative choice of state space regions. For trajectory-following tasks with long optimization horizons and complex reference trajectories, the importance of distinct initial states diminishes.

Incremental schemes in online operation frequently cannot be used to reset the plant to the selected initial state—the complete learning session is then reduced to one example with a single initial state.

6.5 TRAINING METHODS

After having committed to a particular controller structure and represented this structure by a particular neural network with free parameters $\mathbf{w}$, one selects a training method that finds parameter values optimal for the controller performance. The basic options for doing this can be classified along the following lines: (1) The training scheme can be incremental or batch, or (2) the optimization may be explicitly closed-loop or be transformed into the open-loop form.

The reader acquainted with the neurocontrol literature may wonder about our abridged treatment of analytic gradient computation, which is otherwise granted much space in other publications. The reason is that, for general, complex, and realistic control problems, this author recommends using batch-oriented training procedures without an analytic gradient.

6.5.1 Open-Loop Training Methods

Under specified conditions, control theory provides useful statements about controllers with guaranteed properties. One such property is that, by decoupling, any state of a measurable variable can be reached in a number of sampling periods equal to the relative degree of the measurable variable. An arbitrary goal state vector can be reached within n sampling periods. Narendra and Mukhopadhyay [104] have proposed training a controller of type (6.41) or (6.41) by randomly selected pairs $(\mathbf{z}_t, \mathbf{w}_t)$. This method uses the single-step, static cost function of Section 6.4.2.

The cost function is the squared error

$$\|\mathbf{w}_t^* - \mathbf{y}_{t+1}^*\|^2 \qquad (6.126)$$

with $\mathbf{y}_{t+1}^*$ being the vector of $y_{i,t+d_i}$ and d_i the relative degree of the ith measurable variable. Assuming that neural networks are capable of representing an arbitrary mapping with arbitrary precision (which is theoretically guaranteed if the number of neurons is not limited), the error can be reduced to zero.

The training method consists in propagating the error through the fixed input–output model

$$\mathbf{y}_{t+1}^* = f(\mathbf{y}_t, \ldots, \mathbf{y}_{t-d_o}, \mathbf{u}_t, \ldots, \mathbf{u}_{t-d_o}) \qquad (6.127)$$

and applying static backpropagation formulas to controller (6.41) or (6.41).

Furthermore, for a plant model linear in action variables

$$\mathbf{z}_{t+1} = f(\mathbf{z}_t) + h(\mathbf{z}_t)\mathbf{u}_t \qquad (6.128)$$

the input–output form

$$\mathbf{y}_{t+1}^* = k(\mathbf{y}_t, \ldots, \mathbf{y}_{t-d_o}, \mathbf{u}_{t-1}, \ldots, \mathbf{u}_{t-d_o}) + L(\mathbf{y}_t, \ldots, \mathbf{y}_{t-d_o}, \mathbf{u}_{t-1}, \ldots, \mathbf{u}_{t-d_o})\mathbf{u}_t \qquad (6.129)$$

is linear in current action $\mathbf{u}_t$. This allows us to formulate the controller explicitly in the form of *plant inversion*:

$$\mathbf{u}_t = L^{-1}(\mathbf{y}_{t+1}^* - k) \qquad (6.130)$$

A neurocontroller can be trained by random samples of mapping (6.130) with $\mathbf{y}_{t+1}^* = \mathbf{w}_t^*$.

This appealing and theoretically well-founded approach suggests the question of whether the credit assignment problem of Chapter 2 really exists. It is true that under the assumptions of (1) an absolutely accurate model, (2) absolutely accurate measurements, (3) known relative degrees, and most important, (4) unbounded action variables, any measurable state can be reached within a few sampling periods. Unfortunately, taken literally, this is an idealistic picture. Even for accurate models, constraints on action variables prevent us from reaching the goal state in an arbitrarily short time. (Theoretically this time is arbitrarily short because the selected sampling period can be arbitrarily short and the delays are, in discrete control theory, measured in sampling periods.) For example, the spacecraft control described by [54] works with a model of fourth order, but the time necessary for stabilization is thousands of sampling periods. Besides the model imprecisions, the reason is that thrusters for positioning the spacecraft are capable of producing only very limited torques.

If the firm ground of theoretical assumptions is left, a more complex description of closed-loop behavior than through "real output equals desired output" becomes inevitable: Trajectory describing terms such as overshoot avoidance, stabilization within a tolerance, and so on, will be used. The open-loop approaches do not provide means for such specifications.

Note 6.13 It has to be pointed out that a plant inversion does not imply closed-loop stability. More exactly, pure algebraic inversion would be stable for an exact model, but without any stability margin for model inaccuracy. In practical terms, stability is reached by exploiting the degrees of freedom that remain free after doing the inversion. In principle, this is possible to reach by using a sufficiently large neural network. The problem is that the inversion scheme does not provide immediate means for training stability or even for recognizing instability—the rudimentary training horizon equal to the relative degree of the plant is too short to make instability observable in practice. It would have to be combined with some dedicated method for reaching stability. For some proposals, see Chapter 9.

By contrast, closed-loop schemes with longer training horizons make possible training for stabilization by directly penalizing unstable behavior.

In summary, the open-loop scheme is appropriate if the theoretical assumptions apply with high accuracy and if the deviations from the goal state remain very small. This can be the case for some trajectory-following tasks in which a realistic trajectory can be precomputed. The definite advantage of this approach is then in using the static backpropagation in the standard form. Otherwise, closed-loop methods are preferable.

6.5.2 Closed-Loop Training Methods

The task of a controller training algorithm is to minimize the cost function (6.125):

$$C_g = \sum_{j=1}^{J} \sum_{t=0}^{T-1} c_t(\mathbf{y}_{t+1}, \mathbf{u}_t, \mathbf{w}_t, \theta, \zeta) \rho_i(\mathbf{y}_{e_0}, \mathbf{w}, \zeta, \theta) \qquad (6.131)$$

with vectors $\mathbf{y}_{t+1}$ and $\mathbf{u}_t$ computed by recursive simulation of plant model (6.116)

$$\mathbf{y}_{t+1} = F(\mathbf{y}_{e_t}, \mathbf{u}_{e_t}, \theta, \zeta) \qquad (6.132)$$

and controller (6.117)

$$\mathbf{u}_t = G(\mathbf{y}_{e_t}, \mathbf{u}_{e_t}, \mathbf{w}_t, \theta, \omega_g) \qquad (6.133)$$

The algorithms are different for incremental schemes, on the one hand, and batch schemes, on the other. The basic steps for incremental and batch schemes are presented in Sections 6.5.3 and 6.5.4, respectively. Analytic gradient computation is treated as a separate topic in Section 6.5.5.

Note 6.14 Besides the stability of the closed loop, the stability of the controller-training algorithm is an important topic. An unstable controller training algorithm may lead to model quality deterioration, instead of improvement. Theoretical results for general nonlinear systems are still not available, but the principles presented in Section 6.5.4 can be conceptually applied: Static (i.e., open-loop) incremental schemes are more stable than dynamic (closed-loop) ones. Batch schemes can suffer from instability only if *local* optimization algorithms are applied iteratively (for the goals of adaptiveness, see Chapter 8), using the controller from the last iteration as the initial solution.

6.5.3 Incremental Schemes

The general incremental scheme for controller training using neural networks is given in Figure 6.1. Dashed lines represent sampling of triples ($\mathbf{w}$, $\mathbf{u}$, and $\mathbf{d}$). The parameter change computation block contains a fixed or incrementally changing plant model neural network.

As in incremental plant identification, incremental controller training schemes change plant parameters after each sample, or after a small number of samples. In this way, the information contained in the sample is incorporated into the weight change, and the sample is discarded.

The algorithm consists of the following steps, iteratively performed at each sampling period t:

Step 1 The measurement vector $\mathbf{d}$ is sampled.
Step 2 The current controller, represented by a neural network, is called, using as input the extended measurement vector at time t, $\mathbf{d}_{e_t}$ (in static versions), or the extended model output vector $\mathbf{y}_{e_t}$ (in dynamic versions).

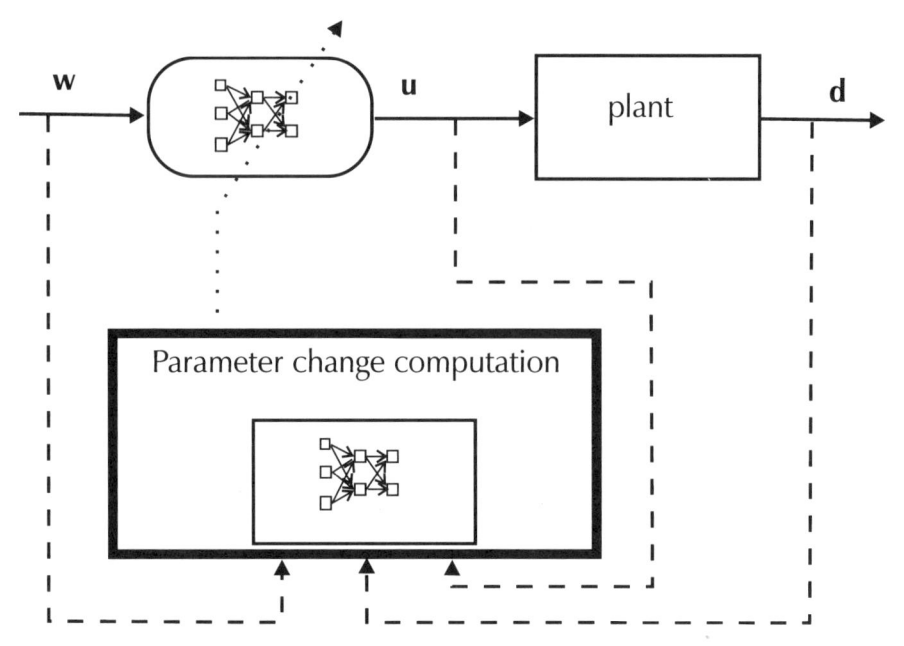

Figure 6.1 Incremental scheme for controller training.

Step 3 The current model is simulated from time t to $t+1$, using as input the extended measurement vector at time t, that is, $\mathbf{d}_{e_t}$ (in static versions), or the extended output vector $\mathbf{y}_{e_t}$ (in dynamic versions).

Step 4 The error term c_{t+1} is evaluated.

Step 5 The change of controller parameters is computed. This is usually done with help of gradient of c_{t+1} with regard to the controller parameter vector:

$$\Delta\omega_f = -\alpha \nabla_{\omega_g} c_{t+1} \qquad (6.134)$$

with α being a constant or variable learning rate. A more sophisticated way to use the gradient is that of Kalman training (Section 4.4).

Step 6 A new parameter vector is computed.

6.5.4 Batch Schemes

As in plant identification, batch procedures first collect data of sufficient quantity and then perform the minimization of criterion (6.131) summed over all measurements. The structure of a typical batch controller training is given in Figure 6.2. A batch procedure usually contains a reference trajectory generating block, using either sampled reference values from a real system or hypothetic reference trajectories (i.e., the dashed line from vector **w** to the reference trajectory block is optional). Vectors **u** and **d** are produced by simulation within the optimization block according to the requirements of the optimization algorithm used. Obviously, if artificial trajectories of reference vectors **w** are used (which is frequently preferable to using sampled reference vectors), the communication between the neurocontroller training algorithm and the real system is unidirectional, via passing optimal parameter values to the embedded neurocontroller. However, this concerns only the neurocontroller training algorithm: As in the incremental scheme shown in Figure 6.1, the plant model neural network is usually used (here, within the optimization block), which has been gained from measurements sampled on the real system.

The steps of a batch procedure are not necessarily iterative:

Step 1 One or more measurement time series of appropriate length are collected.

Step 2 A procedure for evaluating the cost function (5.99), that is, as a sum over the time series length, is made available. The cost function can imply static or dynamic computation. An example for static computation is the algorithm of Section 6.5.1.

Step 3 If necessary, a procedure for gradient computation is made available.

Step 4 An optimization algorithm using both procedures is started.

Each training example is defined by the initial state $\mathbf{y}_{e_0,j}$, reference trajectory matrix $\mathbf{W}_j$ (usually computed from a few reference trajectory parameters such as frequencies, amplitudes, or step sizes), nonobservable parameter vector ζ_j,

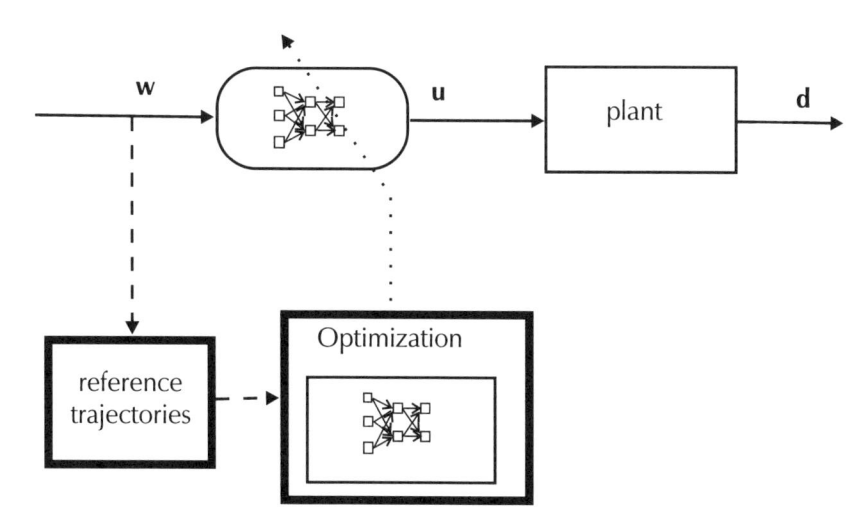

Figure 6.2 Batch scheme for controller training.

200 CONTROLLER TRAINING

and observable parameter vector θ_j. The cost function can then be written as

$$C_g = \sum_{j=1}^{J} \sum_{t=0}^{T_j-1} c_t(\mathbf{y}_{t+1}, \mathbf{u}_t, \mathbf{w}_t, \theta, \zeta) \rho_i(\mathbf{y}_{e_0}, \mathbf{w}, \zeta, \theta) \qquad (6.135)$$
$$= C_g(\mathbf{y}_{e_{0,1}}, \ldots, \mathbf{y}_{e_{0,J}}, \mathbf{W}_1, \ldots, \mathbf{W}_J, \zeta_1, \ldots, \zeta_J, \theta_0, \ldots, \theta_J, \boldsymbol{\omega}_g)$$

Neither computed outputs $\mathbf{y}_{j,t}$ nor the control actions $\mathbf{u}_{j,t}$ occur, since they are a function of initial state $\mathbf{y}_{ej,0}$ and of the reference trajectory $\mathbf{W}_j$.

6.5.5 Gradient Computation

Numeric Gradient With a batch scheme, the numeric gradient can be used. It can be computed as the vector

$$\nabla_{\omega_g} C_g = \left[\frac{C_{gi} - C_{g0}}{\Delta_i} \right] \qquad (6.136)$$

with

$$C_{gi} = C_g(\mathbf{y}_{e_{0,1}}, \ldots, \mathbf{y}_{e_{0,J}}, \mathbf{W}_1, \ldots, \mathbf{W}_J, \zeta_1, \ldots, \zeta_J, \theta_0, \ldots, \theta_J,$$
$$[\omega_1, \ldots, \omega_i + \Delta_i, \ldots, \omega_o]) \qquad (6.137)$$

and

$$C_{g0} = C_g(\mathbf{y}_{e_{0,1}}, \ldots, \mathbf{y}_{e_{0,J}}, \mathbf{W}_1, \ldots, \mathbf{W}_J, \zeta_1, \ldots, \zeta_J, \theta_0, \ldots, \theta_J, [\omega_1, \ldots, \omega_i, \ldots, \omega_o]) \qquad (6.138)$$

(for more, see Section 4.1.3). For recommended values of Δ_i, see Eqs. (4.20) and (4.21). The numeric gradient is computationally more expensive, but both its advantages (discussed in Section 5.5.3) are even more important for controller training than for plant identification. The dynamic simulation of the closed loop is even more apt to round-off and truncation errors than it is the case for an open loop. So the danger of inconsistency between cost function values and gradients is even more serious.

Also the simplicity argument gains importance in view of a large number of alternative controller structures, each of which requires the derivation of dedicated gradient formulas.

Analytic Gradient The cost function in batch schemes (6.125) is a sum of terms c_t. So the gradient can be computed as a sum of gradients of c_t:

$$\nabla C_g = \sum_{j=1}^{J} \sum_{t=1}^{T_J} \nabla c_t(\mathbf{z}_{k,t}, \mathbf{d}_{k,t}) \qquad (6.139)$$

In incremental schemes, the gradient is computed for every c_t and used to perform parameter change.

A gradient of partial error c_t is a vector of derivatives

$$\frac{\partial c_t}{\partial \omega_{jk}} \qquad (6.140)$$

Because c_t is a chain of nested functions of ω (the index f with omega will be omitted in this subsection), the derivatives are to be computed backward through this chain. Parts of this computation can be done by backpropagation algorithm ([126, 146].

With static computation (the algorithm of Narendra and Mukhopadhyay [104] of Section 6.5.1 being a special case), the chain of nested function is the following:

1. The output of a neurocontroller neural network is a function of network parameters ω.
2. The controller output $\mathbf{u}_t$ is a function of the output of neural network. If a discrete-time model is used and directly represented by a neural network, this function is trivial. With numeric derivatives or gain vector networks, the concrete form of the mapping is to be taken into account.
3. Partial cost term c_t is a function of controller output $\mathbf{u}_t$.
4. Model output $\mathbf{y}_t$ is a function of controller output $\mathbf{u}_{t-1}$
5. Partial cost term c_t is also a function of model output $\mathbf{y}_t$.

The computation of gradient is done in an inverse order:

1. The derivative of c_t with regard to model output is computed.
2. The derivative of c_t with regard to model input (i.e., with regard to controller output via the model) is computed.
3. The derivative of c_t with regard to controller output (directly) is computed.
4. The derivatives of c_t with regard to controller output, direct and via the model, are added to get the complete derivative of c_t with regard to controller output.
5. The derivative of c_t with regard to network output is computed.
6. The derivative of c_t with regard to network weights is computed.

202 CONTROLLER TRAINING

For illustration, let us consider the simplest case where the network output is identical to the model output vector. The backpropagation scheme for gradient computation consists of repeated application of a chain rule for computing derivatives. The recursion goes backward (i.e., from output layer toward input layer) over network layers.

For example, with a perceptron with input vector α, output vector γ and a single hidden-layer vector $\mathbf{h}$, the derivatives $\partial c_t / \partial \gamma_j$ of function c_t with regard to network vector γ are to be computed. Then, the procedure of Section 5.5.3, that is, formulas (5.109) to (5.114), is directly applicable.

The derivatives $\partial c_t / \partial \gamma_j$ reflect two nested mappings: the plant model mapping and mapping of network output to control action:

$$\frac{\mathrm{d}c_t}{\mathrm{d}\gamma_j} = \left(\frac{\mathrm{d}c_t}{\mathrm{d}\mathbf{y}}\frac{\mathrm{d}\mathbf{y}}{\mathrm{d}\mathbf{u}} + \frac{\mathrm{d}c_t}{\mathrm{d}\mathbf{u}}\right)\frac{\mathrm{d}\mathbf{u}}{\mathrm{d}\gamma_j} \tag{6.141}$$

The vectors $\mathrm{d}c_t/\mathrm{d}\mathbf{y}$ and $\mathrm{d}c_t/\mathrm{d}\mathbf{u}$ result from the concrete form of cost function. If this function does not contain terms with action variables, the latter vector is zero. The matrix of partial derivatives $\mathrm{d}\mathbf{y}/\mathrm{d}\mathbf{u}$ is the open-loop Jacobian matrix of the model. The matrix $\mathrm{d}\mathbf{u}/\mathrm{d}\gamma_j$ is identity matrix if the controller output is equal to network output.

The *dynamic scheme* differs from the static scheme by introducing a recursive dependence of model outputs on previous model outputs.

1. The output of neurocontroller network is a function of network parameters ω and of previous model outputs.
2. The neurocontroller output $\mathbf{u}_t$ is a function of the output of neurocontroller network and of previous model outputs.
3. The model output $\mathbf{y}_t$ is a function of the output of neurocontroller network and of previous model outputs.
4. The partial cost function c_t is a function of model output $\mathbf{y}_t$.

The computation of gradient is done in an inverse order:

1. The derivative of c_t with regard to model output at time t is computed.
2. The derivative of c_t with regard to controller output at time t is computed.
3. The derivatives of $c_s, s \geq t$ with regard to controller output at time t are computed and summed up. For this, derivatives of the closed-loop model output $\mathbf{y}_s, s > t$ with regard to model output $\mathbf{y}_t$ as well as those of controller output $\mathbf{u}_s$ with regard to model output $\mathbf{y}_t$ are used.
4. The summed derivative of $c_s, s \geq t$ with regard to network output is computed.

5. The summed derivative of $c_s, s \geq t$ with regard to network weights is computed.

Let us consider only the situation with observability index 1. If the open-loop Jacobian matrix of the plant is $\mathbf{J}_t = [\mathbf{K}_t | \mathbf{L}_t]$ ($\mathbf{K}_t = [\partial y_{t+1,i}/\partial y_{t,j}]$ and $\mathbf{L}_t = [\partial y_{t+1,i}/\partial u_{t,j}]$), and the Jacobian matrix of the controller is $\mathbf{H}_t = [\partial u_{t,i}/\partial y_{t,j}]$, then the closed-loop Jacobian matrix is $\mathbf{M}_t = [\partial y_{t+1,i}/\partial y_{t,j}] = \mathbf{K}_t + \mathbf{L}_t \mathbf{H}_t$.

Note 6.15 Although formally similar, $\mathbf{K}_t$ is the matrix of partial derivatives in the open loop, whereas $\mathbf{M}_t$ is the matrix of partial derivatives in the closed loop. The latter partial derivatives account for the additional influence of $\mathbf{y}_t$ upon $\mathbf{z}_{t+1}$ via $\mathbf{u}_t$.

All Jacobian matrices $\mathbf{K}_t$, $\mathbf{L}_t$, and $\mathbf{H}_t$ can be determined analytically at any time point t in a straightforward way by applying the chain rule to the plant-model network (matrices $\mathbf{K}_t$, $\mathbf{L}_t$) or the controller network (matrix $\mathbf{H}_t$).

The row vectors partial derivatives of objective function terms c_t (6.141) are $\mathbf{f}_t = dc_t/d\mathbf{y}_{t+1}$ and $\mathbf{g}_t = dc_t/d\mathbf{u}_t$. They can be determined from the specified objective function by analytic differentiation. The objective function gradient can be written as

$$\frac{dc_t(\mathbf{y}_{t+1}, \mathbf{u}_t, \theta)}{d\omega_g} = \sum_{s=0}^{t} \frac{dc_t(\mathbf{y}_{t+1}, \mathbf{u}_t, \theta)}{d\mathbf{u}_s} \frac{d\mathbf{u}_s}{d\omega_g}$$

$$= \left[\sum_{s=0}^{t-1} \mathbf{P}_t \mathbf{R}_{st} \mathbf{L}_s \frac{d\mathbf{u}_s}{d\omega_g} + \mathbf{Q}_t \frac{d\mathbf{u}_t}{d\omega_g} \right] \tag{6.142}$$

with $\mathbf{R}_{st} = (\prod_{r=s+1}^{t-1} \mathbf{M}_r)$ (product from the left), $\mathbf{P}_t = \mathbf{f}_t \mathbf{M}_t + \mathbf{g}_t \mathbf{H}_t$, $\mathbf{Q}_t = \mathbf{f}_t \mathbf{L}_t + \mathbf{g}_t$. The term $d\mathbf{u}_t/d\omega_g$ depends on the form of the controller. It is received by analogous static backpropagation formulas.

6.6 NEUROCONTROL WITH ANALYTIC MODELS

The research in neural network control has so far almost entirely focused on methods of controller training working with models identified from data. The idea of model-free control is certainly very attractive; taking an arbitrary plant and letting the controller design itself automatically is the dream of every control engineer. However, there are situations in which design methods using an analytic plant model are preferable: For example, some plants cannot be put in operation before the controller design (satellites, aircraft, power plants, etc.). Even more plants cannot be operated in a way necessary for complete *nonlinear* identification, that is, bringing them to all (possibly catastrophic) working points.

These simple constraints are frequently overcome in research studies by simulating the plant. But this is a contradiction with the goal of model-free control, since a simulation always uses a plant model. But, obviously, if a plant model is available, it should enter the design procedure directly. The indirect way of plant model → simulation → generating data → using the data for adaptive identification is connected with a certain loss of information. What is lost with certainty is the manpower and computing time connected with generating data, determining the structure of the plant model, and identifying the model.

For such cases, the possibility of training a neurocontroller with an arbitrary analytic model of general form (6.116) should be mentioned. What has been said about cost function, training examples, and optimization algorithms can be directly applied to this case. The bias toward using the numeric, and not analytic, gradient is even greater because cumbersome computations of application-specific Jacobian matrices can be avoided.

With analytic plant models for neurocontroller network controller training, what is then the main benefit of using neural network controllers at all? This is an important question. Indeed, most control engineers would object that, with a plant model, controllers can be designed by classical analytic methods. The simple answer is that neurocontrol methods are more general than most analytic nonlinear control methods. Even for linear plants, the assumption of unconstrained control action, which makes applicable the analytic linear-quadratic approach for optimal control, is always violated. An example of this is presented in Chapter 14. In nonlinear problems the neurocontrol approaches make almost no assumption about the properties of the plant and impose almost no constraints on the cost function.

6.7 USING NEURAL NETWORK MODELS FOR ANALYTIC CONTROLLER DESIGN

The use of neural-network-based plant models in classical controller design should be briefly addressed.

If the plant output depends on control action $\mathbf{u}_t$ only linearly, with nonlinear dependence on plant state or output, the plant inversion can be performed explicitly. Freund [37] inferred explicit formulas for a continuous-item decoupling model of the form

$$\mathbf{y}^* = f_1(\mathbf{z}) + f_2(\mathbf{z})\mathbf{u} \qquad (6.143)$$

The proposal of Narendra and Mukhopadhyay [104] is based on a discrete-time decoupling neural network model of the form

$$\mathbf{y}_{t+1}^* = f_1(\mathbf{y}_t, \mathbf{y}_{t-1}, \ldots, \mathbf{y}_{t-n_o+1}, \mathbf{u}_{t-1}, \ldots, \mathbf{u}_{t-n_o+1})$$
$$+ f_2(\mathbf{y}_t, \mathbf{y}_{t-1}, \ldots, \mathbf{y}_{t-n_o+1}, \mathbf{u}_{t-1}, \ldots, \mathbf{u}_{t-n_o+1}) \mathbf{u}_t \quad (6.144)$$

This model is sufficiently general for many practical problems. For given reference vector $\mathbf{w}_{t+1}^*$, the required control action can be computed as

$$\mathbf{u}_t = f_2(\mathbf{y}_t, \mathbf{y}_{t-1}, \ldots, \mathbf{y}_{t-n_o+1}, \mathbf{u}_{t-1}, \ldots, \mathbf{u}_{t-n_o+1})^{-1} \mathbf{w}_{t+1}^*$$
$$- f_1(\mathbf{y}_t, \mathbf{y}_{t-1}, \ldots, \mathbf{y}_{t-n_o+1}, \mathbf{u}_{t-1}, \ldots, \mathbf{u}_{t-n_o+1}) \quad (6.145)$$

if the dimension of vector **y** is equal to that of **u**, and matrix f_2 is nonsingular so that its inversion can be performed. This approach is closely related to the open-loop training method described in Section 6.5.1; with realistic reference trajectory $\mathbf{w}_{t+1}^*$, it is an elegant way to bypass the gradient-training algorithms.

Approaches using dynamic optimization [e.g., 99, 129] are independent from the special plant model form (6.144) and from the knowledge of relative degree. A method addressing numerical aspects has been proposed Kreisselmeier and Birkhölzer [79]. Unfortunately, dynamic optimization works, in the general case, with a discretized state space (n-dimensional grid) whose size grows exponentially with the number of variables.

7

ROBUST NEUROCONTROL

As stated in Section 1.4, robust control addresses the problem of controlling a plant whose behavior is different from that of the plant model. One reason for the difference may be the variations in the structure or parameters of an exact model, on one hand, and those of an approximate model used for controller design, on the other hand. Another reason may be a systematic disturbance from outside, that is, a disturbance that acts in a time-invariant way (or that is changing very slowly with time), so observations on its past effects can be used to improve the control in the future. Finally, there may be a random disturbance, that is, a disturbance that is so random that it does not permit its past effects to be observed and used to improve future control.

It is striking that these goals are identical with those of adaptive control. There is really a fundamental duality of means to reach this goal: to cope with changing conditions, it is always possible either to try to be immune, or robust, against the changes, or to try to use the information about the changes for improving one's performance.

Even in taking such a philosophical position, it is difficult to separate adaptive control from nonadaptive control, and thus also from robust control, because every feedback can be interpreted as using information about environmental changes [103]. Feedback is inherent to almost all control concepts; therefore, every approach committed to the field of robust control is somewhere between the definitions of robustness, on the one hand, and adaptiveness, on the other hand.

As long as only linear systems and controllers are considered, a pragmatic separation has been possible by means of constant or varying linear coefficients: Robust, nonadaptive controllers are those for which parameters are constant, and adaptive ones are those for which parameters are changed by adaptation algorithms.

This separation principle is less meaningful if general, nonlinear controllers are considered. Suppose that a fixed controller is represented as a mapping

$$\mathbf{u}_t = G_f(\mathbf{y}_{e_t}, \mathbf{u}_{e_t}) \qquad (7.1)$$

of measurement history $\mathbf{y}_{e_t}$ and action history $\mathbf{u}_{e_t}$ to the next action $\mathbf{u}_t$.

Suppose, further, that an adaptive controller is represented as a parametrized mapping

$$\mathbf{u}_t = G_p(\mathbf{y}_{e_t}, \mathbf{u}_{e_t}, \omega_g) \qquad (7.2)$$

with ω_g being the vector of parameters to be adapted. Let the plant identification algorithm be represented as a fixed mapping

$$\omega_f = H_1(\mathbf{y}_{E_t}, \mathbf{u}_{E_t}) \qquad (7.3)$$

of measurement history $\mathbf{y}_{E_t}$ and action history $\mathbf{u}_{E_t}$ to plant parameter vector ω_f. (The measurement and action histories sufficient for this task are different, and usually substantially longer, than those used as controller inputs.) The control design algorithm is, in turn, represented by a fixed mapping

$$\omega_g = H_2(\omega_f) \qquad (7.4)$$

of plant parameters to controller parameters. With the help of Eqs. (7.3) and (7.4), Eq. (7.2) can be rewritten as

$$\mathbf{u}_t = G_p\{\mathbf{y}_{e_t}, \mathbf{u}_{e_t}, H_2[H_1(\mathbf{y}_{E_t}, \mathbf{u}_{E_t})]\} = G'_p(\mathbf{y}_{eE_t}, \mathbf{u}_{eE_t}) \qquad (7.5)$$

with $\mathbf{y}_{eE_t}$ and $\mathbf{u}_{eE_t}$ being the unions of corresponding histories indexed by e and E.

Obviously, adaptive controller (7.2) with variable parameters can be viewed as a fixed function of measurement and action histories, that is, as a structural equivalent to fixed, nonadaptive controller (7.1). As long as mappings G_f of Eq. (7.1) and G_p of Eq. (7.2) are linear, whereas mapping G'_p of Eq. (7.5) is nonlinear, strict distinguishing between fixed form (7.1) and variable form (7.2) is justified. With all mappings being nonlinear, the justification becomes arbitrary.

7.1 CLASSICAL VIEW OF ROBUST CONTROL

As pointed out in Chapter 6, both standard control tasks, stabilization and trajectory following, have two basic facets: stability and equilibrium point definition.

7.1.1 Robust Stability

The stability problem can be formulated as keeping closed-loop eigenvalues so that they they remain in the stability region (e.g., negative real parts of eigenvalues for continuous time) for all alternative configurations of plant model parameters and other robustness requirements. If all alternative configurations (or, at least, their convex envelope) represent a finite set, they constitute a set of inequalities for eigenvalues and thus also a set of inequalities for controller parameters.

Sometimes, robustness requirements are not specified in terms of concrete plant parameter variations. Then, a pragmatic approach is to require that the eigenvalues be sufficiently deep in the stability region, for example, that their real parts are less than -0.5. A penalty-function-based method for this has been proposed by Konigorski [78]: A goal region for eigenvalues is defined, and a function containing penalties for eigenvalues outside this region is minimized by a gradient method.

A popular frequency domain approach, the H_∞ approach, consists in controller design under a requirement of infinity norm of the closed-loop transfer-function gain (i.e., the transfer-function maximum over all frequencies and phases) being in the stability region.

Let us now consider the case of an external disturbance. One possibility is to view the disturbance as a closed dynamic system with equation

$$\dot{\mathbf{z}}_d = \mathbf{A}_d \mathbf{z}_d \qquad (7.6)$$

and to extend the plant state equation using Eq. (5.14)

$$\dot{\mathbf{z}} = \mathbf{A}\mathbf{z} + \mathbf{A}_e \mathbf{z}_d + \mathbf{B}\mathbf{u} \qquad (7.7)$$

Then, the extended plant

$$\begin{bmatrix} \dot{\mathbf{z}} \\ \mathbf{z}_d \end{bmatrix} = \begin{bmatrix} \mathbf{A} & \mathbf{A}_e \\ -\mathbf{0} & \mathbf{A}_d \end{bmatrix} \begin{bmatrix} \mathbf{z} \\ \mathbf{z}_e \end{bmatrix} + \begin{bmatrix} \mathbf{B} \\ \mathbf{0} \end{bmatrix} \mathbf{u} \qquad (7.8)$$

can be analyzed. This extended plant may be observable but is with certainty not controllable. This is intuitively obvious by taking into account that disturbance is modeled as a closed system—neither the action nor the state influences the disturbance. Another way to see this is by considering a state controller

$$\mathbf{u} = -\mathbf{R}\mathbf{z} - \mathbf{R}_d \mathbf{z}_d \qquad (7.9)$$

Then, the closed-loop behavior is described by

$$\begin{bmatrix} \dot{z} \\ \dot{z}_d \end{bmatrix} = \begin{bmatrix} A & A_e \\ 0 & A_d \end{bmatrix} \begin{bmatrix} z \\ z_e \end{bmatrix} - \begin{bmatrix} B \\ 0 \end{bmatrix} [R \quad R_d] \begin{bmatrix} z \\ z_e \end{bmatrix}$$
$$= \begin{bmatrix} A - BR & A_e - BR_d \\ 0 & A_d \end{bmatrix} \begin{bmatrix} z \\ z_e \end{bmatrix} \quad (7.10)$$

The eigenvalues of the system matrix

$$\begin{bmatrix} A - BR & A_e - BR_d \\ 0 & A_d \end{bmatrix} \quad (7.11)$$

can be factorized [31] into those of matrix $A - BR$ (which can be influenced by choosing an appropriate feedback matrix R) and those of matrix A_d (which cannot be influenced).

This can be summarized in the following way:

A plant with external disturbance can be stabilized only if the disturbance is, itself, a stable system. Then, the disturbance has no influence on the state feedback design; a stable controller for the plant without disturbance is stable also for the plant with disturbance.

The condition of disturbance being a stable system is often trivially satisfied. A constant disturbance is a good example. Also a white noise disturbance can be counted in the stable class.

Note that this statement concerned complete state feedback. It will not be necessarily valid if an input–output model is used. A condition for a controller design based on an input–output model to be equivalent to that based on a state model is that the complete state of the plant be observable. If the measurements are independent of the disturbance, the applicability of the input–output model is not affected.

However, if the measurements are not completely independent of the disturbance, the observability must be extended to the *observability of the disturbance*. This is usually satisfied for constant disturbances but not for white noise or similar random disturbances. White noise really negatively affects the observability of disturbed plant. In particular, it leads to inaccuracy of measurements of higher-order derivatives, as discussed in Section 5.3.2, and restricts the applicability of input–output models with higher observability indexes. In commonsense terms, it can be said that for a disturbance to be harmless, it has to persist a time sufficiently long to make it observable with help of time series measurements.

7.1.2 Robust Closed-Loop Equilibrium Points

A linear closed loop with controller (6.13)

CLASSICAL VIEW OF ROBUST CONTROL 211

$$\mathbf{u} = -\mathbf{Rz} + \mathbf{v} \tag{7.12}$$

is stable only if the equilibrium-point-defining controller component **v** is defined as a product of the prefilter matrix (6.11) and the reference state **w**, that is,

$$\mathbf{v} = \left[\mathbf{C}(\mathbf{BR} - \mathbf{A})^{-1}\mathbf{B}\right]^{-1}\mathbf{w} \tag{7.13}$$

This can be seen by substituting Eq. (7.13) into Eq. (6.14):

$$\mathbf{0} = (\mathbf{A} - \mathbf{BR})\mathbf{z} + \mathbf{B}\left[\mathbf{C}(\mathbf{BR} - \mathbf{A})^{-1}\mathbf{B}\right]^{-1}\mathbf{w}$$
$$\mathbf{w} = \mathbf{Cz} \tag{7.14}$$

The first equation of (7.14) can then be written as

$$(\mathbf{BR} - \mathbf{A})\mathbf{z} = \mathbf{B}\left[\mathbf{C}(\mathbf{BR} - \mathbf{A})^{-1}\mathbf{B}\right]^{-1}\mathbf{w} \tag{7.15}$$

Multiplying both sides of this equation by $\mathbf{C}(\mathbf{BR} - \mathbf{A})^{-1}$ results in

$$\mathbf{Cz} = \mathbf{y} = \mathbf{w} \tag{7.16}$$

It is clear that with prefilter matrix (7.13), equation (7.16) is satisfied exactly only if model matrices **A**, **B**, and **C** are perfectly consistent with reality, that is, if the model is exact. Otherwise, the equilibrium point with the property $\mathbf{y} = \mathbf{w}$ does not exist, and the closed loop cannot ever be stabilized at this point.

As mentioned in Section 6.1, this problem can be solved by a controller concept using error integration of the form

$$\dot{\mathbf{e}} = \mathbf{w} - \mathbf{y} = \mathbf{w} - \mathbf{Cz} \tag{7.17}$$

The plant formally extended by the state variable **e** can be stabilized by the controller

$$\mathbf{u} = -\mathbf{Rz} + \mathbf{Se} \tag{7.18}$$

without prefilter matrix **M**. To figure out the equilibrium point of this extended plant, controller (7.18) is substituted to the extended plant (6.23). The extended closed loop is then described by the equation

$$\begin{bmatrix} \dot{\mathbf{z}} \\ \mathbf{e} \end{bmatrix} = \begin{bmatrix} \mathbf{A} - \mathbf{BR} & \mathbf{BS} \\ -\mathbf{C} & \mathbf{0} \end{bmatrix} \begin{bmatrix} \mathbf{z} \\ \mathbf{e} \end{bmatrix} + \begin{bmatrix} \mathbf{0} \\ \mathbf{I} \end{bmatrix} \mathbf{w} \tag{7.19}$$

The equilibrium point is defined by setting the left side of the equation to the zero vector. The part of the equation corresponding to $\dot{\mathbf{e}} = \mathbf{0}$ obviously defines the equilibrium point at $\mathbf{y} = \mathbf{Cz} = \mathbf{w}$. So, it is sufficient that the part of the equation corresponding to $\dot{\mathbf{z}} = \mathbf{0}$ be consistent with this. It can be reformulated as

$$(\mathbf{BR} - \mathbf{A})\mathbf{z} = \mathbf{BSe} \qquad (7.20)$$

Mutliplying this equation by $\mathbf{C}(\mathbf{BR} - \mathbf{A})^{-1}$, we get

$$\mathbf{Cz} = \mathbf{y} = \mathbf{C}(\mathbf{BR} - \mathbf{A})^{-1}\mathbf{BSe} \qquad (7.21)$$

Obviously, there is a vector **e** consistent with $\mathbf{y} = \mathbf{w}$ (for arbitrary **w**) and Eq. (7.21) whenever the matrix $\mathbf{C}(\mathbf{BR} - \mathbf{A})^{-1}\mathbf{B}$ as well as the matrix **S** are nonsingular. The latter matrix can always be chosen to be nonsingular. The former matrix is the inversion of the prefilter matrix of Eqs. (6.11) and (7.13), so the conditions for existence of the equilibrium point are identical with those for state feedback without error integration: The reference state must be consistent with the plant.

The closed-loop equilibrium point is retained even if the plant state equation (6.1) is modified by an additive offset term β:

$$\dot{\mathbf{z}} = \mathbf{Az} + \mathbf{Bu} + \beta \qquad (7.22)$$

Then, the equilibrium condition (7.20) becomes

$$(\mathbf{BR} - \mathbf{A})\mathbf{z} = \mathbf{BSe} + \beta \qquad (7.23)$$

Mutliplying again this modified equation, by $\mathbf{C}(\mathbf{BR} - \mathbf{A})^{-1}$, we get

$$\mathbf{y} - \mathbf{C}\beta = \mathbf{C}(\mathbf{BR} - \mathbf{A})^{-1}\mathbf{BSe} \qquad (7.24)$$

Once more, for any **w** substituted for **y**, a vector **e** can be found at which the extended plant will stabilize. This extended plant equilibrium point is consistent with $\mathbf{y} = \mathbf{w}$.

Note 7.1 It should be pointed out that the equilibrium point values of **e** defined by Eqs. (7.21) and (7.24) need never be computed. Rather, the closed loop will automatically *converge* to these values.

The additive offset β subsumes

1. An offset of the state or output variables, that is, a measurement offset such as a sensor with the wrong zero calibration
2. An action offset, or active systematic disturbance, such as lateral wind affecting a bike rider

For nonconstant systematic disturbances, these results can be generalized in the following way:

1. Controller input including error integral is necessary if stability under constant additive disturbance is to be reached.
2. Controller input including the integral of error integral is necessary for stability under a linearly changing additive disturbance (i.e., a ramp) to be reached.
3. Controller input including the $(n-1)$th error integral is necessary if stability under additive disturbance which is an nth-order polynomial of time is to be reached.

The output feedback with action integration (6.32) of the form

$$\dot{\mathbf{u}} = \mathbf{M}(\mathbf{w} - \mathbf{y}) \tag{7.25}$$

is another possibility to guarantee the closed-loop equilibrium point at $\mathbf{y} = \mathbf{w}$ with additive constant disturbance as well as inaccurate model. A sufficient condition for existence of the equilibrium point with a given $\mathbf{w}$ is that equation

$$\begin{bmatrix} \dot{\mathbf{0}} \\ \mathbf{w} \end{bmatrix} = \begin{bmatrix} \mathbf{A} & \mathbf{B} \\ \mathbf{C} & \mathbf{0} \end{bmatrix} \begin{bmatrix} \mathbf{z} \\ \mathbf{u} \end{bmatrix} \tag{7.26}$$

have a solution. In turn, a sufficient (but not necessary) condition for this is that matrix

$$\begin{bmatrix} \mathbf{A} & \mathbf{B} \\ \mathbf{C} & \mathbf{0} \end{bmatrix} \tag{7.27}$$

be nonsingular.

How these concepts can be generalized to input–output models can be derived in the following way: Suppose that a linear plant is described alternatively as

$$\ddot{\mathbf{z}} = \mathbf{A}\dot{\mathbf{z}} + \mathbf{B}\dot{\mathbf{u}} \tag{7.28}$$

or in the first-order form

$$\begin{aligned} \dot{\mathbf{z}}_1 &= \mathbf{A}\mathbf{z}_1 + \mathbf{B}\mathbf{u}_1 \\ \dot{\mathbf{z}}_0 &= \mathbf{z}_1 \\ \dot{\mathbf{u}}_0 &= \mathbf{u}_1 \end{aligned} \tag{7.29}$$

The controller for this extended plant model is

$$\dot{\mathbf{u}} = \mathbf{u}_1 = -\mathbf{R}_0 \mathbf{z}_0 - \mathbf{R}_1 \mathbf{z}_1 - \mathbf{S}\mathbf{u}_0 + \mathbf{M}\mathbf{w} \tag{7.30}$$

If the state is observable, the state and its derivative can be substituted by the higher-order derivatives of the measurements and actions. What results is the analogy to the input–output controller whose various forms have been discussed in Chapter 6, with observability index increased by 1, that is, $d_o + 1$.

The point with this controller form is that it is independent from the size of an additive constant disturbance, which disappears from the model by differentiation underlying the form (7.28). With a fixed (i.e., independent from the unknown disturbance) prefilter matrix **M** [analogous to Eq. (6.11) but applied to plant model (7.29)], the closed loop has an equilibrium point at $\mathbf{y} = \mathbf{w}$ under the conditions discussed above and in Chapter 6.

Obviously, differentiating the model i times, an additive disturbance that is a $(i - 1)$th-order polynomial function of time can be neutralized.

Closed-loop equilibrium points at $\mathbf{y} = \mathbf{w}$ under external disturbance can be provided for by (1) integrating error or (2) giving the controller additional higher-order derivatives.

Note 7.2 The equilibrium point existence considered here has to do with the existence of a disturbance-independent prefilter matrix, that is, with the existence of fixed, disturbance-independent controller mapping with which the closed loop converges to $\mathbf{y} = \mathbf{w}$.

This topic is different from constructing neurocontrollers with equilibrium points *at the coordinate origin* of controller arguments, as discussed in Chapter 6. Whenever the existence of disturbance-independent controller mapping is guaranteed, a special neurocontroller form with an equilibrium point at the coordinate origin can be inferred.

7.2 ROBUSTNESS ASPECTS OF NEUROCONTROL

In general control design, the solution proceeds in two steps. First, an appropriate controller structure is chosen. Then, free parameters of this structure are computed. In neurocontroller design, the second step is done by training, that is, by numeric optimization.

In the particular case of robust control, a controller structure is chosen that can potentially be robust against some class of disturbances or plant variations. The controller parameters are then determined taking into account concrete instances, ranges of disturbances, or plant variations. This principle is reflected in two aspects of robust neurocontroller training.

1. A neurocontroller structure is determined that is sufficiently general to subsume mapping classes necessary to satisfy the given robustness requirements.

2. The neurocontroller is trained so that these requirements are really satisfied. The means to enable the training algorithm to find the solution satisfying these requirements is the cost function. The most important part of the cost function that incorporates robustness requirements is the set of training examples.

These two aspects, neurocontroller structure and training examples, are discussed in Sections 7.3 and 7.4, respectively.

7.3 NEUROCONTROLLER STRUCTURE

As shown in Section 7.1, controller structure modifications improving the robustness consist of adding two types of controller inputs: error integrals and higher-order derivatives. Both extensions can be summarized as increasing the order of the controller. Both can be applied to neurocontrollers in a straightforward way—they are subsumed by the neurocontroller structures discussed throughout Chapter 6.

7.3.1 Plant Variations and Systematic Disturbances

An appropriate neurocontroller structure is important for the existence of required equilibrium points under (1) plant variations and (2) systematic disturbances. Input–output controllers of an order corresponding to only an undisturbed plant will exhibit stationary offsets from the reference value, independently from the training. These offsets can be trained away if appropriate neurocontroller structures with increased order are used. The sufficient order increase is typically by one, to account for nearby stationary-plant parameter variations, systematic disturbances, or measurement offsets.

The robustness against systematic disturbances with regard to closed-loop stability cannot be affected by modifying neurocontroller structure for reasons discussed in relation to the extended model (7.8). In the usual case of stable disturbance (the prominent example of which is a constant offset), the conditions for stability of the closed loop are the same as they would be without such a disturbance.

By contrast, stability with plant variations sometimes can be improved by increasing the controller order because plant variations affect closed-loop stability. If observable, they can be used to produce a control action consistent with the modified plant. This principle is related to the controller concept described by Eq. (7.5)—it is a *hyperstate* controller in the grey zone between adaptive and robust control.

7.3.2 Random Disturbances

There are hardly any structural means to improve the control with random disturbances, such as discretization errors. Because most random disturbances can be viewed as sequences of steps or pulses with zero mean value, they do not affect the equilibrium point's existence. However, the equilibrium point cannot be reached with any accuracy whenever (which is often) the time between the pulses is substantially shorter than the time necessary to reach the reference state from the disturbed state. Strictly speaking, random disturbances with a mean of zero do not affect the stability of linear systems. Nevertheless, a fast sequence of such random pulses may bring the closed loop outside of the domain for which it has been trained and thus cause instability. A formal way to investigate such cases is to use the bounded-input–bounded-output (BIBO) stability [see 103].

In addition to their own harmful effects, random disturbances have a negative impact on higher-order controllers used for systematic disturbances because of the inaccuracy of higher-order derivatives (see Section 5.3.2). The most efficient solution to this problem is smoothing or filtering. This can be done whenever the sampling rate sufficiently exceeds the rate required by the sampling theorem.

7.3.3 Measurable Disturbances

Throughout this chapter, it is assumed that disturbance and plant variations are not directly measurable. Should, by chance, the contrary be the case, the disturbances are simply added to the controller input list. For example, if the lateral wind is the disturbance and it can be measured, it becomes an additional input for the controller.

For controllers with defined closed-loop equilibrium points such as Eqs. (6.86) and (6.87), it is usually sufficient to include the disturbance in the input list of the feedforward component φ. (This is analogous to the classical disturbance compensation proposed by Johnson [66].) The reason for this is that, as explained in Section 7.1, the presence of a (stable) disturbance affects the equilibrium point but not the stability of the closed loop. In this controller structure, it is the feedforward component alone that defines the equilibrium point.

7.4 TRAINING EXAMPLES

What *exactly* the controller is to be robust against is specified in the cost function or, more precisely, in the training examples representing the distribution over which the cost function is to be integrated.

First, let us recall the definitions entering the cost function. The plant model is, by Eq. (6.116),

$$\mathbf{y}_{t+1} = F(\mathbf{y}_{e_t}, \mathbf{u}_{e_t}, \theta, \zeta) \qquad (7.31)$$

with the following semantics:

1. $\mathbf{y}_{e_t}$ is the output feedback to the model, including all delayed outputs.
2. $\mathbf{y}_{t+1}$ is the next model output.
3. $\mathbf{u}_{e_t}$ is the vector of genuine action variables (i.e., outputs of the neurocontroller controller), including their delayed values.
4. θ is the vector of all measurable plant inputs or parameters that are not control actions.
5. ζ is the vector of all nonmeasurable plant inputs or parameters.

Both measurable and nonmeasurable plant parameters represent complete alternative plant models, an approach used by Feldkamp and Puskorius [29].

The controller definition, by Eq. (6.117), is

$$\mathbf{u}_t = G(\mathbf{y}_{e_t}, \mathbf{u}_{e_t}, \mathbf{w}_t, \theta, \omega_g) \qquad (7.32)$$

where $\mathbf{u}_t$ is the actual action vector, $\mathbf{w}_t$ is the generalized reference vector, and ω_g is the vector or free parameters of the neurocontroller network.

The cost function is minimized as in Eq. (6.118), that is,

$$C_g = \int_\theta \int_\zeta \int_\mathbf{w} \int_{\mathbf{y}_{e0}} \sum_{t=0}^{T-1} c_t(\mathbf{y}_{t+1}, \mathbf{u}_t, \mathbf{w}_t, \theta, \zeta) \rho(\mathbf{y}_{e0}, \mathbf{w}, \zeta, \theta) d\mathbf{y}_{e0} d\mathbf{w} d\zeta d\theta \qquad (7.33)$$

with integration over:

All values of measurable parameters θ. (More generally, all *sequences* of θ_t might be considered, with a slight notation modification.)
All values of nonmeasurable parameters ζ.
All sequences of reference vectors $\mathbf{w}_t$.
All initial states $\mathbf{y}_{e0}$.

In real computations, the integrals are replaced by summing all the training examples indexed by $j = 1, \ldots, J$:

$$C_g = \sum_{j=1}^{J} \sum_{t=0}^{T-1} c_t(\mathbf{y}_{t+1}, \mathbf{u}_t, \mathbf{w}_t, \theta, \zeta) \rho_i(\mathbf{y}_{e0}, \mathbf{w}, \zeta, \theta) \qquad (7.34)$$

with ρ_i being the weight of individual examples. So each training example consists of specifications of initial state $\mathbf{y}_{e0}$, reference trajectory $\mathbf{w}$, measurable parameters θ, and nonmeasurable parameters ζ.

Most robustness requirements are specified by a range of parameters or disturbances under which the neurocontroller is to operate. A straightforward way to specify this range is by a set of training examples reflecting the range, its convex envelope, or at least its boundary points. Weighting individual training examples reflects the distribution within the range.

Note 7.3 The performance for cases from the convex envelope of a set is not completely representative for the performance on the whole set. Even for linear systems, the internal points of the set may theoretically exist that are more critical from the robustness viewpoint than points from the complex envelope.

However, a pragmatic, and rather unmathematical, assumption can be made that complex envelope points are more critical than internal ones. Since only a limited number of points can be considered training examples, envelope points are certainly good candidates. If points other than those with which the controller has been trained turn out to be critical, they are included in the training set and retraining is started. Such a procedure is in the spirit of the adaptive data accumulation of Section 8.5.

The case of robustness expressed by less concrete means such as by frequency domain constraints is briefly addressed in Section 7.4.4.

7.4.1 Plant Parameter Variations

There are two alternative approaches to accounting for plant parameter variations. The first consists in collecting data for all plant variations and identifying one or more models for these data sets. The other approach is based on introducing deliberate changes in parameters of identified models.

Plant Variations Identified from Data   The discussion in the preceding sections of this chapter has shown that the plant model can be viewed alternatively as general model including all plant variations or as a set or continuum of special models, each assigned to a variation.

In both cases, a set of measurements has to be collected. Let us index these measurement sets by i. Each of these data sets should satisfy the requirements of Section 5.6; that is, each of them must be sufficient for identifying a complete model. The plant model argument list is extended by index i. Index i may extend either the vector of measurable parameters θ or the vector of nonmeasurable parameters ζ, depending on whether it is possible during real-time control to figure out which past alternative for which the data have been collected is active. For example, if the plant is a test bench on which various types of workpieces are tested, each workpiece type can be assigned an index i and identified in a separate model F_i. Then, index i extends the vector of measurable parameters θ, since, in a future real-time control environment, it will be known which of the workpiece types is currently tested. By contrast, if a wastewater treatment plant is operated under various environmental conditions such as weather or nonmeasurable contaminant concentrations, the conditions cannot be reliably indexed. Then, index i extends the vector of nonmeasurable parameters ζ. Similarly, the controller (7.32) may receive the model index i as additional input.

Formally, in the special model method, the model $F(\cdot, i)$ is set equal to the model F_i separately identified with the help of the ith data set:

$$\mathbf{y}_{t+1} = F(\mathbf{y}_{e_t}, \mathbf{u}_{e_t}, i) = F_i(\mathbf{y}_{e_t}, \mathbf{u}_{e_t}) \qquad (7.35)$$

By contrast, in the general model approach, the mapping $F(\cdot, i)$ is identified at once.

The reasons for choosing the method of special models are mostly numeric. It is easier to find a model for a particular variation than for all variations simultaneously. Identifying a set of special models instead of a single general model amounts to a decomposition of the problem, which is the preferred way of handling high complexity.

Deliberate Variations of the Identified Model Sometimes, it is difficult to collect data from all situations that may appear. Even if this were possible, there would be no certainty that the data collected would really represent all relevant variations. Then, the identified plant model must be modified for robust neurocontroller training.

Theoretically, it should be possible to vary the plant model parameters such as the neural network weights directly. There are two arguments against this (1): The number of such parameters is too large (tens to hundreds). Since each variation corresponds to a training example, this would produce a huge number of training examples. (2) Since neural network parameters cannot be meaningfully interpreted, it would be difficult to define variation ranges for individual parameters.

A preferable alternative is to consider variations of arguments of plant model mappings. For example, the model arguments $\mathbf{y}_{e_t}$ and $\mathbf{u}_{e_t}$ can be multiplied by factors varying around the value 1 corresponding to the original model. Formally, this amounts to the multiplication of both vectors with diagonal matrices $\mathbf{D}_y$ and $\mathbf{D}_u$, respectively:

$$\mathbf{y}_{t+1} = F_{\text{mod}}(\mathbf{y}_{e_t}, \mathbf{u}_{e_t}, \mathbf{d}_y, \mathbf{d}_u) = F(\mathbf{D}_y \mathbf{y}_{e_t}, \mathbf{D}_u \mathbf{u}_{e_t}) \qquad (7.36)$$

Here, the diagonal elements of matrices $\mathbf{D}_y$ and $\mathbf{D}_u$ are equal to the elements of vectors $\mathbf{d}_y$ and $\mathbf{d}_u$.

The diagonal elements of both matrices are, formally, elements of nonmeasurable parameter vector ζ. The training set has to contain a representative choice of combinations of such factors. It is mostly sufficient to consider the expected lower and upper bounds of individual factors, for example, factors 0.7 and 1.5 for each input variable. Variations around one for factors concerning feedback of plant outputs have to be substantially smaller—the feedback will amplify the effects in an exponential way. A full Cartesian product of all

220 ROBUST NEUROCONTROL

factors usually makes the training set too large. A set of training examples, each with only one factor having a different value, is a feasible alternative.

7.4.2 Systematic Disturbances

Although a systematic disturbance can generally have the form of an arbitrary dynamic system, for example, of a closed system (7.6), the usual requirement is robustness against a constant systematic disturbance. Between stable systematic disturbances (and it is only the stable ones for which stabilization of the closed-loop is possible; see Section 7.1), the constant ones (more exactly, those with constant offset from zero) are particularly harmful in being persistent. This persistency leads to permanent stationary offsets from reference values. The most general case is the offset on both output and input:

$$\mathbf{y}_{t+1} = F(\mathbf{y}_{e_t} + \mathbf{E}_y \mathbf{o}_y, \mathbf{u}_{e_t} + \mathbf{E}_u \mathbf{o}_u) - \mathbf{o}_y \tag{7.37}$$

The elements of offset vectors $\mathbf{o}_y$ and $\mathbf{o}_u$ corresponding to various delays of the same variable are set to the same value. This is formally implemented as a multiplication by matrices $\mathbf{E}_{y,u}$ consisting of a row of h unit submatrices of dimension q and p, respectively, with h being the length of the history considered.

In practice, an input offset is frequently sufficient to emulate the expected disturbances. The neurocontroller training set then has to contain various offset vectors $\mathbf{o}_u$.

In any case, such a training set is helpful only if an appropriate neurocontroller structure (i.e., with error integration or order increase) is chosen.

7.4.3 Random Disturbances

The stability of a closed loop for random disturbances in the form of pulses or steps is implicitly solved by solving the stabilization task itself. An informal assumption underlying this viewpoint is that individual disturbances are small enough, or infrequent enough, that stabilization can take place, at least on average, before the next disturbance becomes active. However, if a controller is designed for a certain subset of state space of a nonlinear plant, and if such disturbances are very frequent, they may cause either the plant or the controller to diverge faster from this working subspace than the stabilization can be accomplished.

The first case of a plant diverging through random disturbances may easily become out of reach of the control design—the plant may be thrown into some disastrous state corresponding, for example, to material damage, without the possibility of trading this off by any sequence of control actions.

The second and much more frequent case of a diverging controller is caused by disturbances in the form of measurement errors. It is not the plant that is really disturbed but rather the information about the plant that the controller receives. If this information loss exceeds a certain limit, the plant may become unobservable, making stabilization impossible—the feedback is no longer a complete state feedback.

As long as the information loss is within some bounds, closed-loop stabilization is feasible. However, the information loss has its cost. For example, the attainable speed of stabilization may be reduced, in comparison with the case of exact measurements. It is then the task of neurocontroller training to find the right compromise that is both stable and as fast as possible for a given intensity of disturbance.

This is the reason why these disturbances should be included in the training examples. The controller is then trained so that it receives, instead of measurement vectors $\mathbf{y}_t$, vectors $\mathbf{y}_t + \mathbf{r}(\rho)$, with $\mathbf{r}(\rho)$ being a vector generated by a random process parametrized by amplitudes or variances ρ.

Note 7.4 It is advantageous for the optimization in batch schemes if the disturbance generation is only pseudorandom, with an identical initial value of random generator seed. Then, a cost function value is deterministic for given neurocontroller parameters.

Note 7.5 The optimization task with random disturbance modeling is more difficult to solve than without disturbances because the disturbance makes the cost function nonsmooth. It is advisable to train first from scratch without disturbance and then retrain with the disturbance.

7.4.4 Frequency–Domain Specifications

Alternatively to the time domain, robustness requirements, as well as other requirements on control, can be formulated in the frequency domain. For example, the cost function may have the form

$$C_g = \sum_{f=1}^{f_{\max}} \|Y(f)\|^{\infty} \tag{7.38}$$

that is, minimizing the infinity norm (the maximum) of the plant output Fourier transform. To evaluate $F(f)$ numerically, the closed loop can be excited, say, by a sinusoidal reference trajectory of normed amplitude, for example,

$$\mathbf{w}_t = \mathbf{W}_0 + \sin(2\pi t/T)\mathbf{W}_a \tag{7.39}$$

The discrete Fourier transform is applied to the time series of plant outputs **y**. Then, the cost function has the form

$$C_g = \sum_{f=1}^{f_{\max}} \left\| \sum_{t=0}^{T-1} \mathbf{y}_t e^{2\pi i f t/T} \right\|^{\infty} \tag{7.40}$$

This cost function is an instance of a straightforward and easily implementable generalization of cost function of type (7.34). The only modification is that a weighted sum of now complex-valued terms c_t is subject to norm computation.

This makes obvious that methods such as H_∞ can be applied in a straightforward way. The problem with this approach is more theoretical than practical. Fourier transform and other frequency–domain considerations are, strictly speaking, not directly applicable to nonlinear systems. What can be done is to view the nonlinear plant as a set of linearized plants around various working points. The working points are then defined by the reference offsets $\mathbf{W}_0$ of Eq. (7.39), and the cost function can be a weighted sum of partial cost functions (7.40).

To what extent this will allow us to formulate meaningful robustness requirements in the frequency domain remains an open question. It will certainly depend on the personal preferences of the neurocontroller developer.

8

LEARNING AND ADAPTIVENESS IN NEUROCONTROL

The idea of adaptive control has its origin in the vision of systems that need not be accurately designed but can improve their performance by acting in the environment, observing the consequences of these actions, and optimizing their structure and parameters. Such behavior can be observed among biological organisms, although it is an open question how much of their capabilities can be attributed to adaptiveness.

Careful design of a control system is expensive in time and money and requires highly skilled personnel. However, accurate design is possible only if the situation in which the control system is going to act is accurately specified. But this is not the case in most applications. A partial solution to the problem of accurate specification is provided by methods from the field of robust control design. However, there is a trade off between robustness (at least robustness in a narrow sense) and control quality. A more robust controller is usually slower or less stable, and vice versa. So the commercial motivation for developing adaptive control comes from the need for an expensive design of high-performance controllers working under varying conditions.

8.1 ADAPTIVENESS, LEARNING, AND GENERALIZATION

The terms adaptiveness and learning have various, frequently overlapping, meanings. The most usual distinction [e.g., 10] is with help of forgetting. *Adaptation* is viewed as a process in which a system changes to perform better in a new situation. By contrast, *learning* is a process, the goal of which is for the system to perform better in both new and old situations. This concept of learning has a close relationship to that of *generalization*, which is summarizing the experience from old and new situations and building a representation that accounts for both old and new simultaneously. Generalization frequently implies a certain economy of representation—the simultaneous representation of old and new cases should not consist in a simple coexistence of both representations. Economy of representation is closely connected with interpolation and extrapolation capabilities—if the representation is the most economical from those alternative representations that do not lose information, it must make use of the features that are common to as many individual cases as possible.

If one were free to choose between such adaptiveness and learning, learning is obviously the more attractive concept. Good performance in both new and old situations is attractive simply because it is not known when the old ones come back; whenever they do, the system is fit to master them. Even if the environment exhibited a definite trend in which old situations would never come back, an adaptive system has to have good answers to questions, such as, when is a situation new and no longer old, and what makes a situation different from a system state.

These questions are less critical for learning. If the system is to perform equally well in new and old situations, their distinction is irrelevant. This makes also the distinction between situation, or context, and state uninteresting for the system; information about both is simply input for the decisions made by the system.

On the other hand, it is clear that learning is much more difficult. This means that mappings acquired by learning are much more complex than those used in adaptation. Consequently, the search after mappings to be used in learning takes place in a substantially larger function space, or, if the mappings are represented by parametrized functions, in a substantially larger parameter space. A good illustration of this is linear adaptive control applied to nonlinear plants. There are many successful applications of linear adaptive control [e.g., 8] even to such nonlinear systems for which a general invariant linear controller cannot be found. In such cases, the alternatives are to (1) continuously search in the small space of a linear controller parameter (and get a controller that is specific for a given working point and may be totally useless in other working points) or (2) find a universal controller for all working points (or situations), which necessitates a search in a large space of nonlinear controller (e.g., neurocontroller) parameters.

The question of generalization, and particularly of representational economy, is also relevant to neurocontrol. Local representations based on the partitioning of state space, such as those using radial basis functions (Section 3.1.3), are additively collecting information from new and old situations. The nonadditive aspect arises only if new and old information is conflicting. Then, some strategy for resolving this conflict (e.g., forgetting by decay factors) is required. In other words, there is no difference between learning and adaptation as long as there are no conflicts in data about one state space point or region. Note that this conflict is related to the scope of information provided. Suppose that the behavior of a plant depends on temperature. The controller required for low temperature is different from that required for high temperature. If the temperature measurement is not available, the information from the high-temperature working point appears to be in conflict with that from the low-temperature working point, since both working points cannot be distinguished without temperature measurement. With temperature measurements, the information from low- and high-temperature working points is assigned to different points of the extended state space and is thus not conflicting.

Global representation (as with help of a multilayer perceptron, see Section 3.1.2) is really global only if it does not provide enough resources (e.g., hidden units) for local encoding of individual cases. It is the lack of resources that enforces global representation. Then, the globality of representation is closely related to generalization. In global representations, information from different working points may appear to be conflicting (for given resources) as long as the current representation makes no correct generalization over the different working points. These conflicts can be resolved only by trying different generalizations, that is, by searching the parameter space. In other words, conflicts are an indispensable part of building a global representation. This may be computationally expensive, but the expense is rewarded by the generalization capability.

Computationally expensive conflict resolution is omnipresent with global representations. This makes global representations less recommendable for fast, real-time adaptation. The real domain of global representations is where learning and generalization is required.

8.2 PRAGMATIC CHARACTERISTICS OF ADAPTIVE CONTROL

Obviously, the discussion of adaptiveness as a phenomenon would go beyond the scope of this chapter. An excellent overview of the adaptive control definitions is given by Narendra and Annaswamy [103]. Instead, we focus on a pragmatic view of what physical features a control system should possess to be called adaptive.

1. The design of an adaptive control system must be based on measured data. This criterion excludes control systems designed by methods based on the knowledge of the system. It is clear that some problem-specific definitions such as saying which variables can be measured do not offend this principle, but measured data must contribute the most information for the design.

2. The design must be performed automatically. This criterion ensures that all essential problem-dependent choices take place in an algorithmic way, and are not made by the designer.

3. The design must be performed in real time. This criterion obviously depends on the following:

The sampling rates used in the application.

The lifecycle of the application and of the changes to which it has to be adapted.

The computational platform used.

Although this criterion seems to be relative and application-dependent, it corresponds to the view of the control system user. Control of wastewater treatment plant can be implemented in an adaptive way, rather than by the roll control of a military aircraft, because there is much more time to react to environmental changes by optimizing the controller. Using a powerful parallel computer makes adaptiveness feasible for applications in which a microcontroller is just able to perform PID control.

4. An adaptive control cannot be easily distinguished from feedback control, in general, and robust control, in particular, as the preceding discussion has shown (see Chapter 7). Feedback control action is computed from measured data; it is computed automatically and in real time. While, by definition, adaptive controllers have to contain parameters whose values are changing with the adaptation, the parameters can also be viewed as the state of the controller that is evolving under the influence of the data. To show what makes the difference from the user's point of view, we extend this definition by requiring that control parameters be changed by a separate adaptation process or algorithm, typically taking place in a time frame some orders of magnitude slower than the control itself.

8.3 WHEN IS ADAPTIVENESS REALLY NEEDED?

Before the search after the most appropriate adaptive or learning scheme is started, it should be realized that the fundamental alternatives are adaptation or no adaptation.

The discussions of Chapter 7 and Section 8.1 have made clear that there is always a dichotomy between changing the controller during its operation to perform better in particular situations and designing the controller in such a general way that it will perform well in all situations. The latter alternative appears at first glance to be too ambitious and thus is frequently assumed to be infeasible. Since every comparison is relative, this hypothesis is usually based on the assumption that the first alternative is easier. However, the remaining sections of this chapter will show that this is not the case. Some of the topics are just lists of unsolved problems.

One approach, designing the controller in advance, requires addressing directly some problems that are less obvious (but not less serious) in adaptive schemes. They include providing sampled measurements that appropriately cover the state space (see Section 5.6) and using optimization methods that are powerful enough to find a good general neurocontroller (Chapter 4). The lower visibility of these problems in adaptive schemes makes them a bit perilous. Some additional tasks include:

1. Selecting the appropriate measurements out of those spontaneously arising during the real-time operation.
2. Providing for appropriate (at first, not too fast) forgetting in order to weight present and past situations correctly.
3. Implementing sufficiently powerful optimization methods in a real-time target configuration.
4. Providing for, and supervising, the stability of adaptation.
5. Defining emergency procedures in case of adaptation failure.
6. Supervising the operation closely until the quality of control has been proved.

As this list suggests, the use of online adaptive schemes should be considered only if the measurements required for offline training plant parameters cannot be received in advance.

Note 8.1 A particularly bad argument in favor of adaptive neurocontrol has been inherited, by superficial considerations, from the classical linear adaptive control theory. There, adaptive control is sometimes (or maybe mostly) motivated by plant nonlinearities—the plant is linearized around the current working point. This linearization is then adapted for other working points. Such adaptation is useful in the linear domain, but not if both the plant model and the controller are nonlinear. The reasons for this are of a computational nature. A linear plant model and controller can frequently be determined from several, or several tens of, consecutive samples. The adaptation must thus be made sufficiently fast to follow the changing working point.

By contrast, for nontrivial nonlinear controllers, hundreds, thousands, or tens of thousands of samples are required to reach good performance. It is much easier to design the controller directly for all working points.

It can be said that there are substantially fewer nonlinear control applications for which online adaptiveness makes sense with today's methods and computational platforms than is the case for linear applications.

8.4 INCREMENTAL AND BATCH SCHEMES FOR ONLINE ADAPTIVENESS

It is the incremental schemes that are usually seen in close relationship with adaptation. Indeed, the primary form of incremental schemes is online adaptive, such as we have seen in the algorithms of Sections 5.5.1 and 6.5.3 which together constitute an adaptive neurocontroller training scheme. This is illustrated in figure 8.1, where the dashed-dotted connection line between the plant

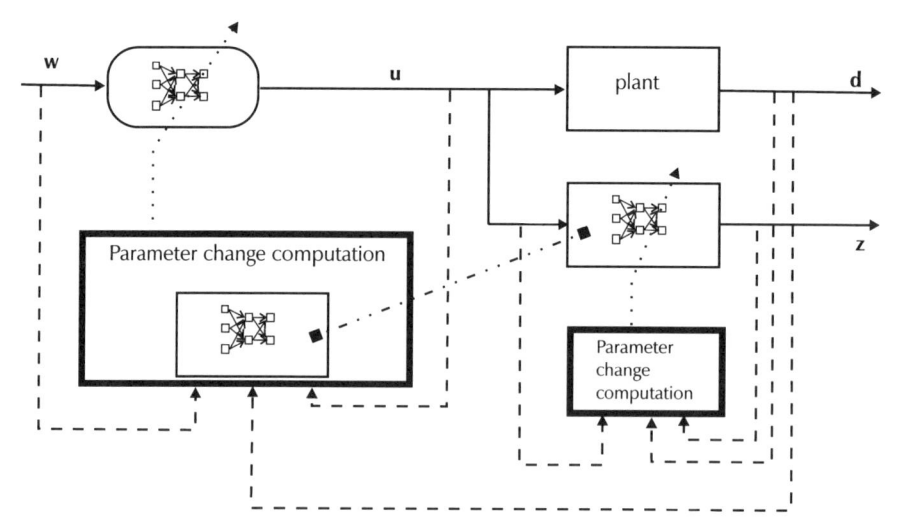

Figure 8.1 Adaptive incremental scheme.

model network symbols in the parameter change computation block and in the right column represents the identity between both. In fact, nonadaptive incremental schemes (Note 4.8) are mere emulations of adaptive ones: The sampling of measurements is substituted by systematic or random presentation of measurements from a data collection sampled in advance.

By contrast, batch schemes are usually considered to be committed to off-line, nonadaptive operation. However, it is equally possible to apply the batch algorithms of Sections 5.5.2 and 6.5.4 in an adaptive configuration running parallel to the plant in operation. The structure and basic components of such an algorithm are presented in Figure 8.2.

There are some components additional to those of the basic batch neurocontroller training algorithm. They are necessary for updating the training sets, both for plant identification and neurocontroller training. In particular, they include the data accumulation components deciding which data will be added to the training sets and the forgetting components deciding which data will be deleted from the training sets. Note that the data accumulation block for controller training set receives samples of reference values **w**, control actions **u**, and plant outputs **d**, in contrast to the sole **w** input into the reference trajectory generation block of the basic batch neurocontroller training scheme of Figure 6.2. The reason for this is that samples of **u** and **d** are used for evaluating which reference trajectories are to be included in the controller training set.

Why are such components required for batch schemes and not for incremental schemes? The answer is that for incremental algorithms, the same decisions have to be made, but they are usually intermingled with the learning algorithm itself, for example, in the form of trivial forgetting implemented by decay factors, or solved by ignoring them, for example, if *all* sampled data are used for learning.

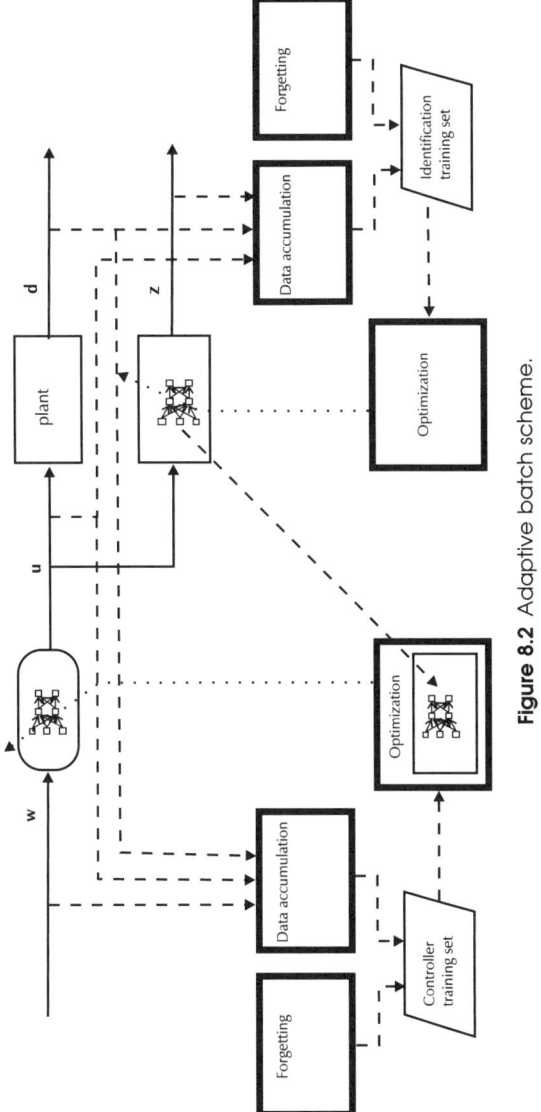

Figure 8.2 Adaptive batch scheme.

The problems of data accumulation and forgetting are briefly discussed in Sections 8.5 and 8.6.

8.5 DATA ACCUMULATION

The discussion of Section 5.6 has shown that the composition of the sampled measurements is important for success in plant identification. The crucial requirements of sampled measurements are sufficient variability of data (i.e., sufficient excitation of the plant) and sufficient coverage of the state space. These requirements set up the necessary conditions and suggest that the more data that are available, the better, since more measurements raise the probability of some among them satisfying the requirements.

However, this is not the whole truth. Besides the natural requirement of data quantity not exceeding some bounds resulting from available computing resources, the distribution of data within the training set is important. If completely exact model identification were possible, it would make no difference whether the training set consists of (1) 1000 samples satisfying the excitation and coverage requirements alone or (2) additionally 100,000 irrelevant, poorly exciting measurements (except for the computing time being 101 times longer in the latter case). In both cases, the minimum error function would be zero.

In reality, exact model identification with neural networks of limited size and optimization algorithms using limited resources is impossible. This means that the error function value cannot be reduced to zero. But also data that do not contribute to better excitation and coverage do contribute to the error function, although the improvement of their error is of less utility than that of relevant data. This makes clear that the distribution of sampled measurements makes a difference, along with good excitation and coverage.

In iterative adaptive schemes, the requirements above have to be reformulated in order to relate the newly collected data to the data already seen (i.e., for batch schemes, data already contained in the training set). The best way may be to assess the additional information content of the data and accept only data whose contribution is sufficient. Fortunately, we can easily do this by comparing the data and the model forecast. If the model forecast for a measurement series is poor, it is certain that the measurements will improve the model if included in the training set. Then, the model has obviously not been sufficiently trained for the state space region; either the data about the region have not occurred so far, or the excitation of the plant has not been sufficient.

The line of argument for including a measurement series into the training set is as follows: Consider a class of functions for functional approximation (e.g., for a given neural network structure with free parameters) parametrized by vector ω. Let us denote the fit (i.e., the measure of difference between the plant model and the plant) at an extended state space point $\mathbf{m}$ [e.g., the value of the cost function C_f described by Eq. (5.103) for a single training example $\mathbf{D}$] as $c(\omega, \mathbf{m})$. Here, the extended state space point consists of an initial state and

DATA ACCUMULATION 231

action or series of actions if a multistep forecast is used. A fit over a *region* M_r of extended state space M can be defined as a supremum

$$C(\varpi, M_r) = \sup_{\mathbf{m} \in M_r} c(\varpi, \mathbf{m}) \tag{8.1}$$

or as an integral

$$C(\varpi, M_r) = \int_{\mathbf{m} \in M_r} c(\varpi, \mathbf{m}) d\mathbf{m} \tag{8.2}$$

both being instances of the general criterion

$$C(\varpi, M_r) = \sqrt[i]{\int_{\mathbf{m} \in M_r} c(\varpi, \mathbf{m})^i d\mathbf{m}} \tag{8.3}$$

with $i = \infty$ (8.1) and $i = 1$ (8.2). The goal of identification is to find the vector $\varpi(M)$ for which the fit over the whole extended state space M is the best, that is, for which

$$C[\varpi(M), M] = \min_{\varpi} C(\varpi, M) \tag{8.4}$$

Also for every region $M_r \subset M$, the best fit

$$C[\varpi(M_r), M_r] = \min_{\varpi} C(\varpi, M_r) \tag{8.5}$$

can be defined.

The supremum-based definition (8.1) has the property that for every sequence of regions M_i such that $M_1 \subset M_2 \ldots \subset M$, the sequence $C[\varpi(M_i), M_i]$ is nondecreasing. The reason for this is that the supremum-based criterion is equal to the worst fit over the region. Obviously, even with the best model, the worst fit over a region M_j that is a superset of M_i can only be worse than the worst fit over M_i.

Let us now investigate what happens if the plant model optimal for region M_i [i.e., the model with parameters $\varpi(M_i)$] is used to evaluate the fit over a new region (possibly consisting of a single extended state) M_k. If

$$C[\varpi(M_i), M_k] > C[\varpi(M_i), M_i] \tag{8.6}$$

it can be concluded that

$$M_k \not\subset M_i \tag{8.7}$$

What does this mean for accumulation of relevant training data? Suppose that the plant identification is done with the help of an optimization algorithm capable of finding the optimum for a given training set, that is, of finding $\omega(M_i)$ for the training set covering (in some pragmatic sense) the region M_i of the extended state space. Then, with a new sample m_i representing subregion N_i, a forecast is done in the form of an evaluation of $C[\omega(N_i), N_i]$. If the condition

$$C[\omega(M_i), N_i] > C[\omega(M_i), M_i] \qquad (8.8)$$

is satisfied, the sample comes obviously from outside the region M_i. Consequently, adding m_i to the training set M_i will improve the coverage of the extended state space. Because the extended state space contains a sequence of plant inputs, its improved coverage subsumes the excitation improvement, that is, a plant model trained with M_i containing only insufficiently excited input sequences will exhibit poor fit for the improved excitation input sequences.

To summarize, the adaptive algorithm will consist of the following steps:

1. Initialize ω randomly, $C_{\max}$ to ∞, and the training set P to the empty set.
2. Perform iteratively the following (potentially infinite, i.e., perpetually adaptive) loop:
   a. Take measurement sample Q.
   b. If $C(\omega, Q) > C(\omega, P)$, then $P \leftarrow P \cup Q$ (extending the training set), $\omega \leftarrow \omega(P)$ (optimization).

The criterion (8.2) and generally all criteria (8.3) with $i \neq \infty$ lack the advantageous properties seen in Eq. (8.7) ensured by Eq. (8.8). A large region of extended state space may contain a small difficult subregion in which the average fit is substantially worse than in the large region. On the other hand, the supremum-based criterion may depend on the single worst point of extended state space, all other points are irrelevant. It is also nonsmooth in terms of plant model parameters (in the points where the maximum shifts from one sample to another) and thus difficult for optimization algorithms. A good compromise is the general criterion (8.3) with a relatively high power such as $i = 4$. Such a criterion behaves as an approximate supremum while smooth and leaves no training example out of consideration.

The assumption that an optimization algorithm always finds an exact global optimum is certainly exaggerated. To account for this, a slight modification of the acceptance procedure is advisable: Samples should be accepted only for inclusion into the training set if the inequality (8.8) is satisfied with some margin.

With incremental schemes, the principles to be applied are similar. Incremental schemes do not maintain an explicit training set. Instead, it is assumed that the information of the previous training examples is incorporated into the current solution. Then, instead of extending a training set by data accepted after some criterion such as Eq. (8.8), only the accepted data are used for the incremental training step (whereas those not accepted are not used).

So far, training the plant model has been discussed. For training the neurocontroller, the procedure is analogous. The only difference is in the definition of the extended state space. In plant model identification the extended state space consists of the initial state (or initial output state) and an input series. For neurocontroller training, it consists of the initial state and a vector characterizing the control requirements, for example, reference states or sequences.

8.6 FORGETTING

A process complementary to data accumulation is data forgetting. The goals of accumulating training data are to make the plant model or the neurocontroller in some sense consistent with additional data. In other words, the plant model or neurocontroller is made dependent on additional data. The function of forgetting is to make data independent of earlier data on which they had been dependent.

There are two fundamentally different reasons for forgetting:

1. *Economy* The training set size remains bounded.
2. *Time variance* Old data lose their validity.

8.6.1 Forgetting to Keep the Training Set Bounded

The economical reason for forgetting training data has to do with the complementary process of data accumulation. It explicitly occurs only with adaptive batch schemes. If the training set is to be kept at an approximately constant size, for additional training samples some other samples have to be deleted from the training set. In doing that, the goals of accumulation must be kept in mind. The goal of accumulating training data is to improve the coverage of the extended state space. Since, after deleting some data, the coverage can only deteriorate, it is important to provide for deleting the data that decrease the coverage less than the newly incoming data increase it. So a principle complement to that of accepting data with the highest information content (relative to the actual training set) is to delete data with the least information content.

One way to implement these principles is to delete training samples from densely populated regions of extended state space.

There are several methods to determine the density of coverage in individual regions. The simplest (and the most widespread) is by using some distance metrics [15]; for example,

$$d(\mathbf{z}_i, \mathbf{z}_j) = (\mathbf{z}_i - \mathbf{z}_j)^T \mathbf{C}^{-1}(\mathbf{z}_i - \mathbf{z}_j) \tag{8.9}$$

where $\mathbf{C}$ is the covariance matrix for extended space vectors $\mathbf{z}$. This method performs a linear rescaling of individual elements of the extended state space vector.

A serious shortcoming of the distance metrics method is that it does not account for different dynamic properties of the plant in different regions. There might be regions in which the plant requires a higher density of coverage than in other regions. One way to account for this is by some metrics of variability of the mapping in different regions. One such measure is *curvature*, which is used for determining the required coverage density by Sing and Wong [132]. For a function f of vector $\mathbf{z}$, curvature is a vector of second derivatives

$$C(\mathbf{z}) = \left[\frac{\partial^2 f}{\partial z_i^2}\right] \tag{8.10}$$

The density along the ith axis can then be made proportional to the absolute value of the ith element of curvature. For multivariable models, the mapping f is multidimensional (corresponding to the dimension of output vector $\mathbf{y}$). Then, some appropriate aggregate (e.g., the maximum) of curvatures with regard to individual output variables has to be used.

However, the use of the first derivative, or of derivatives higher than two, would be equally justified. This is also the main problem with such metrics—the decision in favor of a particular derivative as a measure of irregularity is rather arbitrary. (If the mapping would be known to be a ith-order polynomial, the ith derivative would be the right choice, but the decision in the general nonlinear case is difficult.) Another shortcoming is that the required density along a particular axis is difficult to attain if the dimensionality of the mapping is (as usual) such that it is impossible for complexity reasons to construct a training set as a Cartesian product.

Then, the best approach is a decision-theoretical one. The decision to delete the ith training example is to be assigned some value of loss. According to supremum criterion (8.1) or its approximation, it is best to delete, from the training set P, the training example Q_i for which the fit over the training set without example Q_i, $C[\omega(P \backslash Q_i), P \backslash Q_i]$, is the worst. Then, the difficult examples determining the value of fit are sure to have been retained. However, the evaluation of $C[\omega(P \backslash Q_i), P \backslash Q_i]$ requires performing optimizations for all training examples Q_i which are successively deleted. A computationally less expensive alternative calls for the help of local gradients. Suppose that the numerical form of the general fit measure (8.3) for training set P consisting of examples Q_i with parameter set $\omega(P)$ optimal for P is

$$C(\omega, P) = \sqrt[k]{\sum_i c(\omega, Q_i)^k} \qquad (8.11)$$

Then, setting $C = C(\omega, P)$ and $c_i = c(\omega, Q_i)$, the gradient of C is, for $k \neq 1$,

$$\nabla_\omega C = \frac{d\sqrt[k]{(\sum_i c_i)^k}}{d\omega} = \frac{C^{1-k}}{k} \sum_i k c_i^{k-1} \nabla_\omega c_i = \frac{\sum_i c_i^{k-1} \nabla_\omega c_i}{C^{k-1}} \qquad (8.12)$$

After deleting the jth example, the minimum is lower. The extent of the decrease of this minimum can be locally assessed by the norm of gradient

$$\nabla_\omega C_j = \frac{\sum_{i \neq j} c_i^{k-1} \nabla_\omega c_i}{C_j^{k-1}} \qquad (8.13)$$

with $C_j = \sum_{i \neq j} c_i$. For a relatively large number of training examples, $C_j \approx C$. Then, since in the current optimum $\nabla_\omega C = 0$, Eq. (8.13) is proportional to

$$\nabla_\omega C_j = -c_j^{k-1} \nabla_\omega c_j \qquad (8.14)$$

This is why the norm of $c_j^{k-1} \nabla_\omega c_j$ is a local measure of the contribution of the jth example to the coverage of extended state space.

Note 8.2 Equation (8.14) can also be applied if the jth example is given a weight h in the sum of Eq. (8.11), and if the change of gradient $\nabla_\omega C$ (i.e., the derivative $d\nabla_\omega C/dh$) is considered. This corresponds to a gradual deletion of the jth training example.

The local measure can be used to determine which training example should be deleted. Namely, it is that for which $\|c_j^{k-1} \nabla_\omega c_j\|$ is a minimum.

8.6.2 Forgetting for Time Variance Reasons

The second reason for forgetting the training data is that the plant is genuinely time-variant and past data lose their validity.

Time variance is a very difficult and hardly addressed topic in the research so far. Its discussion is beyond the scope of this book. We confine ourselves to listing some aspects of the problem.

1. The rate with which the past data lose their validity must be determined quantitatively.

2. The notion of data gradually losing their validity with time is probably an oversimplification of reality. In many cases, the time variance of the plant takes the form of discrete structural breakdowns. These breakdowns may make most past data obsolete and may be followed by stable phases in which forgetting is nothing more than information loss. This view is supported by chaos theory [110, 130].
3. Forgetting is complementary, and contradictory, to data accumulation. It should be decided whether making data more up to date by forgetting has, in each particular case, higher value than what is lost by decreased coverage. This decision is influenced by the extent to which the past and the current data are contradictory. In the case of contradiction, acceptance of new data must lead to deleting the old data to keep the training set free of contradiction. For less contradictory data, there is a free space for various decisions from other viewpoints. To be able to do this, a quantitative contradiction measure applicable to measurements of dynamic systems has to be formulated and embedded into the decision algorithm.

The only currently widespread form of data forgetting, using decay factors for weighting between new and old data, hardly provides answers to these questions—it is only a fixed or variable exponential forgetting rate. Using such decay factors with the goal of capturing the time variance, the aspects listed above are to be kept in mind, and expectations must be kept low correspondingly.

Note 8.3 Decay factors sometimes are used also for a completely different goal—for improving generalization [e.g., 65]. They implement a process of simple selection of relevant parameters and may help to decrease the number of active network parameters. This can be successfully applied to the time-invariant systems.

8.7 GLOBAL AND LOCAL OPTIMIZATION METHODS

The discussion of Chapter 4 has shown that for nontrivial neurocontrol applications, the use of global optimization algorithms is advisable. There are several reasons for doing this also in iterative adaptive schemes:

1. Local optimization algorithms with random starting points are frequently not sufficient to solve the problem. With a starting point given by the preceding solution, the chance for success with local optimization is better. This is the case if (and only if!) the next global optimum is in the same attractor as the past one. This relationship may occur frequently, but it cannot be logically

chained infinitely. A solution received with a bad coverage of the extended state space can be arbitrarily far from the optimum for a good coverage case.

2. It was stated in Section 5.5.4 that instability of adaptation is a potential property of incremental schemes, whereas batch schemes do not suffer from it. With batch schemes, in contrast to incremental schemes, the state of parameters after the last optimization iteration is not transferred to the next iteration—the optimization is then not dynamic, and thus potentially unstable. However, this is true only if optimization takes place from scratch. Taking the solution of the last iteration as a starting point for local optimization would violate this assumption.

3. The data accumulation and forgetting algorithms of Sections 8.5 and 8.6 remain well-founded only if global optima are postulated in each phase.

These problems are not only theoretical, as illustrated by the computational experiment described in the next section. In deciding between fast local optimization and slow global optimization, a pragmatic solution is to perform global optimization at least every fixed number of iterations, and local optimizations in the intermediary iterations.

8.8 COMPUTATIONAL EXPERIMENT

Since the implementation of online adaptive control systems is, for security and acceptance reasons, some orders of magnitude more difficult than that of nonadaptive neurocontrol systems, no case study concerned with an online adaptive solution of a difficult real industrial problem is presented in this book. However, to show some properties of batch adaptive schemes, the results of an experimental laboratory device are given in this section.

The experimental device is a manipulator sketched in Figure 8.3. The manipulator has a step alternating-current engine, an ultrasonic distance sensor measuring the position of the load, and an resistance-based angle sensor measuring the angular deviation of the rope from the vertical. The goal is to bring the load from one position to another as quickly as possible without overshoot. The action variable is the acceleration of the step engine.

The simplest reasonable model of this device is fourth order: The state variables are the load position s and its derivative $\dot{s}$, as well as the angle ω and its derivative $\dot{\omega}$. Although the order of the plant is relatively low, and the nonlinearities are not very dramatical, some properties of the plant and of the sensors make its control harder than expected:

1. The ultrasonic sensor works with a sampling rate of 50 ms, while control has to take place substantially faster, in about 10 ms.

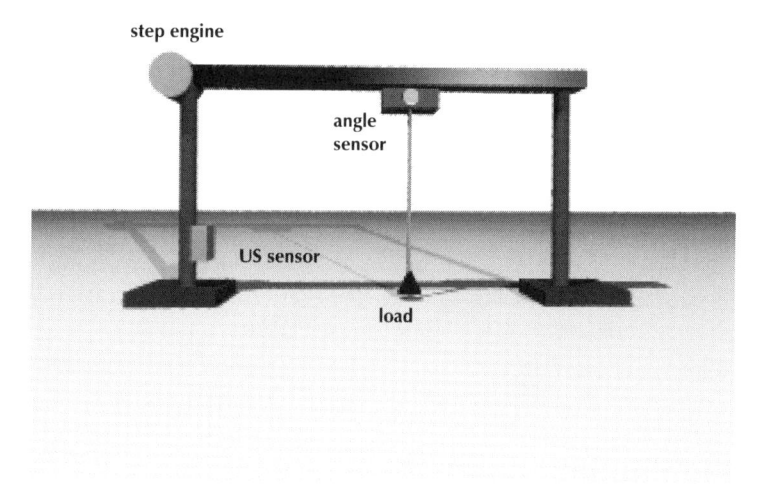

Figure 8.3 Structure of the experimental manipulator.

2. The angle is related to the profile on which the load is suspended. Since the profile is not completely straight (with the time, it becomes still less straight), there are measurement offsets up to 1 degree.
3. The step engine introduces considerable discontinuities into the movement.
4. The load together with the suspension rope constitute a double-pendulum system with the movements of the second pendulum being practically unobservable with the sensory equipment given.

An adaptive online scheme has been implemented, with the following algorithmic modules: (1) The data accumulation module uses the acceptance rule (8.8) and the model fit measure (8.3) with $i = 4$ and yields a good approximation of the supremum-based fit measure (8.1). (2) The batch plant identification module uses a multistep criterion. (3) The final module is the batch neurocontroller training module. For plant identification and neurocontroller modules, global (multilevel) and local (variable metrics) algorithms are available.

The length of the experiment has been limited so that no data-forgetting component had to be implemented.

The development of neurocontroller performance is presented in Table 8.1. The simulation performance (the value of the neurocontroller training cost function) with the plant model received (from the data accumulated so far) in the same iteration, given in the second column, remains similar throughout the experiment. By contrast, the performance on the real plant (the third column) gradually improves, and finally converges to the simulation performance. This is a sign that the plant model gains forecasting capability. For illustration, neurocontroller performance of iterations 3, 6, and 9 is given in Figure 8.4.

COMPUTATIONAL EXPERIMENT **239**

Table 8.1 Neurocontroller performance

Controller Iteration No.	Performance		Rejected Samples
	Simulation	Real Plant	
0	0.115	1.044	—
1	0.101	5.187	0
2	0.123	2.636	0
3	0.086	9.612	0
4	0.087	1.693	2
5	0.038	5.060	1
6	0.059	0.097	9
7	0.058	0.105	42
8	0.044	0.901	58
9	0.045	0.046	114

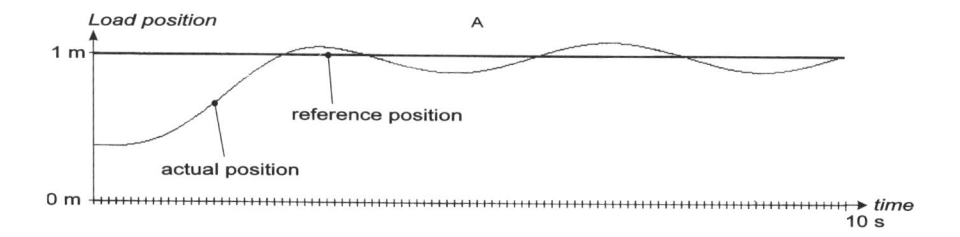

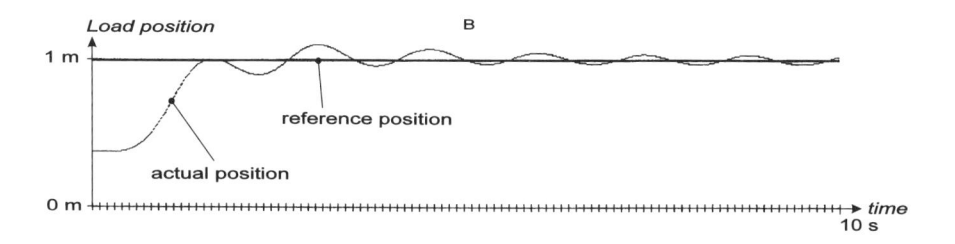

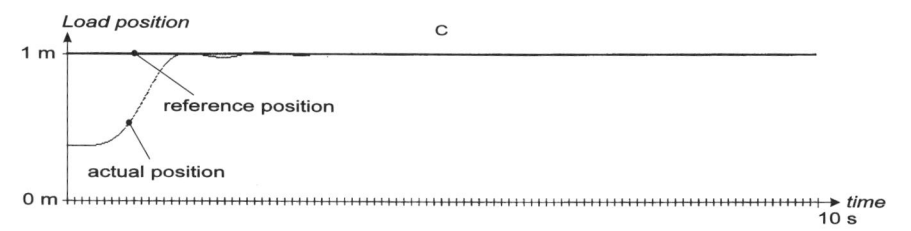

Figure 8.4 Neurocontroller performance—iterations 3, 6, and 9.

The gradual improvement of the model can also be seen in the last column of Table 8.1, containing the numbers of rejected time series (each of 200 sampling period length) until the last of them has been accepted as a training example improving the state space coverage. High numbers of rejected samples in iterations 7 through 9 show the improved forecasting capability of the plant model—a large number of unseen time series are forecast correctly.

Another experiment concerned the performance of global and local optimization for plant identification. In Table 8.2 and Figure 8.5 the successive values of the multistep forecast errors (maximum errors over the training set) are presented. In the zeroth iteration, global optimization has been used in both cases to provide for identical initial conditions.

Global optimization provides a series of roughly monotonically growing errors approaching an upper bound of approximately 0.23. The fluctuations are due to the supremum criterion approximation. The model has been monotonically improving, as confirmed by the performance of the neurocontroller presented above.

Local optimization, in addition to large fluctuations at iterations 1 and 5, abruptly deteriorates at the seventh iteration, without later recovery. The values of the normed squared error near unity show that virtually no nontrivial models have been found: The forecast errors are comparable to the variances of forecast variables. This indicates that the optimum for better-covered state space is obviously in a different attractor than has been the case for initial iterations.

It should be pointed out that this deterioration took place with a very good local optimization method: the Broyden–Fletcher–Goldfarb–Shanno variant of the variable metric method. The consequence of this goes far beyond the scope of incremental adaptation:

Table 8.2 Maxima over training examples of square error values for local and global identification

Iteration	Global	Local
0	0.134	0.134
1	0.170	0.462
2	0.170	0.171
3	0.171	0.172
4	0.170	0.170
5	0.179	0.273
6	0.220	0.222
7	0.202	0.745
8	0.216	0.745
9	0.225	0.503

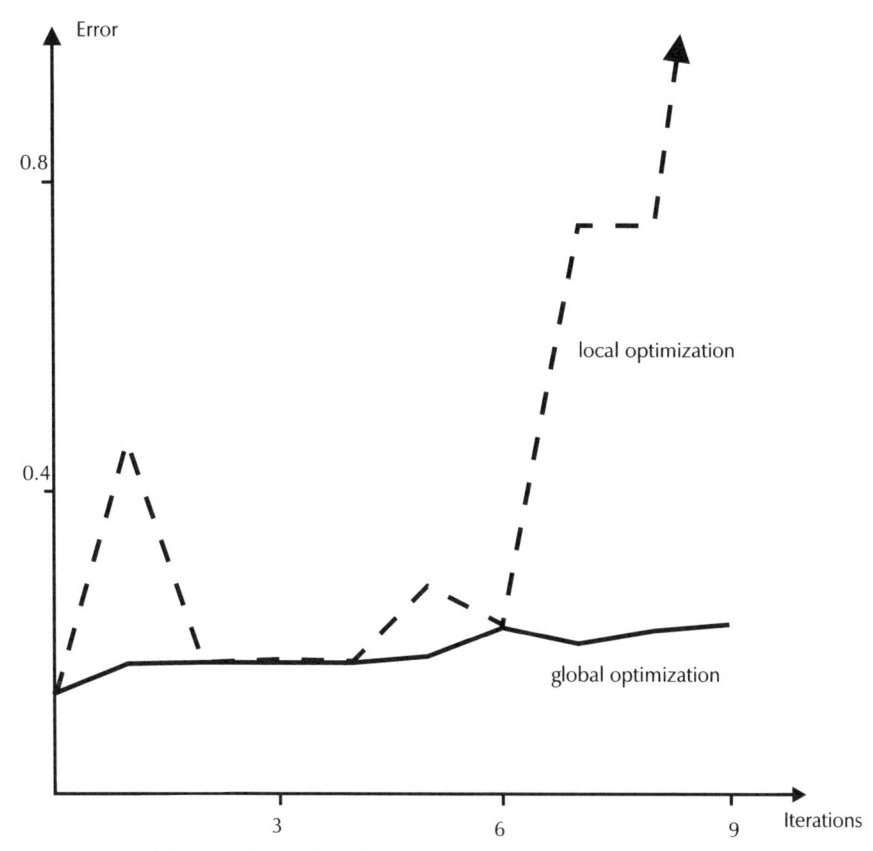

Figure 8.5 Identification by local and global optimization.

Even sophisticated local optimization methods will frequently fail to find not only the global optimum but also any acceptable solution.

It is easy to imagine how less sophisticated algorithms would perform!

9

STABILITY OF NEUROCONTROLLERS

Sooner or later, every neurocontroller developer working on industrial applications is asked the question: Can the stability of neurocontrollers be proved? The simplest, and not false, answer is: Indeed, it can. A more cautious, and also closer to the truth, answer is: It can, the same as the stability of other nonlinear controllers. Most inquirers will be satisfied with such answers, since, in practice, the stability of nonlinear (and usually also linear) controllers is tested by simulations and field tests, rather than relying on the formal proof.

For that rare case where a stability proof must be pursued by the neurocontroller developer, it is good to know what methods are available.

As is given by the nature of neurocontrollers, as well as most nonlinear controllers, the stability proof will be *numeric* rather than analytic. It is important to take into account that the stability proof always requires a *plant model*, and that the proof always *refers* to this plant model. In other words, the proof is valid under the assumption that the plant model is exact. Less positively expressed, whenever the plant model does not exactly match the real plant (and it never does), finding a proof should be interpreted only as increasing the *probability* that the closed loop is stable.

The first point to discuss is what possibilities can prove that an already designed neurocontroller is stable. This is the topic of Section 9.1.

The next step is to design a neurocontroller so that it is provably (under the above-mentioned caveats) stable. Some approaches to this are proposed in Section 9.2.

9.1 PROVING THE STABILITY OF A NEUROCONTROLLER

For nonlinear systems, there are several approaches to proving stability [20, 103]:

Local stability proof by linearization
Lyapunov-function-based stability proof
Input–output stability
Bounded-input–bounded-output (BIBO) stability
Total stability
Hyperstability

The concepts of BIBO stability and total stability are specialized to the analysis of random and systematic disturbances. Input–output stability and hyperstability have strong roots in the frequency domain. It is also not easy to implement purely numeric counterparts of these methods for the case of general nonlinear plants without a known analytic form.

For the most standard cases, the linearization and Lyapunov methods are sufficient. Both methods are well-suited to numeric proofs in time-domain representations. This is why these two methods are considered in this chapter.

9.1.1 Stability Proof by Linearization

To investigate the stability of a closed loop consisting of plant model

$$\dot{\mathbf{z}} = F(\mathbf{z}, \mathbf{u}) \qquad (9.1)$$

and controller

$$\mathbf{u} = G(\mathbf{z}, \mathbf{w}) \qquad (9.2)$$

the closed loop is written in a closed form without external input. Simultaneously, an equilibrium point of the closed loop must be defined around which stability is being investigated.

Suppose that state $\mathbf{z}_w$ and action $\mathbf{u}_w$ are such that

$$\mathbf{0} = F(\mathbf{z}_w, \mathbf{u}_w) \qquad (9.3)$$

and

$$\mathbf{u}_w = G(\mathbf{z}_w, \mathbf{w}) \qquad (9.4)$$

Then, the plant has an equilibrium point at $\mathbf{z} = \mathbf{z}_w$ and $\mathbf{u} = \mathbf{u}_w$. It is certainly desirable that this equilibrium point be the one for which $\mathbf{y} = \mathbf{w}$, but this has no influence on the stability investigation and is not addressed here further. Note that

PROVING THE STABILITY OF A NEUROCONTROLLER 245

$$\dot{\mathbf{z}} = F[\mathbf{z}, G(\mathbf{z}, \mathbf{w})] = H_w(\mathbf{z}) \tag{9.5}$$

and by Taylor expansion around $\mathbf{z}_w$ of Eq. (9.5),

$$\dot{\mathbf{z}} = H_w(\mathbf{z}_w) + \left[\frac{dH_w}{d\mathbf{z}}\right](\mathbf{z} - \mathbf{z}_w) \tag{9.6}$$

with $[d(H_w/d\mathbf{z})]$ being the Jacobian matrix of mapping H_w at the point $\mathbf{z}_w$. For stability, constant terms are irrelevant. A sufficient condition [e.g., 20] for closed-loop stability is that the real parts of all eigenvalues (or of the dominant eigenvalue) of the Jacobian matrix $[H_w/\mathbf{z}]$ be negative.

The Jacobian matrix $[d(H_w/d\mathbf{z})]$ can be constructed easily with help of numeric derivatives. Its jth column is

$$\frac{H_w(\mathbf{z}_w + \mathbf{e}_j \Delta_j) - H_w(\mathbf{z}_w)}{\Delta_j} \tag{9.7}$$

where Δ_j is a small perturbation of the jth element of vector $\mathbf{z}_w$ and $\mathbf{e}_j$ is a vector with the jth element unity and the remaining elements zero. The eigenvalues themselves are then computed by some numeric algorithm such as the QR algorithm in Press et al. [117]. The procedure can be successively applied to an arbitrary number of working points defined by corresponding reference vectors.

So far, a state model (9.1) has been considered. However, this procedure can be applied without change to input–output models. Then, instead of the state vector $\mathbf{z}$, the vector received by the concatenation of extended input and output vectors $\mathbf{y}_i$ for $i = 0, \ldots, n_o - 1$ and $\mathbf{u}_i$ for $i = 0, \ldots, n_o - 2$ is taken. The mapping H_w [with derivative $\dot{\mathbf{z}}_{n_o-1} = \mathbf{z}^{(n_o)}$ on the left-hand side] is extended by trivial integration equations of type $\dot{\mathbf{y}}_{i-1} = \mathbf{y}_i$. The Jacobian matrix and its eigenvalues are then computed for this extended mapping.

9.1.2 Stability Proof Using a Lyapunov Function

A very general tool for investigating the stability of arbitrary nonlinear systems has been proposed by Lyapunov [88, 89] (e.g., see also La Salle and Lefschetz [84], Brogan [20], Föllinger [32], Narendra and Annaswamy [103]). It is based on finding a function $V(\mathbf{x}) = V(\mathbf{z} - \mathbf{z}_w)$ of the plant state transformed so that its origin is in the equilibrium point, with following properties:

1. The function V is positive definite; that is, it is zero if its argument $\mathbf{x} = \mathbf{z} - \mathbf{z}_w$ is a zero vector, and positive elsewhere, and it has a single global minimum for $\mathbf{z} = \mathbf{z}_w$.
2. The function V has nonpositive time derivative; that is, it does not increase with time.

The second condition is sometimes reformulated as

$$\dot{V}(\mathbf{x}) = \frac{\partial V}{\partial \mathbf{x}} \dot{\mathbf{x}} = \frac{\partial V}{\partial \mathbf{z}} \dot{\mathbf{z}} = \frac{\partial V}{\partial \mathbf{z}} H_w(\mathbf{z}) \leq 0 \qquad (9.8)$$

As discussed in Section 2.4.1, this function is closely related to the utility function of control. It expresses, at least in some ordinal sense, the utility or value of each state, from the viewpoint of reaching the reference state, that is, the equilibrium point $\mathbf{x} = \mathbf{0}$.

An important theorem of Lyapunov's stability theory states that if *any* function V with these properties can be found for the given system, the system is asymptotically stable. The region of the state space in which the system is stable is identical with the domain on which the Lyapunov function is defined and on which it possesses these properties. If this is the case for the entire state space, the system is globally asymptotically stable; that is, it will converge to the equilibrium point from all points of the state space.

For every reference vector $\mathbf{w}$, the equilibrium point is different and so is the description of the system. This means that a Lyapunov function can (but need not) be defined formally as a function also of the reference state. If it is, it is more complex, and so is the search for it. On the other hand, if it is not a function of the reference state, no such general Lyapunov function may exist, although a Lyapunov function for each particular reference state does exist (and this would be sufficient for proving stability).

The existence of a Lyapunov function is a sufficient condition for stability, but there is no general method for finding it. There are two basic practical alternatives for search after the Lyapunov function for general nonlinear systems: looking for a parametrized function of a certain type, and constructing the function numerically on a grid in the state space.

Analytical Lyapunov Function For linear plants, a Lyapunov function can be found by using a deterministic algorithm. It has the form of a quadratic function of the state

$$V(\mathbf{x}) = \mathbf{x}^T \mathbf{V} \mathbf{x} \qquad (9.9)$$

This can be attempted also for nonlinear systems. There is a fair chance that such a quadratic function exists, although it has been shown by Fu and Abed [38] that there are singular cases for which a quadratic Lyapunov function does not exist while Lyapunov functions of the third or the fourth order exist.

The search can be formulated as a constraint satisfaction problem, or a minimization problem with penalty terms for constraint violation. The parameters optimized are the elements of matrix $\mathbf{V}$.

The first constraint concerns the positive definiteness of function (9.9). A quadratic function of this form is positive definite iff all principal minors V_1, V_2, ... V_n, that is, all determinants

$$|v_{11}|, \begin{vmatrix} v_{11} & v_{12} \\ v_{21} & v_{22} \end{vmatrix}, \begin{vmatrix} v_{11} & \cdots & v_{1n} \\ & \cdots & \\ v_{n1} & \cdots & v_{nn} \end{vmatrix} \quad (9.10)$$

are positive.

The second constraint concerns the negative definiteness of $\dot{V}$. There is no analytic way to test whether this constraint is satisfied for general nonlinear systems. However, it is sufficient in practice to test it for some representative selection of vectors $\mathbf{x} = \mathbf{z} - \mathbf{z}_w$ such as a grid on the state space Z. Then, the constraints to be satisfied are

$$\mathbf{z}^T(\mathbf{V} + \mathbf{V}^T)\dot{\mathbf{z}} \leq 0 \quad (9.11)$$

for some set $\mathbf{z} \in Z_0 \subset Z$.

The constraint satisfaction problem can be then formulated as a problem of minimization of

$$\min_{\mathbf{V}} \left[w_1 \sum_i s(V_i, \delta_1) + w_2 \sum_{\mathbf{z} \in Z_0} s(\mathbf{z}^T \mathbf{V} \dot{\mathbf{z}}, \delta_2) \right] \quad (9.12)$$

with $s(x, y)$ being $x - y$ if $x > y$ and zero otherwise. The constants $\delta_{1,x}$ represent security margins for constraint satisfaction. Particularly, δ_2 should be set to a positive value to make the second constraint satisfiable also for $\mathbf{z} \notin Z_0$. The constants $w_{1,2}$ are weighting factors. Their role is to bring both summands to a comparable scale.

If the minimum reached by the optimization is zero, the closed loop is stable.

Numeric Lyapunov Function A numeric procedure for Lyapunov function construction is based on the discretization of state space and computing function values for all discrete states. It is closely related to the Ljapuov-based numeric design procedure of Kreisselmeier and Birkhölzer [79], who also provide a theoretically founded discussion of the algorithm.

The algorithm requires a direct use of a discrete-time closed-loop model, for example,

$$\mathbf{z}' = \mathbf{z} + \Delta_t H_w(\mathbf{z}) \quad (9.13)$$

as an analogy to the continuous-time model (9.5). Each state is assigned some direct cost $c(\mathbf{z})$, for example, proportional to its distance from the equilibrium point $\mathbf{z}_w$. For the equilibrium point itself, the cost is zero.

The algorithm is iterative. At the beginning, the function values for all states $\mathbf{z}_i$ are set to ∞. From each state $\mathbf{z}_i$, the next state $\mathbf{z}'_i$ is computed with the help of Eq. (9.13). The value of the Lyapunov function V for state $\mathbf{z}_i$ is updated according to the following rule:

$$V(\mathbf{z}_i) \leftarrow \min \left[c(\mathbf{z}_i) + V(\mathbf{z}'_i), V(\mathbf{z}_i) \right] \quad (9.14)$$

The minimum operator guarantees that the final value of $V(\mathbf{z}_i)$ is approached from above.

If the state space were genuinely discrete, $\mathbf{z}'_i$ would exactly correspond to some discrete state, say, $\mathbf{z}_j$. Then, the space could be represented by a directed graph, and the closed-loop equation (9.13) would define transitions, that is, directed edges from nodes $\mathbf{z}$ to those of corresponding $\mathbf{z}'_i$. The value V of a node would correspond to the sum of costs c assigned to the nodes on the unique path to the node corresponding to the equilibrium state. So, during the iterative application of Eq. (9.14), each node could attain only two values: either ∞ (if the path to the equilibrium point node has not yet been established) or its final value $V(\mathbf{z}_i) = c(\mathbf{z}_i) + V(\mathbf{z}'_i)$. With K state nodes, the labeling would be finished after at most K iterations over all nodes. V would constitute a Lyapunov function iff there were a path from every node to the equilibrium point node.

With a continuous state space, the value $V(\mathbf{z}'_i)$ in Eq. (9.14) can be determined only by interpolation. Interpolation on a rectangular grid can be done by determining into which grid field (an n-dimensional hypercube in an n-dimensional state space) $\mathbf{z}'_i$ falls, and weighting the values of 2^n discrete states in the grid corners. In a two-dimensional case (see Fig. 9.1) with state variables x and y, it first must be determined in which field (i,j), such that $x_{i-1} \leq x < x_i$ and $y_{j-1} \leq y < y_j$, the pair (x,y) falls. Then, the interpolated value is

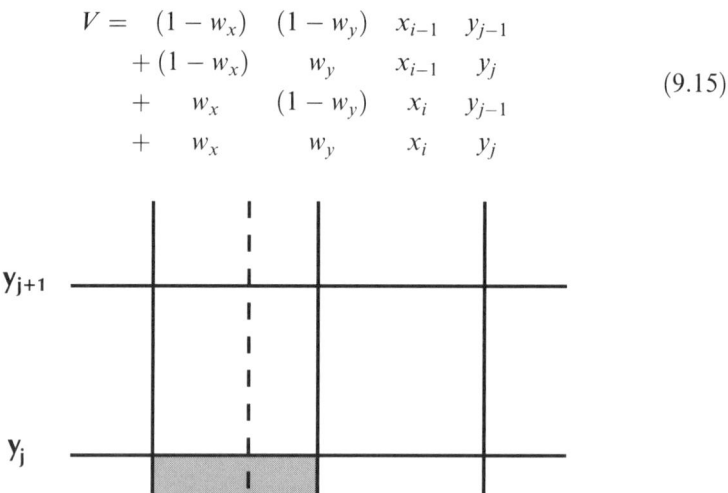

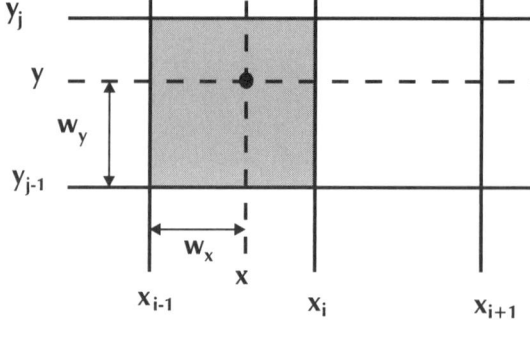

$$\begin{aligned} V = &\;(1-w_x)\;(1-w_y)\;\;x_{i-1}\;\;y_{j-1} \\ &+(1-w_x)\;\;\;\;w_y\;\;\;\;x_{i-1}\;\;y_j \\ &+\;\;\;\;w_x\;\;(1-w_y)\;\;\;x_i\;\;\;y_{j-1} \\ &+\;\;\;\;w_x\;\;\;\;\;w_y\;\;\;\;\;x_i\;\;\;y_j \end{aligned} \qquad (9.15)$$

Figure 9.1 Interpolation on a two-dimensional grid.

with

$$w_x = \frac{x - x_{i-1}}{x_i - x_{i-1}}$$
$$w_y = \frac{y - y_{j-1}}{y_j - y_{j-1}}$$
(9.16)

It is has been shown by Kreisselmcier and Birkhölzer [79] that this algorithm (in a slightly different and expanded form) converges to the solution with a given precision. Then, Eq. (9.14) becomes within this precision

$$V(\mathbf{z}_i) = c(\mathbf{z}_i) + V(\mathbf{z}'_i)$$
(9.17)

Like the case of a genuinely discrete state space, the values V in the grid points represent, after the algorithm has converged to the solution with a stable state, a discretized Lyapunov function whenever $V(\mathbf{z}) < \infty$ is true for all $\mathbf{z}$:

The function is obviously positive-definite, since the costs c are positive and so are the interpolated values $V(\mathbf{z}'_i)$.

Along the closed-loop trajectory, V is descending, as is obvious from Eq. (9.17). The value of V for the next state $V(\mathbf{z}'_i)$ is lower than that of current state $V(\mathbf{z}_i)$ by $c(\mathbf{z})$.

Explicit construction of a Lyapunov function is a very general approach with which it is possible to prove the stability for an arbitrary form of plant and controller. However, the obvious problems connected with the use of this method are discretizing the state space (1) so roughly that the proof is computable with reasonable expense and (2) so finely that the error from interpolating between the discrete values is acceptable. These problems are complementary; it is difficult to find a good compromise for state spaces of dimension higher than, say, three. To prove global stability, the grid has to cover all states that can occur in practice. On the other hand, it has to be sufficiently fine-grained at the proximity of the equilibrium point to discern the stability from, for example, small-sized limit cycles. It is advisable to construct a grid with the step size increasing (e.g., exponentially) with the distance from the equilibrium point. In any case, the interpolation will remain a source of inaccuracy that can easily lead to numeric instability even for closed loops that are otherwise stable. This can be accounted for by defining the zero cost, not only in the exact equilibrium point but also in some neighborhood of it. Of course, the stability proof will then concern the stability of convergence toward the zero cost region and not toward the exact equilibrium point.

Stability under Disturbances A great advantage of the Lyapunov-based stability proof is its flexibility. It can be adapted easily to nondeterministic cases, such as random disturbances, systematic disturbances of unknown size, and plant variations.

The only necessary modification is to consider the worst case of the random spectrum instead of a deterministic system. This can be done because the proof tests whether sufficient conditions for stability are satisfied. If they are satisfied in the worst case, they are satisfied as well in all other cases. The term *worst case* refers to the descent on the Lyapunov function along the system trajectory. So it is necessary to find a function V of state $\mathbf{z}$ (or of the equilibrium-point-centered state $\mathbf{x} = \mathbf{z} - \mathbf{z}_w$) that is (1) positive definite and (2) whose value decreases along the system trajectory whichever random disturbance is considered.

Formally, a closed loop model (9.5) can be viewed as depending on additional parameters ψ that take on values from the set Φ:

$$\dot{\mathbf{z}} = H_w(\mathbf{z}, \psi) \qquad (9.18)$$

The discrete-time variant is

$$\mathbf{z}' = \mathbf{z} + \Delta_t H_w(\mathbf{z}, \psi) \qquad (9.19)$$

The closed-loop model must comprise the model of the particular disturbance parameterized by individual elements of vector ψ. This is done in a similar way as in Section 7.4. For example, measurement errors can be modeled as

$$\dot{\mathbf{z}} = F[\mathbf{z}, G(\mathbf{z} + \psi, \mathbf{w})] = H_w(\mathbf{z}, \psi) \qquad (9.20)$$

with measurement error vector ψ. The domain Ψ of ψ will be a Cartesian product of some the maximum positive and negative values expected in practice.

With such an extended closed-loop model, the descent condition has to be adapted to the worst case consideration. With an analytic Lyapunov function of type (9.9), the constraints (9.11) to be satisfied are

$$\max_{\psi \in \Psi}\left[\mathbf{z}^T(\mathbf{V} + \mathbf{V}^T)\dot{\mathbf{z}}\right] = \max_{\psi \in \Psi}\left[\mathbf{z}^T(\mathbf{V} + \mathbf{V}^T)H_w(\mathbf{z}, \psi)\right] \leq 0 \qquad (9.21)$$

Then, the penalty terms in the optimization formulation (9.12) are changed correspondingly.

With numerical Lyapunov function, the update rule (9.14) is modified to

$$\begin{aligned}V(\mathbf{z}_i) &\leftarrow \min\left\{c(\mathbf{z}_i) + \max_{\psi \in \Psi}\left[V(\mathbf{z}'_i), V(\mathbf{z}_i)\right]\right\} \\ &= \min\left\{c(\mathbf{z}_i) + \max_{\psi \in \Psi}[V(\mathbf{z}_i + \Delta_t H_w(\mathbf{z}, \psi)), V(\mathbf{z}_i)]\right\}\end{aligned} \qquad (9.22)$$

The remaining parts of the algorithms can be applied without change.

Note 9.1 With some types of random disturbances, complete stability in the sense defined here cannot be reached. For example, with random measurement errors, the controller will always make small movements even if the plant is exactly at the equilibrium point; its nonzero inputs will induce an action, unless small deviations are filtered away by some dead-zone filter.

Then, the infeasible should not be attempted. The stability definition can be modified so that remaining in a small region around the reference vector is considered stable. (In fact, this is how "common" stability is defined, in contrast to the asymptotic stability implicitly considered throughout this chapter.) With the analytic Lyapunov function, condition (9.21) is satisfied then only with some limited precision around the equilibrium point. With the numeric Lyapunov function, the zero cost is extended to a small region around the reference state, instead of only in the reference state.

The size of this region can either be defined or be, inversely, determined by trial and error for the size at which the stability proof succeeds.

Note 9.2 The worst case approach is only applied if the disturbance can change from state to state. If the disturbance or plant parameter variation is assumed to be constant over a time sufficient for the closed loop to reach the equilibrium point, it is sufficient to find a Lyapunov function for each type, size, or direction of such a disturbance.

9.2 DESIGNING A STABLE NEUROCONTROLLER

The fact that the neurocontroller design takes place with the help of numeric algorithms, as do all stability proofs proposed in this chapter, suggests the question whether it is possible to combine the training with the stability proof in order to obtain a controller that is provably stable.

This is certainly possible, and the way to do it is straightforward. It consists simply in extending the cost function (6.125) by an appropriately weighted stability constraint penalty term:

$$C_g = \sum_{j=1}^{J} \sum_{t=0}^{T-1} c_t(\mathbf{y}_{t+1}, \mathbf{u}_t, \mathbf{w}_t, \theta, \zeta) \rho_i(\mathbf{y}_{e0}, \mathbf{w}, \zeta, \theta) + w_s S \quad (9.23)$$

The weight should be high enough to enforce exact satisfaction of the constraints by reducing S to zero. (After the optimization, it has to be checked whether this is really the case—the minimum found may not be global, or the weight w_s may be inappropriate to enforce the exact constraint satisfaction.)

For analytic Lyapunov functions, the penalty term was formulated earlier in Eq. (9.12). For the numeric Lyapunov function, the positive definiteness of the function V is given by definition. That the condition of descent along the closed-loop trajectory is not satisfied can be recognized by V being ∞ for some states. Penalizing the sum of all infinite values of V will lead theoretically to a correct formulation of the optimization problem, but one that cannot be solved by any existing optimization algorithm. Substituting ∞ by some very high but constant value also cannot be helpful, since such a penalty term will have a gradient of (almost) zero as long as the controller is not provably stable (which will not be the case during most of the optimization time). The best solution is to initialize V not by infinity but by values depending on the distance from the equilibrium point, though larger than the upper bound V_{max} of ultimate values V for a stable controller. This upper bound can be assessed as the sum of costs c along some long finite realistic path from a state space point distant from the equilibrium point. The concrete form of the penalty term S of Eq. (9.23) then would be

$$S = \sum_{\mathbf{z} \in Z_0} s[V(\mathbf{z}), V_{max}] \qquad (9.24)$$

with Z_0 being the discretized state space. The function $s(x, y)$ is analogous to that of Eq. (9.12). By requiring that $V(\mathbf{z}) \leq V_{max}$, it is quaranteed that there are paths from all states to the equilibrium point. The value V_{max} separates the low values, which represent the cost of the path to the equilibrium point, from those resulting from the initial high values of V.

Also the stability proof based on linearization can be easily reformulated as a constraint satisfaction problem. The penalty term is

$$S = s(\lambda_0, \delta) \qquad (9.25)$$

with λ_0 being the dominant eigenvector of the Jacobian matrix computed, for example, according to Eq. (9.7).

Note 9.3 The values of the cost function extended by such penalty terms can be computed with the help of various algorithms such as the QR algorithm for eigenvalue computation or determinant evaluation algorithms. Although theoretically possible, it may be a hopeless task to try to derive formulas for the analytic gradient of such terms with regard to neurocontroller parameters. Then, the use of the numeric gradient is clearly indicated.

An unfortunate consequence is that the algorithms for a stable neurocontroller design proposed in this section cannot be effectively used in the incremental training mode. This author recommends stability proofs using the methods of Section 9.1 instead.

10

A NEUROCONTROL ALGORITHM—THE EASIEST WAY TO THE GOAL

This book attempts to restrict the number of neurocontroller approaches presented to those for which the immediate industrial application potential is the highest. Yet, there are many alternative choices available. The criteria for selecting among alternatives are mostly theoretical. In practice, there are no reliable choice criteria.

This may be embarrassing for the user who is, as many industrial control engineers certainly are, interested in a quick application with a high chance of success. From this viewpoint, the most useful book would give unambiguous recommendations on how to proceed, at least at the start.

Since this book is meant to be useful, this chapter provides such unambiguous recommendations. Of course, no selection can be entirely free from one's subjective experience, and this chapter's recommendations are no exception. Let me list my objectives in making a selection:

The methods I selected had to be relatively simple to implement. In particular, methods requiring extensive (sometimes even case-specific) mathematical derivations are omitted because error-free software implementation of such methods requires an immense amount of testing, without the possibility of ever being sure that all cases have been really considered.

The methods had to be generally applicable to a broad class of neurocontrol problems and be sufficiently robust to really solve them.

An easy and straightforward encoding of user specifications had to be an important feature.

The requirement profile had to be typical for industrial applications. This profile can be characterized as a free-style solution of a nonlinear optimal control problem, without overemphasizing online adaptiveness.

These objectives have led me to make the following general commitments for the rest of this chapter:

1. *Batch schemes* are used since they do not constrain the choice of optimization method. More exactly, they do not preclude the use of global optimization algorithms nor of arbitrary local algorithms (see Chapter 4).
2. *Numeric gradients* are used (if gradient information is required) since they make the elaboration of analytic gradient formulas obsolete. There is freedom in formulating an arbitrary cost function as well as in using single-step and multistep criteria with arbitrary plant model and neurocontroller structures. Another argument in favor of the numeric gradient is its consistency with the function value call—an inconsistency may lead to poor convergence of the algorithms (see Chapter 4).
3. *Multistep performance criteria* are favored since they reflect the correspondence between simulation and reality to a substantially higher degree than single-step criteria do (see Chapters 5 and 6). It is true that they are more difficult to optimize because of larger differences between good and bad results. (In incremental schemes, they are frequently even unstable, but here only batch schemes are going to be used.) However, after optimization has succeeded in finding a low value for the cost function, it is more probable that the neurocontroller will work in a closed loop than with single-step criteria.
4. Neural networks that build *global representations*, such as the multilayer perceptron, are used since they have better generalization and interpolation capabilities (see Chapter 3). They are also more economical for higher-dimensional tasks; neural networks based on local representations such as RBF networks require exponential growth in size if their advantageous properties are to be preserved.

It is important to point out that the recommendations of this chapter make sense only as a whole. Some choices would be bad if others are omitted. For example, multistep performance criteria cannot be recommended together with incremental schemes [101]. To make this clear, an analogous list of recommendations are proposed in Section 10.4 for incremental online adaptation. The following two sections sketch concrete plant identification (Section 10.1) and neurocontroller training (Section 10.2) algorithms for batch optimization. Section 10.3 discusses briefly which optimization algorithm can be used for both subtasks.

Although the algorithms given in this chapter are easy to implement, many users might prefer using commercial tools for neurocontrol development.

However, the available tools on the market are rather disappointing. General neural-network tools fail to provide algorithms specific for neurocontrol, particularly if serious applications are planned. Nonetheless, some recommendations are given in Section 10.5.

10.1 PLANT IDENTIFICATION ALGORITHM

10.1.1 Cost Function

Let a measurement series be written in the form of two matrices: matrix $\mathbf{D}$ whose columns are the vectors of measured outputs $\mathbf{d}_t$ for $t = 0, \ldots, T$, and matrix $\mathbf{U}$ whose columns are the vectors of inputs $\mathbf{u}_t$ for $t = 0, \ldots, T-1$.

Cost function (5.84) then depends on these measurements as well on the plant model parameter vector ω_f. This cost function is computed according to the following algorithm:

```
C_f(D, U, ω_f):
S ← 0
y_0 ← d_0
y_1 ← d_1
For t = 2 to T:
Begin
    y_t ← F(y_{t-1}, y_{t-2}, u_{t-1}, u_{t-2}, ω_f)
    For i = 1 to q:  S ← S + (y_{i,t} - d_{i,t})² / V_i
End
Return S
```

For all samples in the measurement series, model forecasts $\mathbf{y}_t$ are computed, using the past forecasts of outputs and past inputs. To be concrete, the model is committed to the observability index two, which is both sufficient and feasible in the most applications (see Section 5.3.2). The first forecast computed is for $t = 2$, since the measured outputs and inputs of two past sampling periods are required. For observability index n_o, the samples from n_o past sampling periods would be required.

The forecasts are evaluated using quadratic error terms normed to unity variances, that is, divided by variances V_i of individual output variables. The sum of quadratic error terms over all forecast periods and all output variables is the value of the cost function.

Generally, the optimization goes over multiple measurement series. The function to be optimized is then computed as

$$C_F(\mathbf{D}_1, \ldots, \mathbf{D}_K, \mathbf{U}_1, \ldots, \mathbf{U}_K, \omega_f) \leftarrow \sum_k C_f(\mathbf{D}_k, \mathbf{U}_k, \omega_f)$$

Possible variants of this procedure will consist of (1) considering models of an observability index different from two (e.g., reduced to one for high-dimensional problems) and (2) taking other than quadratic error terms (which is scarcely advisable).

10.1.2 Plant Model

The operative input–output model (i.e., a model working only with delayed variables, and not with numeric derivatives) for observability index two (5.71) can be implemented as a function

$$F(\mathbf{y}_1, \mathbf{y}_2, \mathbf{u}_1, \mathbf{u}_2, \boldsymbol{\omega}_f) \leftarrow 2\mathbf{y}_1 - \mathbf{y}_2 + \Delta_t^2 P\left(\begin{bmatrix} \mathbf{y}_1 \\ \dfrac{\mathbf{y}_1 - \mathbf{y}_2}{\Delta_t} \\ \mathbf{u}_1 \\ \dfrac{\mathbf{u}_1 - \mathbf{u}_2}{\Delta_t} \end{bmatrix}, \boldsymbol{\omega}_f\right)$$

with arguments $\mathbf{y}_1$, $\mathbf{y}_2$ (outputs delayed by one and two sampling periods), $\mathbf{u}_1$, $\mathbf{u}_2$ (inputs delayed by one and two sampling periods), and parameter vector $\boldsymbol{\omega}_f$. This function uses a subroutine call of the conceptual model (5.70) expressed in terms of numeric time derivatives.

Using gain vector representation, the concatenated vector is both the input vector of neural network P and a factor multiplying the network output:

$$F(\mathbf{y}_1, \mathbf{y}_2, \mathbf{u}_1, \mathbf{u}_2, \boldsymbol{\omega}_f) \leftarrow 2\mathbf{y}_1 - \mathbf{y}_2 + \Delta_t^2 MP\left(\begin{bmatrix} \mathbf{y}_1 \\ \dfrac{\mathbf{y}_1 - \mathbf{y}_2}{\Delta_t} \\ \mathbf{u}_1 \\ \dfrac{\mathbf{u}_1 - \mathbf{u}_2}{\Delta_t} \end{bmatrix}, \boldsymbol{\omega}_f\right)\begin{bmatrix} \mathbf{y}_1 \\ \dfrac{\mathbf{y}_1 - \mathbf{y}_2}{\Delta_t} \\ \mathbf{u}_1 \\ \dfrac{\mathbf{u}_1 - \mathbf{u}_2}{\Delta_t} \end{bmatrix}$$

The operator M transforms the network output vector into a matrix. With plant output vector of dimension q, plant input vector of dimension p, and observability index two, the neural network input vector (i.e., the concatenated vector) has dimension $2(p+q)$. Operator M transforms the neural network output vector of dimension $2(p+q)q$ to a $q \times 2(p+q)$-matrix. This is multiplied by the $2(p+q)$-dimensional concatenated vector from the right to get the numeric second derivative of q-dimensional output forecast vector, returned by the function f. Alternatively, a discrete-time model using time delays directly could be used. However, models using numeric derivatives are preferable for numeric reasons (see Section 5.3.1).

It is up to the user to decide whether to take the direct or the gain vector representation. Direct representation requires fewer neural-network parameters, but the gain vector representation seems to improve the convergence of the optimization (see Sections 3.2.5 and 3.3.1). In both cases, a single network represents the whole plant model. An alternative representation of each state equation by an individual network has been discussed in Section 5.3.3. This alternative should be considered whenever the plant possesses clear causal structure making possible a reduction of potential dependences.

10.1.3 Neural Network

The multilayer perceptron implementation can be directly based on the matrix notation (3.9):

$$P\left(\mathbf{x}, \begin{bmatrix} \mathbf{v}_0 \\ \mathbf{v}_1 \\ \ldots \\ \mathbf{v}_H \\ \mathbf{w}_0 \\ \mathbf{w}_1 \\ \ldots \\ \mathbf{w}_N \end{bmatrix}\right) \leftarrow \mathbf{w}_0 + \mathbf{W}S(\mathbf{v}_0 + \mathbf{V}\vec{x})$$

The parameter vector, which is the second argument of function P, has to be transformed to matrices $\mathbf{V}$ and $\mathbf{W}$ as well as vectors $\mathbf{v}_0$ and $\mathbf{w}_0$. This is symbolized in the argument list of P. Function P uses the vector sigmoid function (3.7):

$S(\mathbf{h})$:

For $i = 1$ to H: $r_i \leftarrow \dfrac{1}{1 + e^{-h_i}}$

Return $\mathbf{r}$

The output vector dimension N is equal to q for direct representation of the conceptual model and to $2(p+q)q$ for gain vector representation. Large numbers of hidden units would lead to prohibitive computational costs. This author has developed all neurocontrol applications with hidden unit number between three and six, depending on the number of network inputs. For k inputs, the rule $H = \min[\max(3, k), 6]$ has been used.

Alternative network types such as RBF networks can be used if they are adequate to the dimension of the plant.

10.2 NEUROCONTROLLER TRAINING ALGORITHM

10.2.1 Cost function

For plant identification, it is reasonable to make a commitment to a certain form of the cost function: A weighted square deviation of the forecast from the measurement is a nearly universal plant model quality criterion.

By contrast, for control goals there is a wide variety to choose from. While stabilization and trajectory-following tasks are preferred in control theory, this is not the case in industrial practice where there are many different constraints and cost criteria to consider in even the most nontrivial applications. Their diversity can be covered only on an abstract level. This was done in Chapters 6, 7, and 9. However, it would not be consistent for this chapter to give algorithms whose primary facets remain unsubstantiated. For this reason the cost function is given in the concrete form of trajectory-following tasks. Stabilization tasks, of course, are subsumed. This concerns only the cost function formulation. The controller implementation procedures are directly applicable to the general case.

Now, let the reference trajectory for a training example be given in the form of a matrix $\mathbf{W}$ whose columns are the reference vectors $\mathbf{w}_t$ for $t = 0, \ldots, T$. The columns of matrix $\mathbf{D}$ describe the initial state of output history. For observability index two, the matrix $\mathbf{D}$ has two columns. The cost function (6.125) for a single training example then depends on this reference trajectory as well as on the neurocontroller parameter vector ω_g. This cost function is computed according to the following algorithm:

$C_g(\mathbf{D}, \mathbf{W}, \omega_g)$:
$S \leftarrow 0$
$\mathbf{y}_0 \leftarrow \mathbf{d}_0$
$\mathbf{y}_1 \leftarrow \mathbf{d}_1$
$\mathbf{u}_0 \leftarrow \mathbf{0}$
For $t = 1$ To $T - 1$:
Begin
   $\mathbf{u}_t \leftarrow G(\mathbf{w}_t, \mathbf{w}_{t-1}, \mathbf{y}_t, \mathbf{y}_{t-1}, \mathbf{u}_{t-1}, \omega_g)$
   $\mathbf{y}_{t+1} \leftarrow F(\mathbf{y}_t, \mathbf{y}_{t-1}, \mathbf{u}_t, \mathbf{u}_{t-1}, \omega_f)$
   For $i = 1$ to q: $S \leftarrow S + (y_{i,t} - w_{i,t})^2 / V_i$
End
Return S

As in the plant identification algorithm, the neurocontroller structure chosen implies observability index two. The relative degree is assumed to be one. The neurocontroller structure used covers cases (6.99) through (6.107). With different assumptions, the neurocontroller structure could be extended by including more past values of $\mathbf{w}$ (for a higher observability index) or more past values of $\mathbf{u}$ (for either a higher observability index or a higher relative degree).

The control errors are evaluated with quadratic error terms normed to unity variances, that is, divided by assumed variances V_i of individual output variables. The sum of quadratic error terms over all control periods and all output variables is the value of the cost function.

The vector of plant parameters ω_f is now fixed to values determined by the plant identification algorithm. Therefore, it does not appear in the argument list of procedure C_g.

Generally, the optimization goes over multiple training examples. The function to be optimized is then computed as

$$C_G(\mathbf{D}_1,\ldots,\mathbf{D}_K,\mathbf{W}_1,\ldots,\mathbf{W}_K,\omega_g) \leftarrow \sum_k C_g(\mathbf{D}_k,\mathbf{W}_k,\omega_g)$$

Possible variants of this procedure consist of the following:

1. Considering models of observability index different from two (e.g., reduced to one for high-dimensional problems)
2. Taking other than quadratic error terms, possibly including cost or constraint penalties depending on $\mathbf{y}_{t+1}$ or $\mathbf{u}_t$
3. Training for robustness against plant variations or disturbances

The last point is of particular importance because it substantially increases the chance that the trained neurocontroller will perform equally well with the real plant as it does with the identified plant model F. The extensions of cost function accounting for robustness are discussed in Chapter 7. For example, training for robustness against measurement errors would require the following modification of procedure C_g:

```
C_g(D,W,α,ω_g):
S ← 0
y_0 ← d_0
y_1 ← d_1
u_0 ← 0
m_0 ← d(y_0,α)
m_1 ← d(y_1,α)
For t = 1 To T − 1:
Begin
    u_t    ← G(w_t, w_{t−1}, m_t, m_{t−1}, u_{t−1}, ω_g)
    y_{t+1} ← F(y_t, y_{t−1}, u_t, u_{t−1}, ω_f)
    For i = 1 to q:  S ← S + (y_{i,t} − d_{i,t})² / V_i
    m_{t+1} ← d(y_{t+1}, α)
End
Return S
```

The vectors $\mathbf{m}_t$ stand for inaccurate measurements of $\mathbf{y}_t$. They are generated by the random disturbance vector function $d(\mathbf{y},\alpha)$ that, for example, adds a random vector from the range $\langle -\alpha, \alpha \rangle$ to the model output $\mathbf{y}$. The specification of α is either constant or a part of training example (a nonobservable disturbance).

10.2.2 Neurocontroller

The function computing the operative form of the neurocontroller (i.e., a procedure working only with delayed variables, and not with numeric

260 A NEUROCONTROL ALGORITHM—THE EASIEST WAY TO THE GOAL

derivatives) encapsulates the various conceptual forms. Probably the most useful conceptual forms are the basic input–output controller (6.85) and the controller with separate feedforward mapping (6.87) and (6.88).

For observability index two, the basic trajectory-following controller can be implemented as

$$G(\mathbf{w}_0, \mathbf{w}_1, \mathbf{y}_0, \mathbf{y}_1, \mathbf{u}_1, \boldsymbol{\omega}_g) \leftarrow \mathbf{u}_1 + \Delta_t P \left(\begin{bmatrix} \mathbf{w}_0 \\ \dfrac{\mathbf{w}_0 - \mathbf{w}_1}{\Delta_t} \\ \mathbf{y}_0 \\ \dfrac{\mathbf{y}_0 - \mathbf{y}_1}{\Delta_t} \\ \mathbf{u}_1 \end{bmatrix}, \boldsymbol{\omega}_g \right)$$

where arguments $\mathbf{w}_0, \mathbf{w}_1$ are reference vectors delayed by one and two sampling periods, $\mathbf{y}_0, \mathbf{y}_1$ are outputs delayed by one and two sampling periods, $\mathbf{u}_1$ is an input delayed by one, and the parameter vector is $\boldsymbol{\omega}_g$. This function calls the conceptual model mapping expressed in terms of numerical time derivatives, and materialized directly by a perceptron P.

The gain vector representation of this controller structure is

$$G(\mathbf{w}_0, \mathbf{w}_1, \mathbf{y}_0, \mathbf{y}_1, \mathbf{u}_1, \boldsymbol{\omega}_g) \leftarrow \mathbf{u}_1 + \Delta_t MP \left(\begin{bmatrix} \mathbf{w}_0 \\ \dfrac{\mathbf{w}_0 - \mathbf{w}_0}{\Delta_t} \\ \mathbf{y}_0 \\ \dfrac{\mathbf{y}_0 - \mathbf{y}_1}{\Delta_t} \\ \mathbf{u}_1 \end{bmatrix}, \boldsymbol{\omega}_g \right) \begin{bmatrix} \mathbf{w}_0 \\ \dfrac{\mathbf{w}_0 - \mathbf{w}_0}{\Delta_t} \\ \mathbf{y}_0 \\ \dfrac{\mathbf{y}_0 - \mathbf{y}_1}{\Delta_t} \\ \mathbf{u}_1 \end{bmatrix}$$

The operator M transforms the network output vector into a matrix. With plant output and reference vectors of dimensions q, and action vector of dimension p, the neural network input vector (i.e., the concatenated vector) has dimension $4q + p$. Operator M transforms the neural network output vector of dimension $(4q + p)p$ to a $(4q + p) \times p$ matrix. This is multiplied by the $(4q + p)$-dimensional concatenated vector from the right to get the numeric first derivative of the p-dimensional action vector returned by the function G.

The controller with separate feedforward mapping of type (6.87) and (6.88) can be implemented only in the gain vector representation

$$G\left(\mathbf{w}_0, \mathbf{w}_1, \mathbf{y}_0, \mathbf{y}_1, \mathbf{u}_1, \begin{bmatrix} \boldsymbol{\omega}_{g1} \\ \boldsymbol{\omega}_{g2} \end{bmatrix}\right):$$

$$\varphi_0 \leftarrow \varphi_1 + \Delta_t MP \left(\begin{bmatrix} \mathbf{w}_0 \\ \dfrac{\mathbf{w} - \mathbf{w}_1}{\Delta_t} \\ \varphi_1 \end{bmatrix}, \boldsymbol{\omega}_{g2} \right) \begin{bmatrix} \mathbf{w}_0 \\ \dfrac{\mathbf{w} - \mathbf{w}_0}{\Delta_t} \\ \varphi_1 \end{bmatrix}$$

$$\mathbf{u}_0 \leftarrow \mathbf{u}_1 + (\varphi_0 - \varphi_1) + \Delta_t MP \left(\begin{bmatrix} \mathbf{y}_0 - \mathbf{w}_0 \\ \frac{(\mathbf{y}_0 - \mathbf{w}_0) - (\mathbf{y}_1 - \mathbf{w}_1)}{\Delta_t} \\ \mathbf{u}_1 - \varphi_1 \\ \mathbf{w}_0 \\ \frac{\mathbf{w}_0 - \mathbf{w}_1}{\Delta_t} \end{bmatrix}, \mathbf{w}_{g1} \right) \begin{bmatrix} \mathbf{y}_0 - \mathbf{w}_0 \\ \frac{(\mathbf{y}_0 - \mathbf{w}_0) - (\mathbf{y}_1 - \mathbf{w}_1)}{\Delta_t} \\ \mathbf{u}_1 - \varphi_1 \end{bmatrix}$$

$\varphi_1 \leftarrow \varphi_0$
Return $\mathbf{u}_0$

The neural network parameter vector ω_g is a concatenation of the parameter vectors $\omega_{g1,2}$ of neural networks representing the feedback and feedforward parts of the controller.

Operator M transforms the neural network output vectors of dimension $(2q+p)p$ to $[(2q+p) \times p]$ matrices. These are multiplied by the $(2q+p)$-dimensional concatenated vectors from the right to get the numeric first derivative of a p-dimensional feedback action vector returned by the function G_1 or the p-dimensional feedforward action vector returned by the function G_2.

Alternatively, discrete-time neurocontrollers directly using time delays could be used. However, controllers using numeric derivatives are preferable for numeric reasons (see Section 5.3.1). The implementation of the neural network used is identical with that for plant identification.

10.3 OPTIMIZATION ALGORITHM

The formulation of plant identification and neurocontroller training in the cost function form is a preparation for an optimization algorithm to be applied. Now, the work has to be done with a numeric optimization algorithm.

A good optimization algorithm is a genuine key to the success. A very brief introduction to the large field of numeric optimization is given in Chapter 4. This section presents a starting point. It is important for good results to use the best algorithms within one's reach.

The algorithm proposed below is rather rudimentary, but it is structured like the more sophisticated ones. It consists of a global optimization procedure that generates starting points for the local algorithms which search the minimum of the particular attractor.

10.3.1 Global Optimization

First, it is important to point out that there is no choice between global and local optimization in nonlinear neurocontrol applications. What is searched for is the global optimum of the cost function. Local optimization algorithms need a starting point. They will find the global optimum with *arbitrary* starting point only if the cost function is convex. In nonlinear neurocontrol applications, the cost function is known not to be convex.

Consequently, some (implicit or explicit) global optimization algorithm has to be used. The most widely used (and the worst) global algorithm is one that generates a single random starting point. This is nothing but a lottery, and so are the results (although there are people who win in a lottery). The chances of winning can be substantially enhanced by taking as many starting points as possible. This is also the task of the global optimization algorithms discussed in Section 4.2. These algorithms are admittedly complex and often not available commercially. The global optimization algorithm recommended next may be trivial and not very efficient, but it is better than nothing. For a P-dimensional parameter vector ω, it generates $2P$ starting points:

```
GLOBAL(C())
c_opt ← ∞
For i = 1 to 2P:
Begin
  For j = 1 to P:
    If          i = 2j − 1:  ω_j ← β_j
    Else If  i = 2j:         ω_j ← −β_j
    Else:                    ω_j ← r(δβ_j)
  (c_L, ω_L) ← LOCAL(C(), ω)
  If c_L < c_opt:  (c_opt, ω_opt) ← (c_L, ω_L)
End
Return ω_opt
```

The function $C()$ is instantiated to $C_F()$ for plant identfication and to $C_G()$ for neurocontroller training. The constant vector β gives the scaling of individual parameters ω. This scaling can be derived from the scaling of plant input and output variables, and from the role the individual elements ω_j play in the neural network P (i.e., to which positions of weight matrices V and W they correspond).

In each initial vector ω, one element ω_j is set either to β_j or to $-\beta_j$. The remaining elements are set to small random numbers from intervals $\langle -\delta\beta_j, \delta\beta_j \rangle$ generated by function $r()$. The randomness is necessary for breaking the symmetry in the initial parameter vector: Identical initializations of weights of several hidden units would be harmful for convergence.

Another simple alternative to this algorithm consists of randomly generated starting points. This is a good approach as long as the number of points generated is relatively large. However, with as few as $2P$ starting points (which are frequently in the proximity of the upper limit from the viewpoint of computational resources), no real uniform statistical coverage can be reached. Then, the deterministic procedure described above leads to a better (equidistant) coverage of the parameter space.

10.3.2 Local Optimization

For local optimization, some of the procedures mentioned in Section 4.1 can be used. Here, the conjugate gradient algorithm will be formulated. Its strength is its low memory requirement (linear with the problem size). On the other hand, the variable-metrics methods frequently converge faster, and Powell's method is more robust, because the gradient information is not used at all. All these algorithms are discussed in detail in Press et al. [117].

The Polak–Ribiere variant of conjugate gradient algorithm based on (4.13) can be implemented as the following procedure:

$LOCAL(C(), \omega_0)$:
$c_{opt} \leftarrow \infty$
$\omega \leftarrow \omega_0$
$c \leftarrow C(\omega_0)$
$\mathbf{g}_0 \leftarrow \nabla(C(), \omega_0)$
$\mathbf{h} \leftarrow \mathbf{g}_0$
For $i = 1$ to I:
Begin
   $(c, \omega) \leftarrow LINE(C(), \mathbf{h}, \omega)$
   If $c < c_{opt} - \epsilon$: $(c_{opt}, \omega_{opt}) \leftarrow (c, \omega)$
   Else   : Return (c_{opt}, ω_{opt})
   $\mathbf{g} \leftarrow \nabla(C(), \omega)$
   $\mathbf{h} \leftarrow \mathbf{g} + \dfrac{(\mathbf{g} - \mathbf{g}_0)^T \mathbf{g}}{\mathbf{g}_0^T \mathbf{g}_0} \mathbf{h}$
   $\mathbf{g}_0 \leftarrow \mathbf{g}$
   $(c_L, \omega_L) \leftarrow LOCAL(C(), \omega)$
End

Two subroutines are called, a gradient procedure $\nabla()$ and the line search procedure $Line()$. The gradient procedure computes the numeric gradient at a given point ω. It simplest variant for P-dimensional parameter vector is

$\nabla(C(), \omega)$:
For $i = 1$ to P: $g_i \leftarrow \dfrac{C(\omega + \mathbf{e}_i \delta) - C(\omega)}{\delta}$
Return $\mathbf{g}$

The line search procedure $Line((C(), \mathbf{h}, c, \omega_0)$ delivers the local minimum of function $C()$ in direction $\mathbf{h}$, starting from point ω_0:

$LINE(C(), \mathbf{h}, \omega_0)$:
$c_{opt} \leftarrow \infty$
$\omega_{opt} \leftarrow \omega_0$
$(a, b, c, c_b) \leftarrow INIT(C(), \mathbf{h}, \omega_0)$

```
For i = 1 To I:
Begin
  If c - b > b - a:
  Begin
    x                ← b + γ(c - b)
    c_x              ← C(ω_0 + xh)
    (a, b, c, d)     ← (a, b, x, c)
    (c_a, c_b, c_c, c_d) ← (c_a, c_b, c_x, c_c)
  End
  Else:
  Begin
    x                ← b - γ(b - a)
    c_x              ← C(ω_0 + xh)
    (a, b, c, d)     ← (a, x, b, c)
    (c_a, c_b, c_c, c_d) ← (c_a, c_x, c_b, c_c)
  End
  If c_x < c_opt   : (c_opt, ω_opt) ← (c_x, ω_0 + xh)
  If |b - c| < ε   : Return (c_opt, ω_opt)
  If c_b > c_c:
  Begin
    (a, b, c)        ← (b, c, d)
    (c_a, c_b, c_c)  ← (c_b, c_c, c_d)
  End
End
```

The golden search constant is $\gamma = (3 - \sqrt{5})/2 \approx 0.38197$.

```
INIT(C(), h, ω_0):
c_0   ← C(ω_0)
x     ← δ
c_x0  ← C(ω_0 + xh)
If c_x0 < c_0:
Begin
  While 2x < BIG:
  Begin
    x   ← 2x
    c_x ← C(ω_0 + xh)
    If c_x > c_x0 : Return (0, x/2, x, c_x0)
    c_x ← c_n
  End
End
Else Return (0, 0, 0, c_x0)
```

It is assumed that the cost function values are not unbounded negative. The exception exit returning $(0, 0, 0, c_{x0})$ will occur if the cost function value for the smallest step δ in the given direction **h** is not lower than that at the starting point. This will stop the calling procedure $LINE()$ in its first iteration.

10.4 APPROPRIATE CHOICES FOR INCREMENTAL ADAPTATION

Implementing a working application with incremental adaptation schemes is substantially more difficult than with batch schemes. Even when successive online adaptation is required, the batch algorithm proposed in Chapter 8 can be used. This is particularly advisable if the control problem is of a dimension higher than two or three.

When, for some reason, an application requires incremental adaptation, then the most promising algorithms, which are completely different from those appropriate for batch schemes, must include the following:

1. Formulas for the analytic gradient valid in each chosen setting. There is hardly any reasonably accurate way to compute the numeric gradient in online incremental schemes.
2. Single-step performance criteria despite their lower reliability with regard to the applicability of the simulation results to the real plant. There are two reasons for this. First, adaptation may become unstable with multi-step criteria (see Section 5.5.4), whereas, for single-step criteria, there are schemes with guaranteed stability [102]. Second, analytic gradient accuracy rapidly decreases with the horizon of the criterion and becomes not very useful for optimization; the sign-reversed gradient may point in a direction opposite to that of the cost function descent.
3. Neural networks with local representations such as RBF networks. Once trained, a local representation is more likely to survive the subsequent training steps presenting data from different points of state space.
4. Second-order training algorithms, such as Kalman training (Section 4.4). Second-order algorithms exhibit superior convergence if implemented correctly.

10.5 TOOLS FOR NEUROCONTROL

The algorithms provided in this chapter are of limited complexity and are certainly relatively easy to implement by anyone with experience in programming numeric procedures.

However, many users prefer to avoid implementations from scratch, using instead public domain or commercial tools. So a book on neurocontrol would not be complete without some hints about which tools are useful for application development.

The preceding chapters have shown that neurocontrol is interdisciplinary, with at least three substantial components:

1. Neural-network algorithms
2. General control support
3. Numeric optimization algorithms

Consequently, a tool for neurocontrol should include all three fields. According to this author's experience, the order of importance of the three components for the success with applications of industrial complexity is the following:

1. Numeric optimization algorithms
2. General control support
3. Neural-network algorithms

This order reversal may seem controversial, but there are several arguments supporting it. First, the lowest rank of the neural-network algorithms does not result from any notion that they are dispensable (then, it would be hardly justified to speak about neurocontrol). Rather, the reason is that pure neural-network algorithms are of the lowest complexity and involve no particular numeric computing risks. They consist basically of certain neural-network implementations (simple matrix operations and nonlinear functions) and the formulas for computing the gradient (trivial if the numeric gradient is used, as noted in this chapter and shown in the preceding ones). If the user prefers to use a critic-based algorithm, several alternative procedures are available, but the problem is more choosing among them than any complexity inherent in them.

Second, the contribution of general control theory is more crucial for neurocontrol. Its second rank is justified by the fact that its influence concerns less the algorithms than the problem formulation itself. This is difficult to support by a tool—rather, the control engineer's competence is the decisive factor.

Finally, the first rank of numeric optimization algorithms results from the fact that inappropriate algorithms from this group can spoil the solution in a disastrous way. This author has experienced a couple of projects in which the only reason for the failure has been the use of an inappropriate numeric optimization algorithm, such as rudimentary gradient descent, for otherwise sound problem formulations. In particular, the importance of numeric optimization algorithms grows with the control problem size. Experience shows that an algorithm that works for hypothetical problems cannot be assumed to work for industrial control problems. The criterion for accepting the neurocontrol solution instead of a classical one is usually that it surpasses all classical solutions, many of which have been developed with considerable expertise and algorithmic expense.

The ranking above also suggests that tools providing neural-network algorithms (even if some them are referred to as neurocontrol algorithms) cannot be recommended. It should be clear that the market, which seemingly abounds in commercial neural-network tools, does not provide much for control. So that the reader does not feel completely alone, some recommendations are made here. But this author has a personal bias for programming algorithms himself and thus does not claim any expertise in commercial tools. This is why the recommendations should be viewed rather as suggestions.

10.5.1 Libraries of Numerical Algorithms

A compromise between programming a neurocontrol application from scratch and using a complete tool is to use commercial or public domain tools for the most critical part, the numeric optimization.

One example is to use the algorithms presented in *Numerical Recipes in C* by Press et al. [117], with sources given on a supplementary floppy disk. This collection of algorithms contains simple but reliable implementations of local optimization algorithms. The local optimization part of the algorithm of this chapter, consisting of the procedures *LOCAL()*, *INIT()*, and *LINE()* has its counterparts in the functions:

1. *frprmn()* (conjugate gradient) or *powell()* (Powell's algorithm) as a local optimization (to be substituted for *LOCAL()*)
2. *linmin()*, *f1dim()*, and *golden()* (Golden section search) or *brent()* (an algorithm using quadratic approximation) for line-search function *LINE()*
3. *mnbrak()* for line-search initial point determination *INIT()*

A simple global optimization algorithm such as the procedure *GLOBAL()* given above as well as neural network functions are to be supplied by the user. Viewing the algorithms of this chapter makes clear that the algorithms remaining for the implementation are those of low complexity.

Using the libraries of numeric procedures such as that of *Numerical Recipes in C* does not free the user from programming. However, a clear advantage is the transparency and flexibility, allowing a fast assessment of which parts of the overall algorithm are critical (or which are to be extended for broader scope of problems) and substituting them with more sophisticated ones.

10.5.2 Mathematic and Engineering Packages

Implementations of local optimization algorithms are also provided by mathematic and engineering packages such as *Mathematica*, *MATLAB*, and *Maple*. An advantage over numeric optimization libraries is that in some of these tools, components for general control application development, graphics representation of results, and neural networks basics are supported.

For example, *MATLAB*'s *Simulink* toolbox provides graphic interface for modeling plants from predefined and user-defined blocks, and their simulation by various numeric integration methods. The *Neural Network Toolbox* of *MATLAB* supports simulation of the usual neural-network types. With both toolboxes, several things still remain to be implemented manually:

1. Global optimization, the most trivial form of which is the procedure *GLOBAL( )* of this chapter
2. Procedures embedding the neural networks into the plant model and the controller [i.e. procedures *F( )* and *G( )*]
3. Procedures implementing cost functions for plant identification and neurocontroller training [i.e., procedures *C_F( )* and *C_G( )*]

These procedures are implemented in the *MATLAB* programming language. It is a matter of taste whether the user prefers this or implementation in a general programming language such as Pascal or C++.

To summarize, the strengths of general mathematic and engineering tools are as follows:

Using predefined procedures for local optimization and neural-network handling

Availability of facilities for graphic presentation of simulation results

Possibility of modeling complex plants with the help of graphic block schemes

However, there are also disadvantages, in particular, if advanced neurocontrol applications are involved: (1) Global optimization procedures are mostly not available; their trivial surrogates may fail to solve difficult problems. (2) Combining modules designed graphically with user-programmed procedures may require thorough knowledge of the tool. The time spent in acquiring this knowledge may be comparable to (or in excess of) that spent implementing the algorithms in general programming languages with the use of mathematic libraries. (3) The comfort of interactive definition of variables and arrays has its cost in resources spent, since memory requirements and execution times can substantially exceed the cost of directly programmed software.

10.5.3 Dedicated Neurocontrol Tools

It is probably the difficult interdisciplinary character of neurocontrol that has led to the lack of genuine neurocontrol tools on the market. There are several proprietary research packages not available to the public. The only tool really specialized to neurocontrol with which the author has had personal experience is *NNcon* of the Czech company *Art of Intelligence* mentioned in the preface. Because this company is not particularly well known in the United States, here is its address: Nad Šárkou 25, CZ-16000 Prague, Czech Republic.

This tool implements closed-loop optimization in a way similar to the methods described in this book. One of its most important features is the availability of sophisticated global optimization methods from the clustering family of algorithms.

Currently, it is written for PC-DOS (an MS-Windows version is under development). Interactive graphic interface is provided only for simulations. This is no severe obstacle when developing applications in the face of the batch character of the neurocontrol development, consisting mostly of rather long training sessions.

The specification of the control problem takes place in a mathematic specification language that provides for the whole spectrum, from a purely analytic specification of plant models over identification of individual free parameters from measurements to a complete identification of neural-network-based plant model from measured data. If the plant exhibits specific symmetries (e.g., the behavior of the left and right wheels of a car), they can be used to reduce the number of free parameters of the model.

The specification language can also be used to model classical controllers for comparative simulations. The optimal neurocontroller is delivered in the form of the C source code. Simulation results can be plotted and compared with classical controllers.

II

CASE STUDIES

11

INTRODUCTORY REMARKS TO CASE STUDIES

The second part of this book is concerned with case studies. All of them share the following three attributes:

1. They have been developed using the methods described in Part I. Moreover, they have been developed using the same (proprietary) *software tool* committed to a subset of these methods.
2. All of them have been tested in the field. At the time the book is being written, an exception is the most complex application, the wastewater treatment plant control, because legal constraints do not allow application of methods whose adequacy has not been proved. To this date, only as simplified controller mimicking the control strategy of the neurocontroller has been tested in the field. Furthermore, a series of very realistic simulations of the neurocontroller have been performed.
3. All of them have a clear product or service potential.

Before developing a neurocontrol application, several fundamental and task-specific choices have to be made. The fundamental choices include the following ones:

Neurocontrol Framework   Among the frameworks presented in Chapter 2, the closed-loop optimization scheme and the critic-based system can be used for general applications.

Batch versus Incremental Training Schemes   The arguments in favor of batch and incremental approaches are discussed in Chapters 5 through 10.

Neural-Network Types Used Although an arbitrary mixture of various network types is conceivable, there are hardly any arguments in favor of a mixture. Moreover, commitment to a single type substantially reduces the complexity of software development. Arguments that can be helpful in making the choice can be found in Chapter 3.

Numeric Optimization Methods Some possible alternatives are suggested in Chapter 4.

The task-specific choices are too many to be exhaustively listed, but the essential ones can be grouped in the following way.

1. Formulation of the plant identification task.
    a. Determining which parts of the plant can be modeled explicitly and which are to be identified from measured time series.
    b. Determining measurable variables.
    c. Making an assumption about the observability index of the plant and a corresponding order of the input–output model.
    d. Describing the known dependency structure of the model.
    e. Collecting measurements in sufficient structure and quantity.
2. Formulation of control optimization task.
    a. Determining controller inputs and outputs for the given observability index of the plant, with possible extensions motivated by robustness requirements (Chapter 7).
    b. Modeling typical reference values and trajectories.
    c. Modeling typical disturbances against which the control is to be robust.
    d. Determining training examples.
    e. Formulating the constraints to be applied.
    f. Formulating the goal of control.
3. Formulation of adaptive loop.
    a. Periodicity of measurement.
    b. Mode for substitution of past measurement series by actual ones (data accumulation and forgetting; see Chapter 8).

The way these three groups of choices are addressed in the case studies has been determined by the origin of case studies. All control problems presented in the case studies have been solved using a proprietary software package TREG (Trainable REGulators) of the Daimler–Benz Corporate Research (some of them have been cross-checked by the *NNcon* tool mentioned in Section 10.5, reaching similar results). As far as the author is informed, this software package is not intended ever to be offered as a commercial product. Therefore, no one should suspect us of doing a disguised advertisement, even if

INTRODUCTORY REMARKS TO CASE STUDIES   275

some descriptive features of this software package are used in the case study presentation. In this author's opinion, this is the best framework for making the description both reasonably formalized and realistic.

During the time TREG has been under development, the experience gained has led to recommendations with regard to the fundamental choices listed above. Some of them are related to the recommendations of Chapter 10. In particular, they are the following:

1. The neurocontrol framework used should be the closed-loop optimization scheme.
2. Batch training schemes can be used for both plant identification and controller training.
3. A multilayer perceptron in two forms can be chosen. The first variant uses the output directly, and the hidden layer consists of modulated sigmoid units. The second variant uses the network output as a gain vector and the usual asymmetric sigmoids in the hidden layer. All case studies in this book use the gain vector representation.
4. Proprietary global numeric optimization methods of the multilevel type have been implemented.

The framework provided by TREG is represented by a specification language that allows problem description to the extent that all task-specific choices are covered. Some features irrelevant to the goals of this book or that are under ongoing research are omitted or simplified.

The case studies chosen do not include online adaptive features. A simple case study concerning an adaptive loop is the manipulator of Section 8.8. This experimental device does not currently represent an industrial application with a clear market potential for a product or service, and is thus not included in the case study part of this book.

The conceptual structure of the software package is given in Figure 11.1.

Finally, we need to address the assessment of neurocontroller performance. Such an assessment usually consists of two parts: (1) simulation results and (2) results of field tests. Even though field tests are performed, it frequently is important to include simulation results. The case studies were developed in a normal industrial environment. This has affected the experimental freedom substantially, in particular, in the following ways:

The experiments are not always documented as in scientific publications. (On the other hand, most publications do not contain the results of real field tests.) In most cases, field test charts are substituted for simulation charts, even when field tests had been successfully performed.

The comparison with classical methods were performed by domain specialists. Their results were frequently verbal or qualitative. Because classical methods were implemented before (sometimes a long time before) the

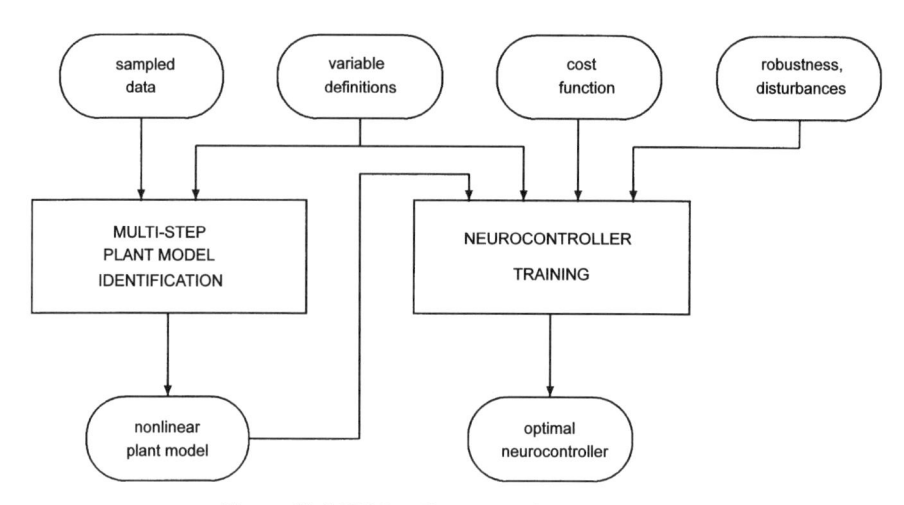

Figure 11.1 TREG software package structure.

corresponding neurocontrol project, their form and documentation could not be adapted to the needs of this book.

Simulations do not suffer from such restrictions. Therefore field tests are usually described by quantitative aggregates or verbally, with reference to simulation charts that resemble the real performance closely.

Another reason for including simulations is that the goal of case studies is not only to present the achievements of this author and his colleagues, but primarily to illustrate by concrete examples the theoretical concepts of Part I of this book. For this goal, the comparative experiments (e.g., comparing various neurocontroller structures) that are helpful are not necessarily a part of application projects, and they would never be allowed on expensive real objects in routine operation, such as test benches or even wastewater treatment plants.

Nevertheless, this author hopes that despite these limitations, the presentation of results is convincing and useful to the reader.

12

ELASTOMERE TEST BENCH CONTROL

This case study is concerned with the dynamic control of an elastomere test bench. Structurally, it is a relatively simple trajectory-following problem. However, strong nonlinearities make it a considerable challenge. Through its low dimensionality and thus acceptable computing times for plant identification and controller training, comparisons between various controller structures have been possible. Therefore, this case study is particularly illustrative.

12.1 TEST BENCHES

Test benches for industrial automation are devices for physical simulation of certain extreme conditions (e.g., loads) to which tested workpieces are exposed. A usual procedure is to prescribe a certain load schedule (i.e., time series of the load), with the goal of reproducing this schedule as closely as possible on the test bench. Load schedules can either be observed (e.g., during a test drive of a car) or deliberately defined to cover the scope of loads that may occur during the lifetime of the workpiece.

In many cases, the load cannot be controlled directly but only via a certain action variable such as the accelerator lever position or a hydraulic valve. So following a given test load schedule represents a control problem. The source of particular difficulties is that the plant to be controlled consists of both the test bench itself and the tested workpiece. The latter is frequently changed, either in a series of specimens of the same type that deviate by production tolerances (e.g., a series of engines of the same type) or in tests of different product types (e.g., engines of various sizes). In the first case, the control has to

278 ELASTOMERE TEST BENCH CONTROL

be very robust to retain stability for a whole series of varying workpieces. In the second case, a complete controller redesign, an expensive manual task, may be necessary.

Available manual expertise is usually too expensive to be justified for such repetitive design. This motivates a strong need for automatic plant identification and controller design directly from measured data.

12.2 ELASTOMERE TEST BENCH

The elastomere test bench consists of a hydraulic press that exerts force on the tested workpiece, an elastomere shock-damping component. The structural scheme is given in Figure 12.1. The press is controlled by a continuous bidirectional valve steering flow into and from the hydraulic pressure cylinder. The objective of control is to keep the force as close as possible to a dynamic load schedule, typically a sequence of sinusoidals of various mean values, frequencies, and amplitudes.

The basic relationship between the action variable, the valve control voltage, and the hydraulic fluid pressure is nonlinearly integrative, with unknown gains, delays, and time-constants. Gain variations of the controlled plant are within a broad ratio of approximately 1:100. Thus, for reaching a given sinus amplitude of the pressure force, the required valve control voltage amplitude in low-pressure regions is one hundred times higher than the voltage amplitude at high-pressure regions.

Concrete values of integrator parameters depend to a great extent on the particular workpiece tested and its instantaneous elastic compression. During the test, considerable changes are possible. These changes are both reversible,

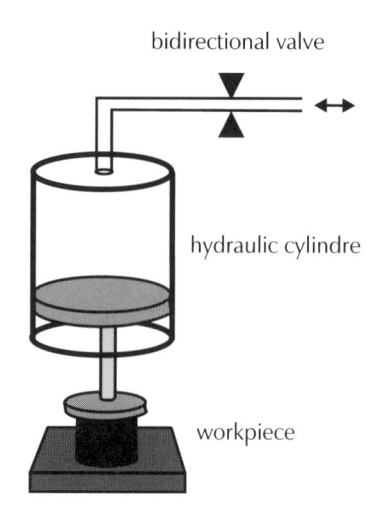

Figure 12.1 Scheme of elastomere test bench.

such as temperature dependencies, and irreversible, such as material damages. The trajectory following controller must be robust against such changes.

The integrative character of the plant makes the feedback loop tend strongly to instability. Every pressure overshoot shifts the plant into a working point with a drastically different behavior. Together with varying workpiece properties, this may lead to critical situations, such as damages on the test bench equipment.

Classical control methods failed to deliver a controller that would remain stable in all working points and simultaneously follow the test schedule with reasonable precision; there is a requirement that a maximum deviation should not exceed 20 percent of the scheduled reference sinusoidal amplitude. The pragmatic solution has been a field of gain-scheduled PID controllers, with specific coefficients for each working point. Individual controllers were activated for individual working points, but this resulted in frequent instability during and after working point transitions. The control design has been purely experimental—the design deficiencies have been recognized only under operation of the test bench. This has necessitated continuous supervision by an operator.

12.3 PLANT MODEL IDENTIFICATION

The only measurable output of the elastomere test bench is the pressure force. There is also a single action variable, the control voltage of bidirectional hydraulic valve. This constitutes a univariable system.

Force is sampled with the help of a 12-bit analog–digital converter in the range of –25 to 25 kN, but its amplitude, most of the time, is below 2 kN. A single-bit difference in discretized values is 50 kN/4096 ≈ 0.0122 kN. The minimum sampling period implementable on the control device is 1 ms. With this sampling period, the values of the first numeric derivative of the force are, on average, 48 kN/s. The smallest discernible derivative corresponding to the one-bit change in sampled values is 0.0122 kN/0.001 s = 12.2 kN. So the average derivative is only four times larger than the resolution limit. Because the oscillations of the plant in unstable closed loops have reached frequencies over 100 Hz, the 1000-Hz sampling rate does not provide a big margin against the rate required by the sampling theorem. A consequence of this is that the resolution cannot be enhanced substantially by smoothing or low-pass filtering, without the danger of filtering out relevant frequency information about plant dynamics. This makes obvious that the maximum-order numeric derivative available is the first-order derivative. This constrains the model to second order. Because TREG is committed to gain vector representation, the form of the model is

```
output
  Force eq_nr: 1
```

```
input
  Voltage
dependences
  DD Force   depends_on:    Force,D Force,Voltage,
                            D Voltage
             gain_depends_on:  Force,D Force,Voltage,
                            D Voltage
```

The symbol D denotes the first derivative of the following variable, the symbol DD the second derivative. The two types of dependences, depends_on and gain_depends_on are related to the use of the gain vector network. For a mapping represented as

$$f(\mathbf{x}_2)\mathbf{x}_1 \qquad (12.1)$$

the list after depends_on corresponds to $\mathbf{x}_1$, whereas that following gain_depends_on corresponds to $\mathbf{x}_2$. The vectors $\mathbf{x}_1$ and $\mathbf{x}_2$ are identical. They might be different if the model were parametrized by additional parameters—these parameters would appear in the vector $\mathbf{x}_2$, that is, in the list following the keyword gain_depends_on. The model is represented by a set of neural networks corresponding each to one model equation. The overall neural-network structure is depicted in Figure 12.2 (for notational rules, see Note 12.1).

Note 12.1 The neural-network structure schemes of this and the following chapters are drawn according to the following principles:

1. The continuous-time symbols for numeric derivatives and integrals are used, but the computations, of course, are performed in discrete time.
2. Feedbacks from output to input (e.g., the feedback of Force in Fig. 12.2) are connected, as usual, with a delay by one sampling period.
3. Gain vector representation is used, represented by multiplications of network output units with corresponding values.
4. The network itself, delimited by a dashed rectangle, is a multilayer perceptron with a single hidden layer. Bold arrows denote completely interconnected layers.

The method implemented is a multistep batch scheme with identification horizon defined by the user. For the elastomer test bench application, the identification horizon has been 200 sampling periods. The model has been trained with 24 sampled time series consisting of approximately sinusoid force trajectories with frequencies from 4 to 15 Hz and amplitudes of approximately 2 kN. The offsets of the sinusoids have been varying between −1 and 15 kN. Combinations of frequencies and offsets have been chosen so that both

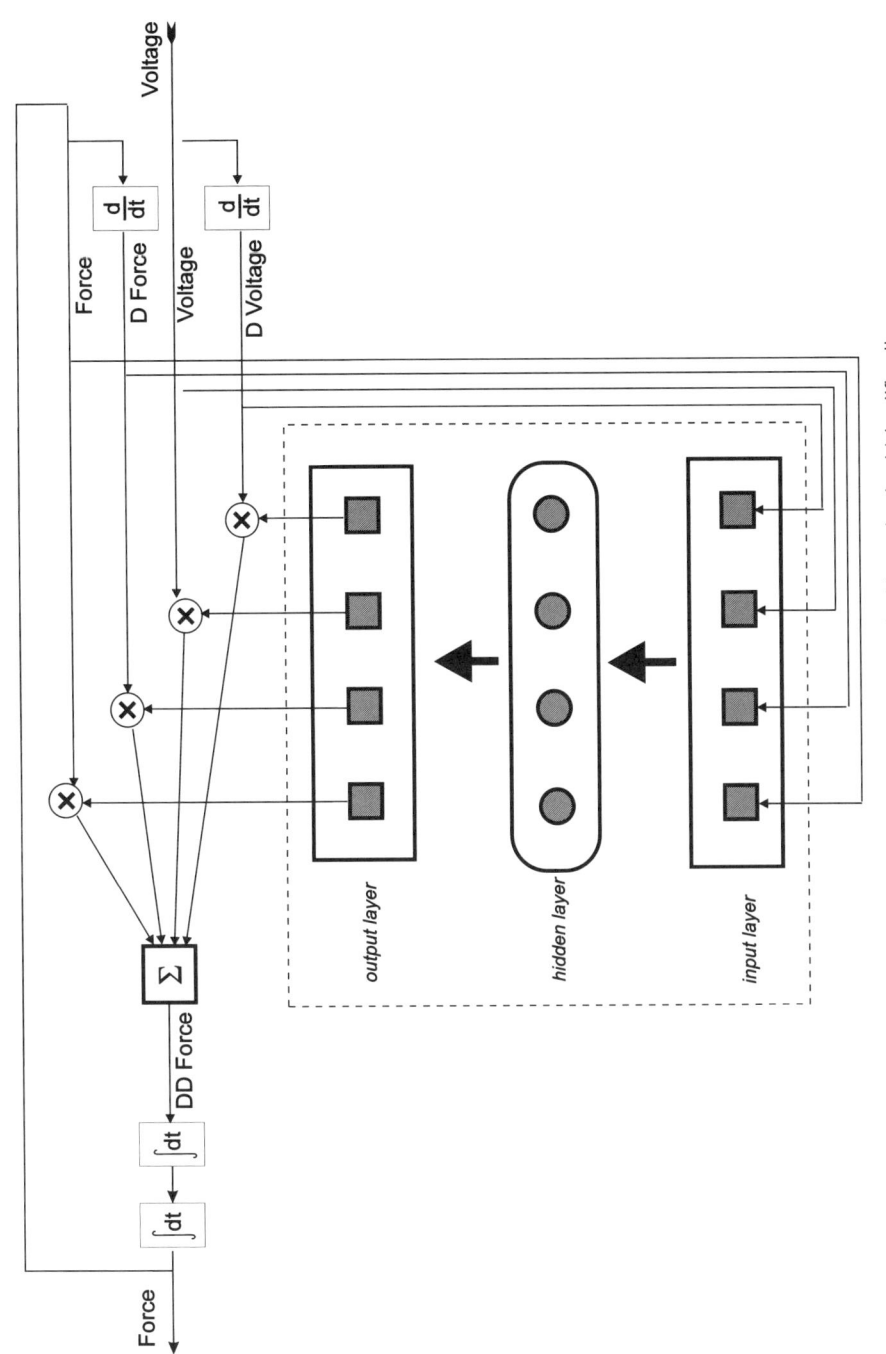

Figure 12.2 Neural network for elastomere test bench plant identification.

parameters are uncorrelated, with, for example, low forces occurring only with high frequencies and high forces with low frequencies. Thus, the model might erroneously infer that force depends on frequency.

Note 12.2 Security reasons have not allowed open-loop excitation of the plant. To identify a plant using data generated in a closed loop is dangerous, because it is possible that the data cover only a subspace instead of the whole state space. This danger is larger, the more the controller does what it is expected to. So it is advisable to use a sufficiently bad controller. Fortunately, this has not been a problem in this application; all controllers available so far have been sufficiently bad.

12.3.1 Results of Plant Identification

The results of plant identification are presented in Figure 12.3. The charts A, B, C, and D show 4 of the 24 sampled sequences of 300 sampling periods each. The bold lines are sampled values of `Force`. The thin lines close to the sampled trajectories represent the 300-step (i.e., using the series model developed by Narenda [101]), open-loop force forecast by the identified model for a given initial state of force and a given trajectory of input variable `Voltage`, represented by the third line. Note that in sample C, the open-loop forecast gradually drifts away from the sampled trajectory. This is not critical because the plant model will be used for training in a closed loop. For closed-loop training, it is important that the fast dynamics in all working points be captured; slow drifting can be compensated easily by feedback.

12.4 CONTROLLER TRAINING

The input–output model gained by identification has been used for training several alternative controller structures. Because the plant model is second order, so are the neurocontroller structures tested. In particular, they are the basic trajectory-following controller (6.105), the controller with error integration (6.106), the trajectory-following controller with a defined equilibrium point (6.101), and the controller with integration of action (6.103).

In addition to the genuine system variables `Force` and `Voltage`, some auxiliary variables are necessary for the following:

1. Representing controller inputs that are not identical with the two system variables, such as numerical derivatives, errors, and error integrals
2. Computing the reference trajectory
3. Representing disturbances and plant variations
4. Computing the cost function

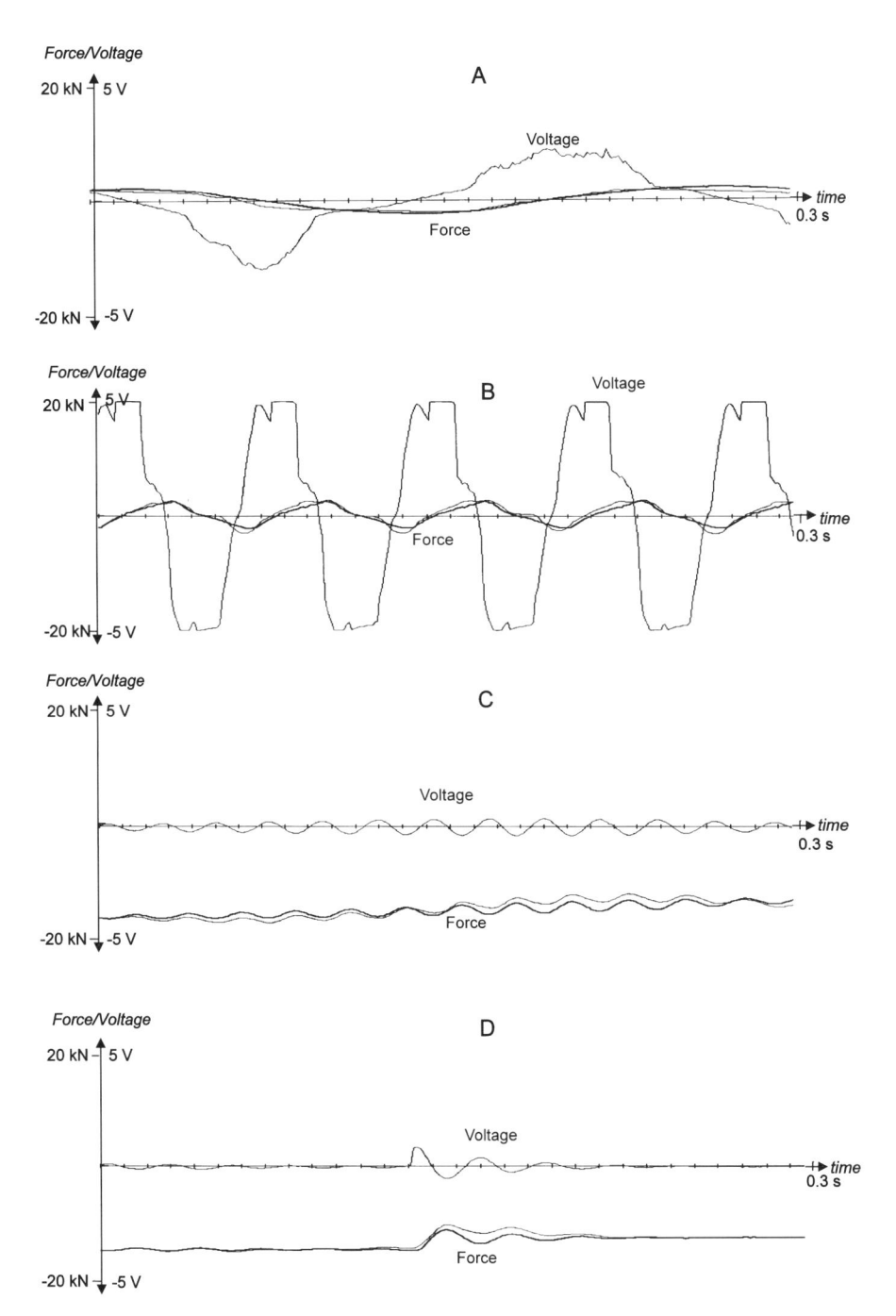

Figure 12.3 Measurement time series and model forecast.

These variables are as follows:

Force	// Modeled force
M_Force	// Measured force, i.e., force
	// including measurement errors
	// and plant variations
R_Force	// Reference force
E_Force	// Force error, i.e., deviation
	// from reference force
iE_Force	// Force error integral
Frequency	// Reference force frequency
ForceLo	// Reference force lower bound
ForceHi	// Reference force upper bound
Voltage	// Valve-control voltage
Voltage1	// Feedback part of valve-control voltage
Voltage2	// Feedforward part of valve-control voltage

There is a set of parameters that receive specific values for each training example. Some of them correspond to the parameters ζ of the model (6.116) in the general scheme of computing the cost function (6.118). They represent measurement errors and plant variations. Others parametrize the reference trajectory.

Force reference trajectory is a sinusoid of given frequency and lower and upper bounds. Because one critical competence required from the controller is to remain stable after transitions between different sinusoidal sequences, such transitions must be explicitly trained. To be able to formulate training examples for transitions, each reference trajectory can be composed from individually specified Segments 1 and 2.

Measurement errors are supposed to be random, uniformly distributed on an interval $\langle$-Dist,Dist$\rangle$. The identified plant model depends on the momentary condition of the plant. More exactly, it is a kind of average model over the varying conditions of the plant, with this variability remaining unmodeled. To simulate the varying conditions, the plant output is modified by factor GainFact. (This is slightly different from the approach proposed in Section 7.4.1, but the effect is similar.)

Frequency1	// Reference frequency for Segment 1
ForceLo1	// Lower bound for Segment 1
ForceHi1	// Upper bound for Segment 1
StepTime	// Time in seconds at which transition between
	// Segment 1 and 2 occurs

```
Frequency2    // Reference frequency for Segment 2
ForceLo2      // Lower bound for Segment 2
ForceHi2      // Upper bound for Segment 2
Dist          // Maximum amplitude of random measurement error
GainFact      // Gain factor for modifying the plant
```

Finally, in the reference trajectory computation, the constant π is used:

```
PI            3.14
```

Model equations, extended for auxiliary variables, are the following

```
n_Force      = nn(1)
n_M_Force    = (n_Force + 2*(rand()-1)*Dist) *GainFact;
n_ForceLo    = (T<StepTime ? ForceLo1 : ForceLo2);
n_ForceHi    = (T<StepTime ? ForceHi1 : ForceHi2);
n_Frequency  = (T<StepTime ? Frequency1 : Frequency2);
n_R_Force    = (n_ForceHi+n_ForceLo)/2
                 + (n_ForceHi-n_ForceLo)/
                   2*cos(2*PI*T*n_Frequency);
n_E_Force    = n_R_Force-n_M_Force;
n_iE_Force   = iE_Force + DT*n_E_Force;
```

Constant DT represents the sampling period in seconds, variable T the elapsed time. Prefix n_ denotes the value at the next sampling period, that is, the value with index $t+1$. The variable values without prefix are understood to be those with index t. The function rand() generates pseudorandom real numbers uniformly distributed on $\langle 0, 1 \rangle$. The function nn(i) establishes the relationship to the result of plant identification. It refers to the ith output variable of the plant model, with dependences declared in the plant identification part of the algorithm.

The cost function represents two goals: exact trajectory following and stability. The accuracy of trajectory following is measured by square deviation from the reference force. From the two means for formulating stability goals proposed in Section 6.4.3, penalties on the squared error progressively growing with time cannot be used—in trajectory following tasks, the accuracy cannot be expected to improve with time down to zero error. This is why the second means proposed is used, the high pass penalty based on higher-order derivatives. Here, the second derivative of the force error is taken. The second derivative is normed by dividing through the second derivative of the reference trajectory, that is, $(2\pi f)^2$ for a reference trajectory with frequency f. So the cost function or, more exactly, the partial cost term c_t of (6.118) is

```
E_Force^2 + (DD E_Force/(2*PI*Frequency)^2)^2;
```

Basic trajectory following controller (6.105) is defined as

```
D Voltage depends_on       : M_Force,D M_Force,Voltage,
                             R_Force,D R_Force
         gain_depends_on:    M_Force,D M_Force,Voltage,
                             R_Force,D R_Force
```

It receives the measurement of the state variable M_Force and its derivative, the action variable Voltage, and the reference force R_Force together with its derivative. The derivative of the action variable is not included because the highest-order derivative of action variable (6.105) is lower by one than that of the measurement. The meaning of symbol D as well as of keywords depends_on and gain_depends_on is analogous to that in identification specification.

The controller with error integration (6.106) receives additionally the error integral:

```
D Voltage    depends_on     : M_Force,D M_Force,Voltage,
                              R_Force,D R_Force,iE_Force
             gain_depends_on: M_Force,D M_Force,Voltage,
                              R_Force,D R_Force,iE_Force
```

Controllers with error integrations are primarily stabilization controllers. However, it is included here for comparison.

A trajectory-following controller with a defined equilibrium point as shown by Eq. (6.101) consists of two dynamic equations for feedback and feedforward parts of the controller:

```
D Voltage1  depends_on      : E_Force,D E_Force,Voltage1
            gain_depends_on: E_Force,D E_Force,Voltage1,
                             R_Force,D R_Force
D Voltage2  depends_on      : R_Force,D R_Force,Voltage2
            gain_depends_on: R_Force,D R_Force,Voltage2
```

Both components are to be added together to receive the controller action Voltage:

```
Voltage = Voltage1+Voltage2
```

Note that for the feedback controller part, the variable list following the keyword gain_depends_on is longer than that following the keyword depends_on. This is due to the form of the controller (6.101). The controller is defined around the equilibrium point R_Force - M_Force = E_Force = 0 and D R_Force - D M_Force = D E_Force = 0 with the help of the gain vector network representation. Additionally, it depends on the reference value `R_Force` and its derivative `D E_Force`, which jointly define this equilibrium point (or, more exactly, equilibrium trajectory).

The controller with integration of action (6.103)

```
D Voltage depends_on      : E_Force,D M_Force
         gain_depends_on: E_Force,D M_Force,R_Force
```

is also defined around an equilibrium point. Because this controller form has been derived as a stabilization controller, `D R_Force` is assumed to be zero and is omitted. The simplicity of this controller, as well as its defined equilibrium point which does not call for training a separate feedforward part, make it suitable for a large range of control tasks. This is why it has been included here despite the fact that it is to be applied preferably to stabilization tasks.

The neural-network structures of these neurocontrollers are given in Figures 12.4 to 12.7 (for notational rules, see Note 12.1).

Each neurocontroller training example consists of an initial state-for-state variable `Force` and parameter vector (`Frequency1, ForceLo1, ForceHi1, StepTime, Frequency2, ForceLo2, ForceHi2, Dist, GainFact`). The last four training examples contain reference frequency, amplitude, and offset change after 0.25 s. The remaining examples represent a single sinusoid.

These training examples have been used for global training from scratch. Note that robustness parameters are set to trivial values in this phase: The measurement error `Dist` is uniformly zero, while the gain factor on the identified plant `GainFact` is set to one. In other words, the controller is first trained for exact measurements and plant corresponding exactly to the identified model.

In the next phase, training has been refined by setting the measurement error to 0.05 kN and the gain factor to 0.5 and 2.0 for each reference trajectory. The robustness against plant variations, trained by varying gain factors, is discussed in Section 12.4.1. Other parameters remain the same.

The reason for two-phase training is that modeling a random measurement error introduces considerable nonsmoothness into the cost function. This makes the optimization substantially less efficient. It is preferable to train first the smooth, nonrobust variant and then to retrain with robustness requirements.

12.4.1 Reduced-Order Controller

The order of the plant model hints at the definition of the order of the controller. The expectations about the controller performance then have firm theoretical foundations. However, the theoretical statements involved formulate sufficient conditions (e.g., under some assumptions, absence of disturbances and model inaccuracies). Consequently, it is possible that controllers of an order lower than the plant order perform sufficiently well and are more

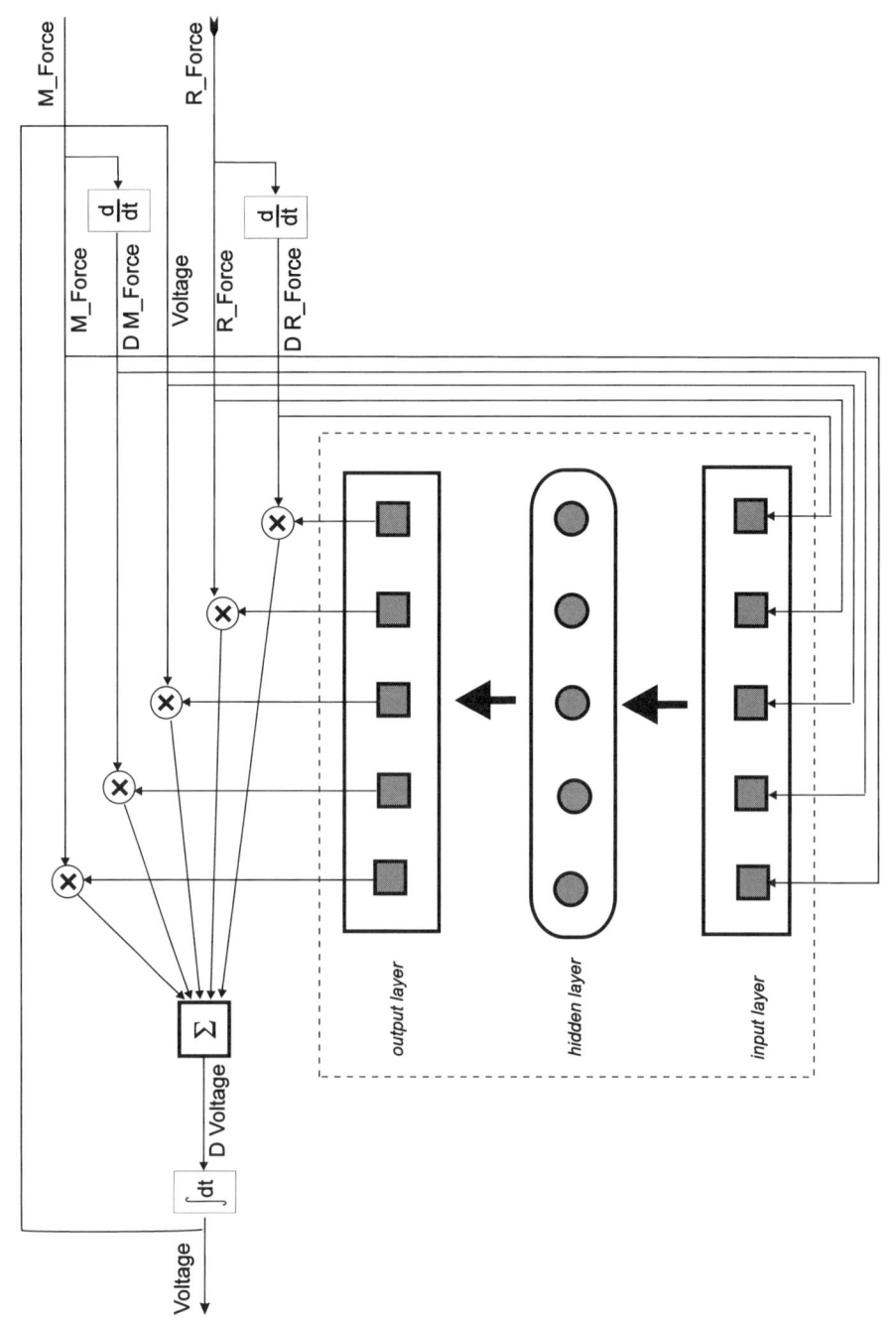

Figure 12.4 Neural network structure for elastomere best bench—basic trajectory-following neurocontroller.

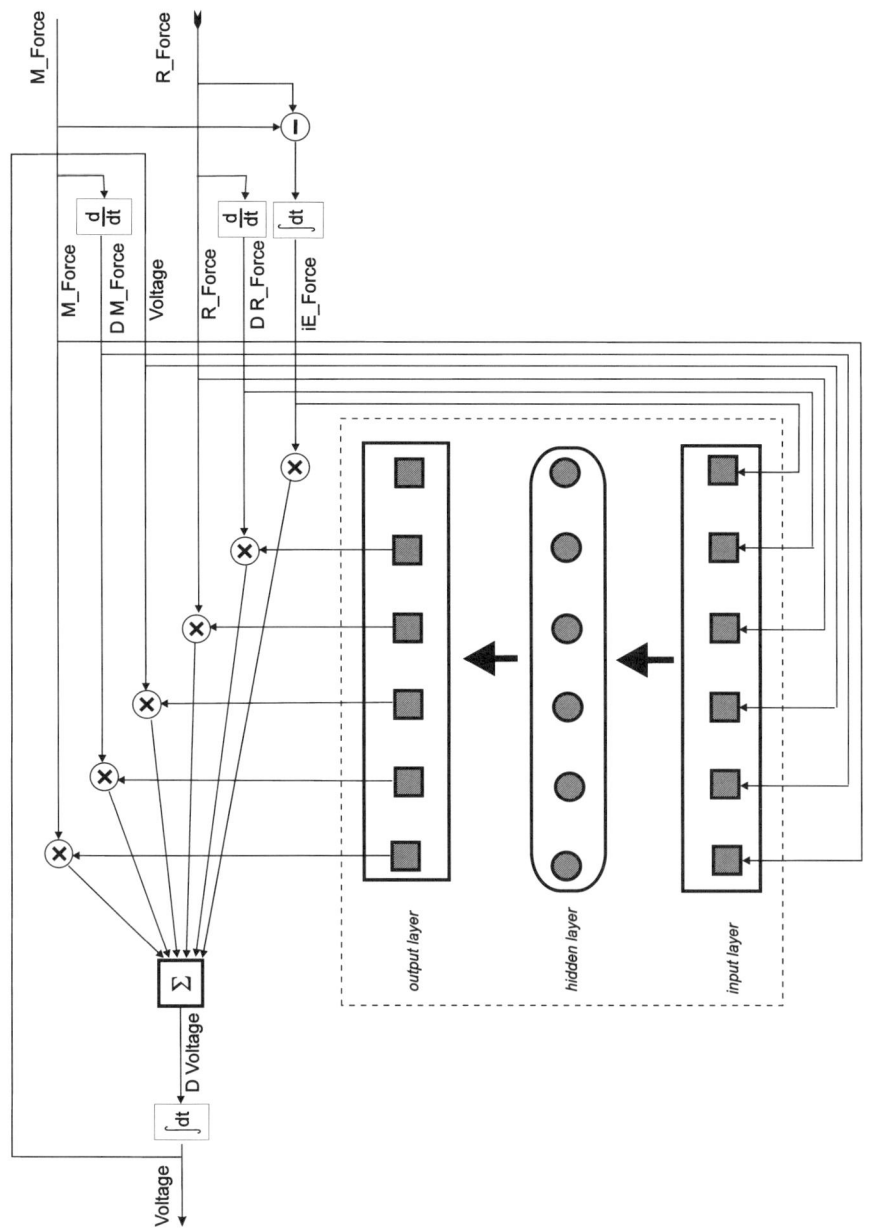

Figure 12.5 Neural-network structure for elastomerre test bench—neurocontroller with error integration.

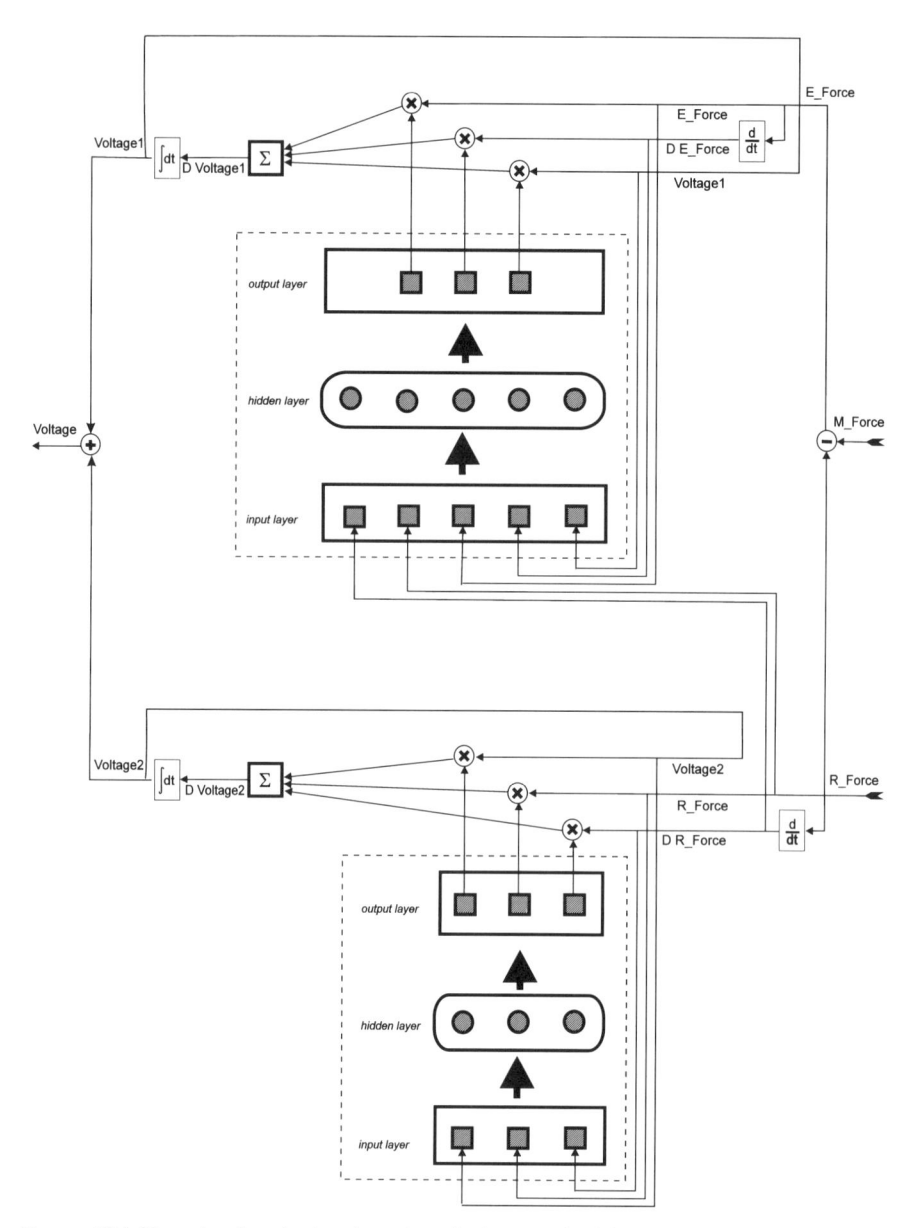

Figure 12.6 Neural-network structure for elastomere test bench—trajectory-following neurocontroller with defined equilibrium point.

computationally tractable because of the lower number of plant parameters. It is frequently worth an experiment to train a reduced-order neurocontroller and to compare its performance with neurocontrollers of order resulting from the plant order.

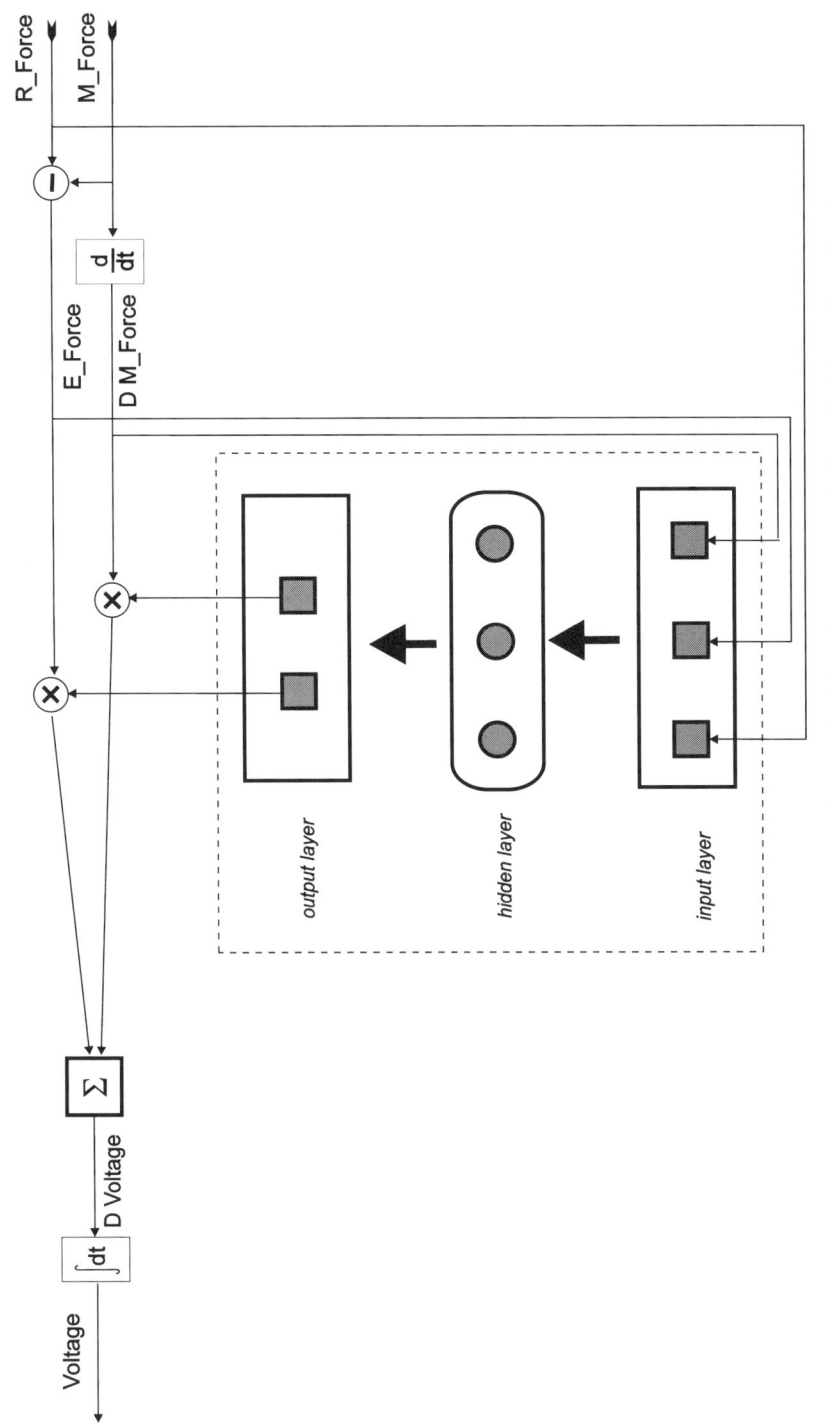

Figure 12.7 Neural-network structure for elastomere test bench—neurocontroller with integration of action.

Such a reduced-order trajectory-following controller with a defined equilibrium point (6.109), corresponding to the assumption of observability index one, is described by the following two dynamic equations for feedback and feedforward parts of the controller:

```
Voltage1 depends_on     : E_Force,D E_Force
         gain_depends_on: E_Force,D E_Force
                          R_Force,D R_Force
Voltage2 depends_on     : R_Force,D R_Force
         gain_depends_on: R_Force,D R_Force
```

Also both components have to be added together to receive the controller action then `Voltage`:

`Voltage = Voltage1+Voltage2`

12.4.2 Results of Neurocontroller Training

The simulated performance of the four controller structures tested is shown in Figures 12.8 through 12.11. The performance of the reduced-order controller is given in Figure 12.13. For each controller, two examples from the training set of Table 12.1 are taken (Charts *A* and *B*) as well as a novel reference trajectory (Chart *C*). The novel trajectory has a frequency that is twice the maximum in the training set and an amplitude that is half the maximum. The motivation for this choice is that both the reference force and its derivative are within the range of training examples, although their combination has not exactly occurred in the training set. The minimum values of cost function reached with individual controllers are listed in Table 12.2.

The basic trajectory-following controller performs well in both examples for which it has been trained. However, it does not cope well with the novel trajectory; the closed-loop trajectory has an amplitude more than twice as large as the reference trajectory, and there are also higher harmonics making the closed-loop trajectory nonsmooth. Its transient phase before reaching the reference trajectory is oscillatory.

The error integration controller is, surprisingly, almost as good as the basic trajectory-following controller, although it has to do without specific trajectory-following inputs. The error integrations seem to give it good robustness at various working points. While the novel-case trajectory exhibits large overshoots, it is smoother than that of the basic trajectory-following controller, and so is its transient phase.

The simplest controller, the one with integration of action, is obviously too slow. It follows the reference trajectory with a substantial lag. Nevertheless, the reference trajectory amplitude is approximately reproduced, at least in the examples it has been trained for. This makes it potentially useful; time lag is

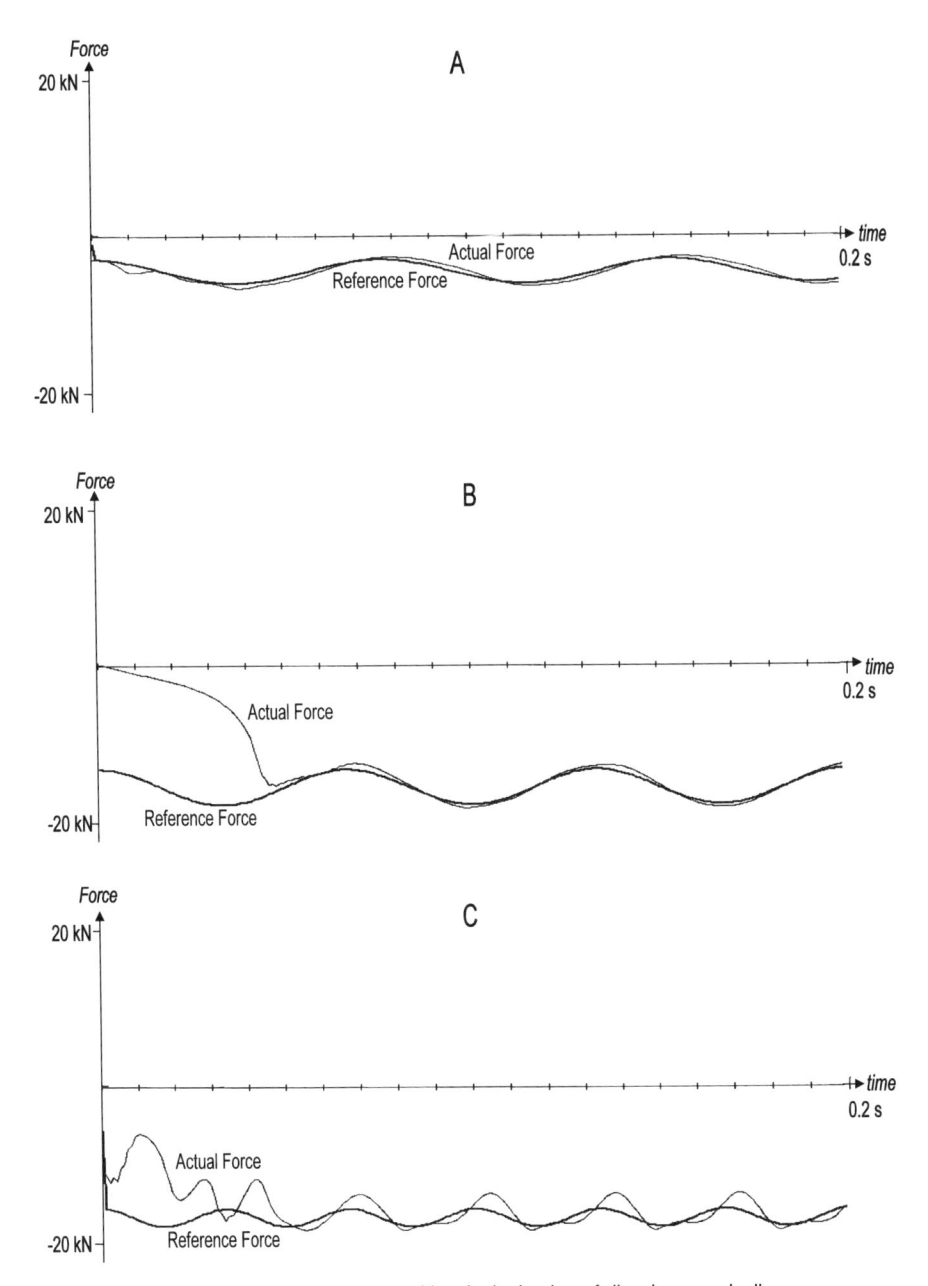

Figure 12.8 Performance of basic trajectory-following controller.

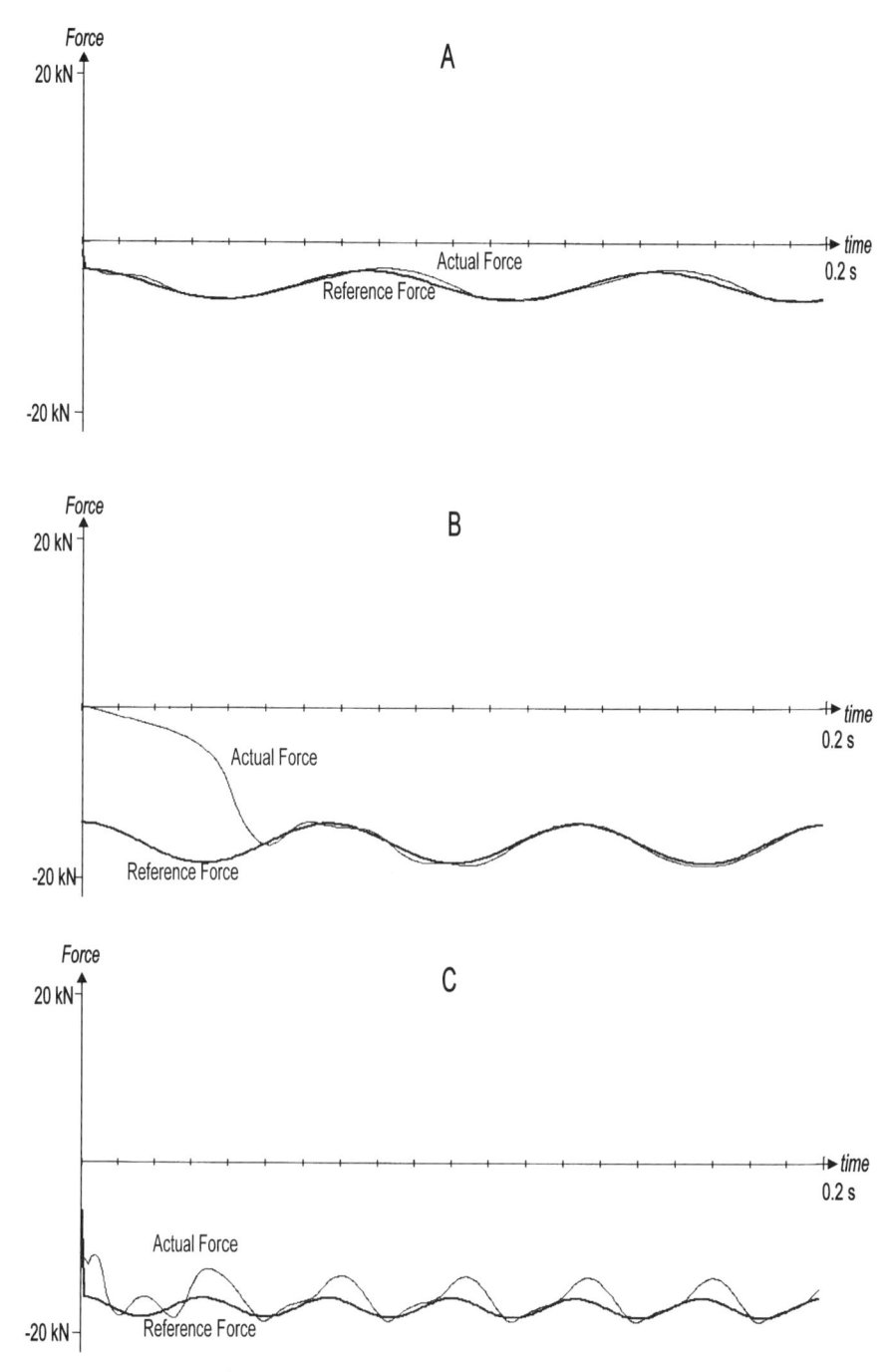

Figure 12.9 Performance of controller with error integration.

CONTROLLER TRAINING 295

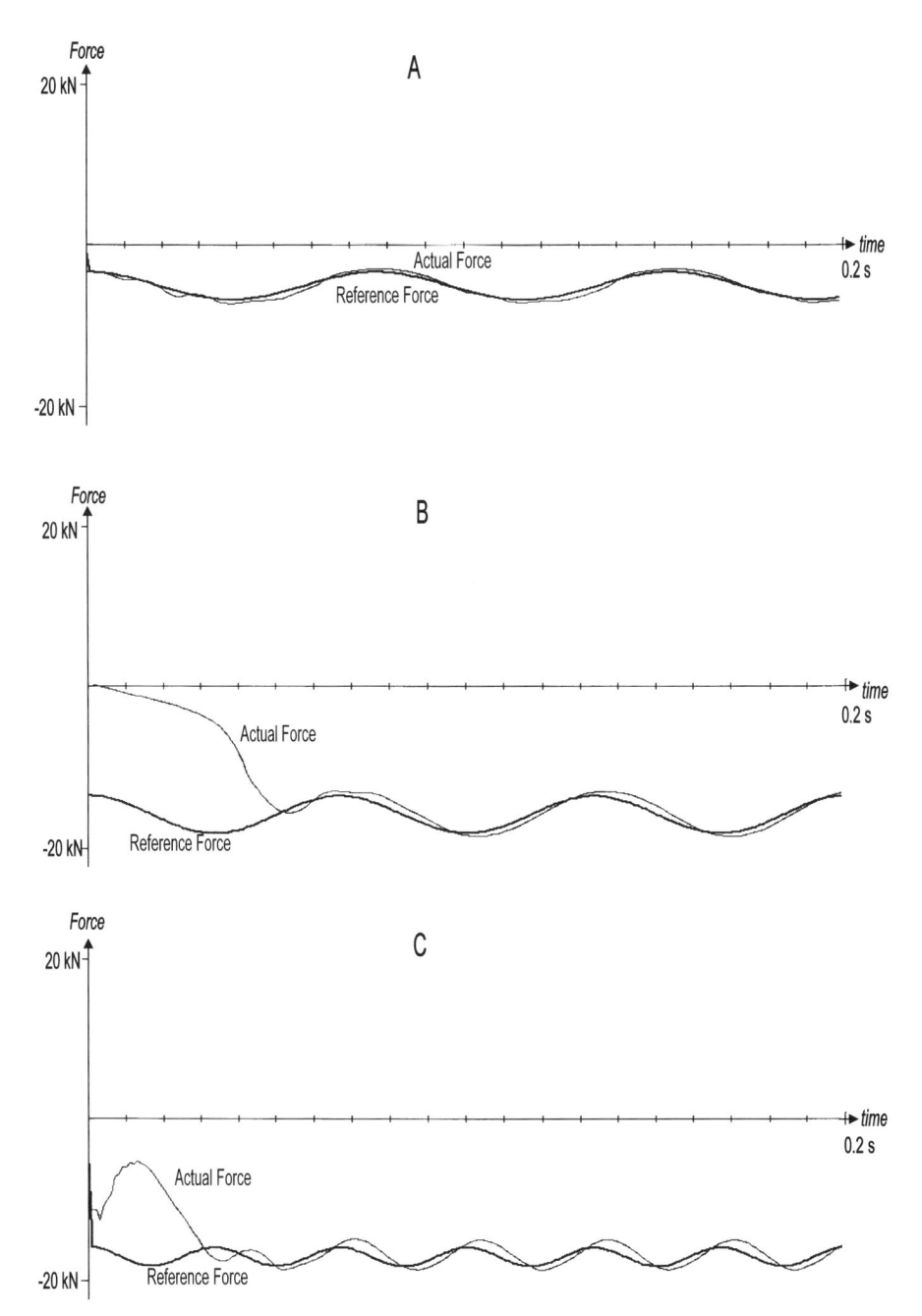

Figure 12.10 Performance of trajectory-following controller with a defined equilibrium point.

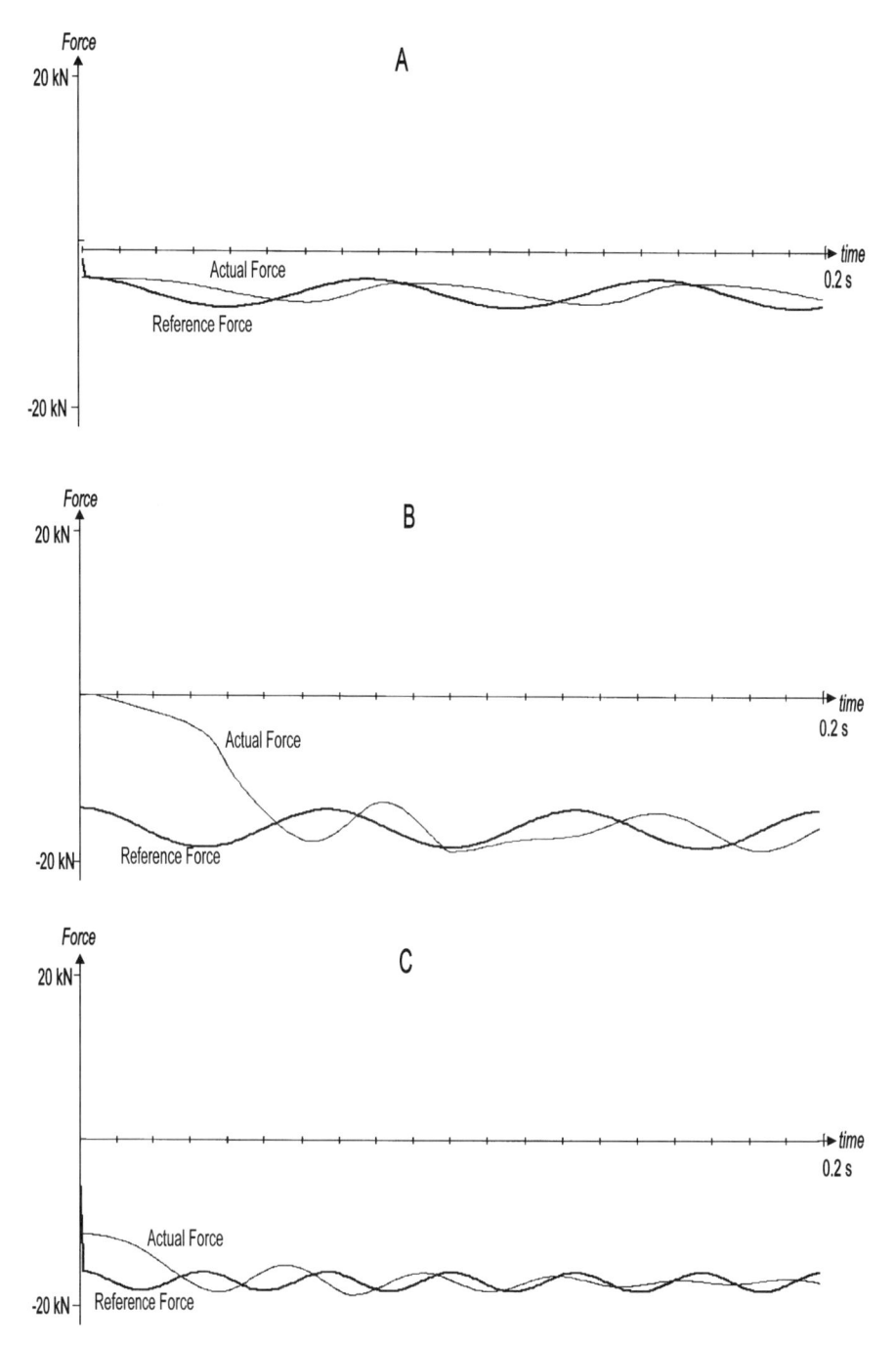

Figure 12.11 Performance of a controller with integration of action.

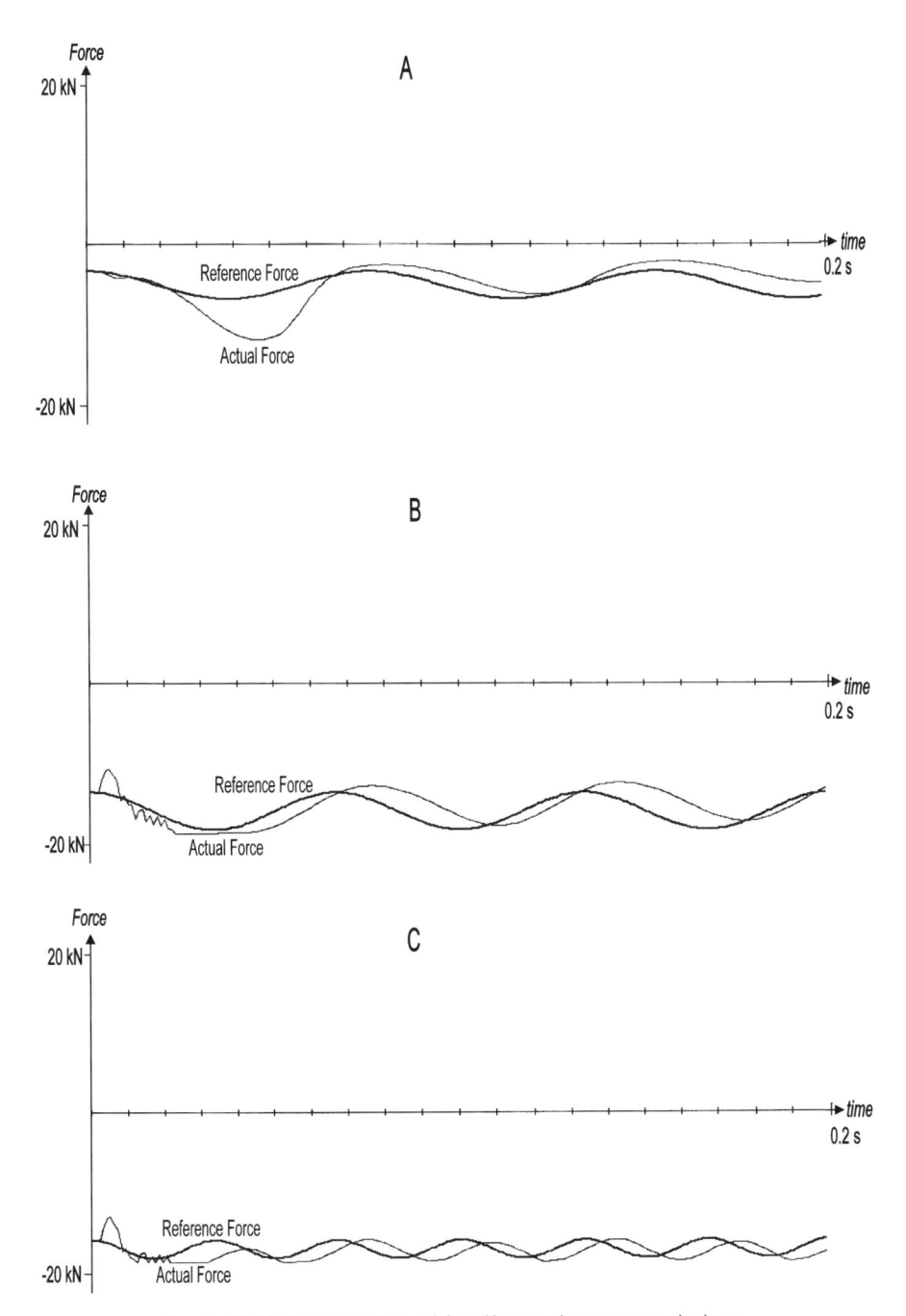

Figure 12.12 Performance of feedforward component alone.

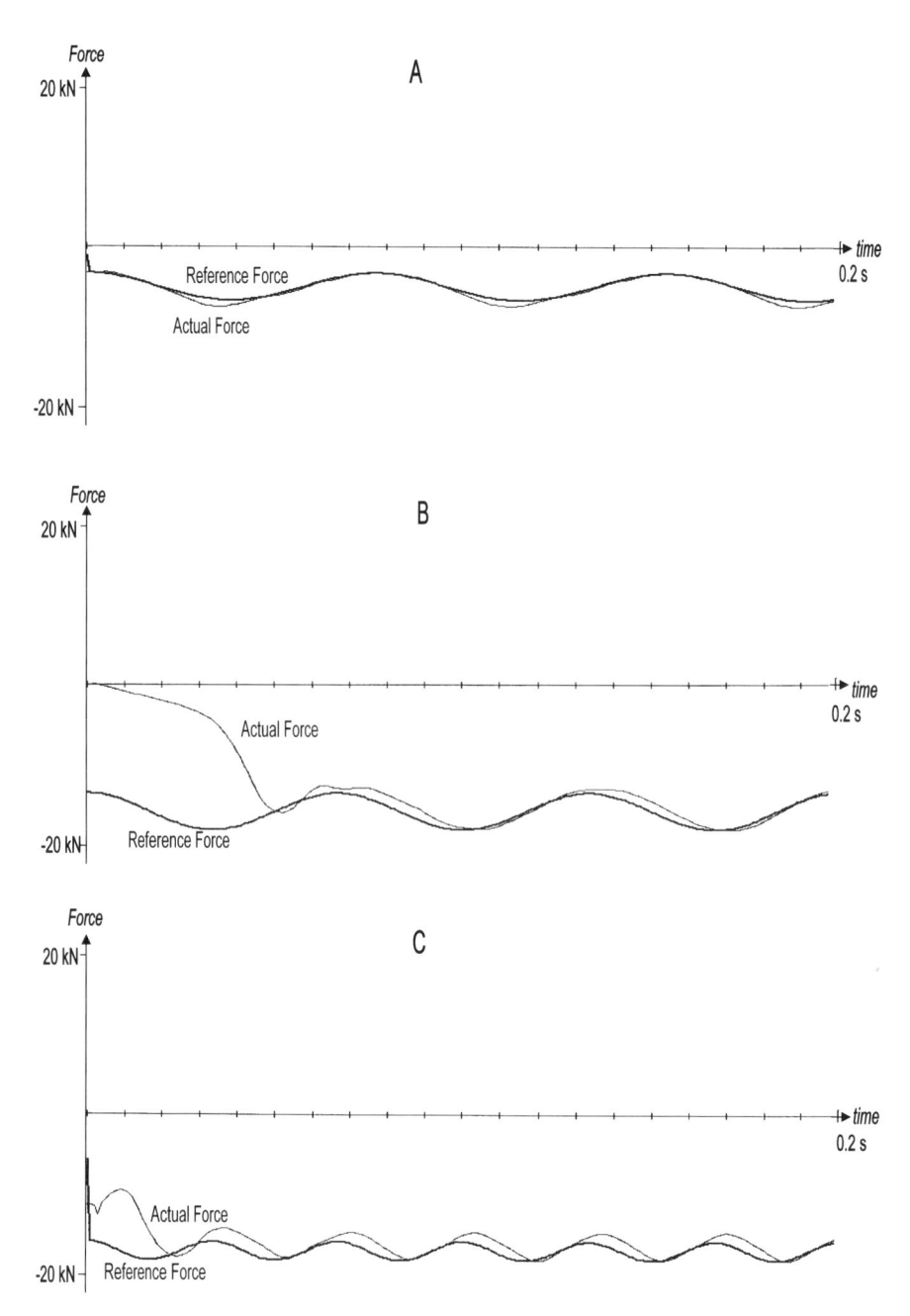

Figure 12.13 Performance of a reduced-order trajectory-following controller with a defined equilibrium point.

Table 12.1 Neurocontroller training examples

Force		Reference Trajectory Parameters						Robustness Parameters	
1.5	1.0	−2.0	1.5	9.9	15.0	−2.0	1.5	0.0	1.0
1.5	4.0	−2.0	1.5	9.9	15.0	−2.0	1.5	0.0	1.0
1.5	15.0	−2.0	1.5	9.9	15.0	−2.0	1.5	0.0	1.0
−1.0	1.0	−6.0	−1.0	9.9	15.0	−6.0	−1.0	0.0	1.0
−1.0	4.0	−6.0	−1.0	9.9	15.0	−6.0	−1.0	0.0	1.0
−1.0	15.0	−6.0	−1.0	9.9	15.0	−6.0	−1.0	0.0	1.0
−5.0	1.0	−14.0	−5.0	9.9	4.0	−14.0	−5.0	0.0	1.0
−5.0	4.0	−14.0	−5.0	9.9	4.0	−14.0	−5.0	0.0	1.0
−5.0	15.0	−14.0	−5.0	9.9	4.0	−14.0	−5.0	0.0	1.0
−12.0	1.0	−16.0	−12.0	9.9	4.0	−16.0	−12.0	0.0	1.0
−12.0	4.0	−16.0	−12.0	9.9	4.0	−16.0	−12.0	0.0	1.0
−12.0	15.0	−16.0	−12.0	9.9	4.0	−16.0	−12.0	0.0	1.0
1.5	15.0	−2.0	1.5	0.25	4.0	−14.0	−5.0	0.0	1.0
−5.0	4.0	−14.0	−5.0	0.25	15.0	−2.0	1.5	0.0	1.0
1.5	15.0	−2.0	1.5	0.25	4.0	−16.0	−12.0	0.0	1.0
−12.0	4.0	−16.0	−12.0	0.25	15.0	−2.0	1.5	0.0	1.0

Table 12.2 The minimum values of cost function reached

Controller Structure	Cost Function Minimum
Basic trajectory-following controller	0.349
Controller with error integration	0.356
Trajectory-following controller with defined equilibrium point	0.332
Controller with integration of action	1.935
Reduced-order controller	0.413

no problem in this test mode. Its strength is its simplicity: Its training time has been the shortest.

The best performing controller at all working points is the trajectory-following controller with a defined equilibrium point. It performs well at all working points, including the novel example. The feedforward component of the controller does a good job, as can be seen in Figure 12.12. Its potential gradual divergence, hardly discernible on the charts within this time horizon, can easily be compensated by the feedback component.

The superiority of the controller with a defined equilibrium point for novel examples can be explained by the fact that, to a certain degree, all trained and novel reference trajectories are shifted into the equilibrum point. A good performance around this equilibrium point is then, to some extent, universal for all combinations. By contrast, a controller without a defined equilibrium point has to learn its performance for all combinatorial working points separately.

The overall performance of a reduced-order controller with a defined equilibrium point is, according to Table 12.2, only slightly worse than for its full-order counterpart. The difference can hardly be discerned from the charts. The strength of this controller is its low number of parameters to be trained (47 instead of 72 for the full-order controller). Without further robustness requirements, this controller structure would probably be preferred for the final implementation.

12.4.3 Robustness Experiments

So far, the neurocontrollers considered have been trained, and their performance evaluated, under the assumptions that the plant model identified from the data is the accurate model of the real plant and that no plant modifications and disturbances will occur in the target setting. Besides random measurement inaccuracies, the most dangerous disturbance type for the elastomere test bench consists of plant variations caused by varying workpiece properties. Both differences between workpiece individuals and time-variant properties of a single workpiece during the test (resulting from material overload) are expected.

To account for this, the neurocontroller has to perform well under gain variations emulated by values of parameter `GainFact` different from one. It has been determined that variations between values 0.5 and 2 are to be considered.

The performance of three most promising neurocontrollers trained without plant model variations (i.e., with `Gainfact` equal to one) is shown in Figures 12.14 (basic trajectory-following controller), 12.15 (full-order controller with a defined equilibrium point) and 12.16 (reduced-order controller with a defined equilibrium point). The gain factor is set to 0.5 (Charts *A*) and 2 (Charts *B*). Obviously, all the controllers presented exhibit serious performance deficiencies at least for one of the two plant modifications. Since plant variations of this order of magnitude can be expected, implementing these controllers in the real-time environment will lead to disaster.

It is one of the magics of neurocontrol that all that needs to be done to solve this problem is to retrain with different training examples. The training set of Table 12.1 is doubled: One copy is provided with `GainFact = 0.5` (last column of the table), the other with `GainFact = 2.0`. Both sets together constitute the new training set.

The performance of all three neurocontrollers after the robustness training is shown in Figures 12.17 through 12.19. The cost function values reached for the training set with gain factor variations are listed in Table 12.3.

The performance of the basic trajectory following controller is unstable. It completely fails in some working points. This once more confirms the hypothesis that it is difficult to find a single mapping that satisfies both the stability requirements and the algebraic equilibrium point requirements at all working

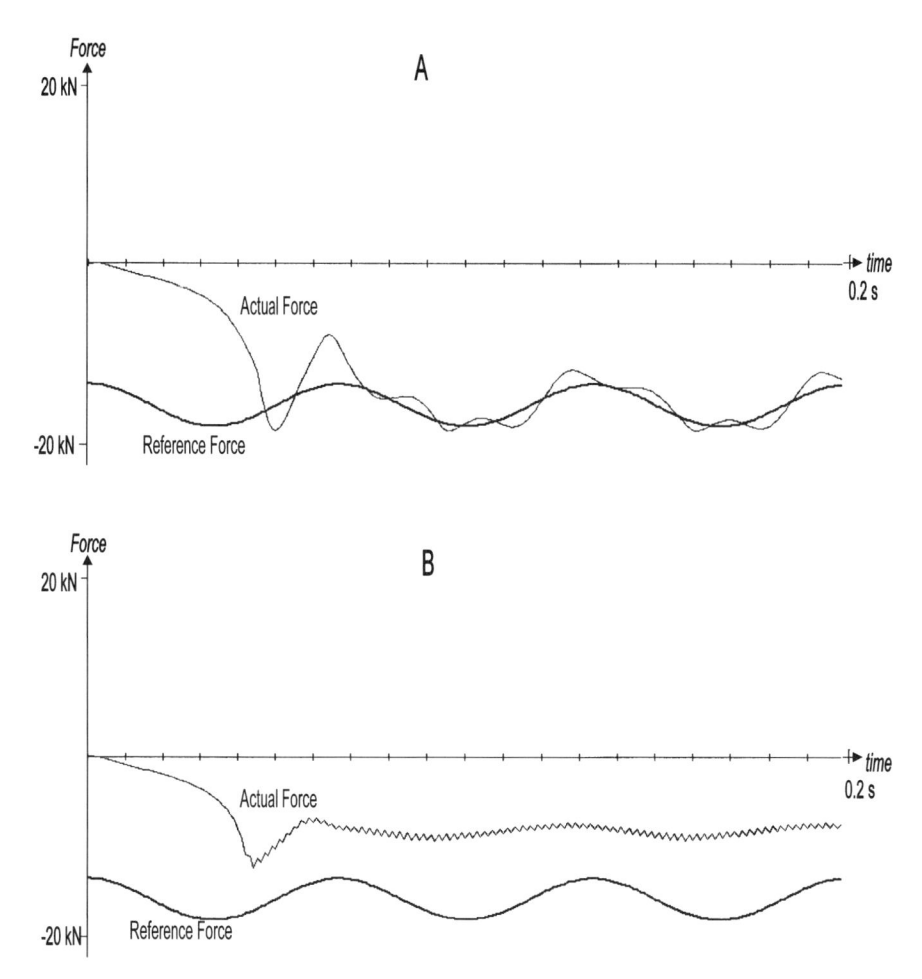

Figure 12.14 Performance of (full-order) basic trajectory-following controller without robustness training.

points simultaneously, even if this is theoretically possible by virtue of functional approximation theorems for neural networks.

The reduced-order controller with an equilibrium point defined by a separate feedforward component has been able to be tuned a little toward improved robustness, but it comes to the limits given by its order. Although its reduced order has been sufficient for the plant without disturbances, it is no longer the case if disturbances are present. Algebraic reasons for this have been discussed in Sections 7.1 and 7.3.

This is why only the full-order controller with a defined equilibrium point has been able to perform satisfactorily. In fact, its order is not sufficient to guarantee the robustness against plant variations—for this, its order would have to exceed the full order. However, it could be observed that the

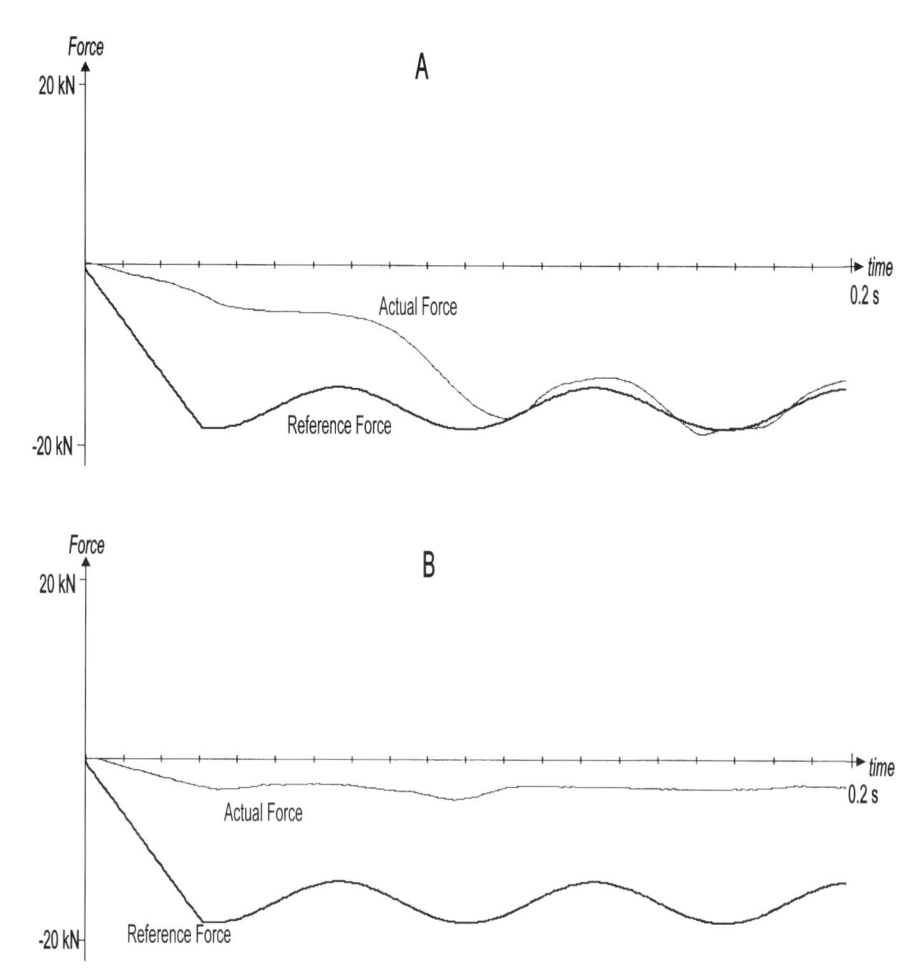

Figure 12.15 Performance of full-order trajectory-following controller with a defined equilibrium point without robustness training.

reduced-order controller is sufficient without plant variations in this specific control task. The full-order neurocontroller exceeds this sufficient order by one and thus has enough potential for robustness.

The algebraic insufficiency of the reduced-order controller becomes more obvious if Figures 12.20 and 12.21, depicting the behavior of feedforward components alone, are observed. (See also the last two rows of Table 12.3.) It is the feedforward components that are to solve the algebraic plant inversion under varying plant parameters; the feedback component is responsible only for a stable convergence after deviations from the reference trajectory. The reduced-order feedforward component (Figure 12.21) makes hardly any contribution to the performance of the reduced-order controller. If this controller works at all, it is only because the closed-loop convergence to the reference

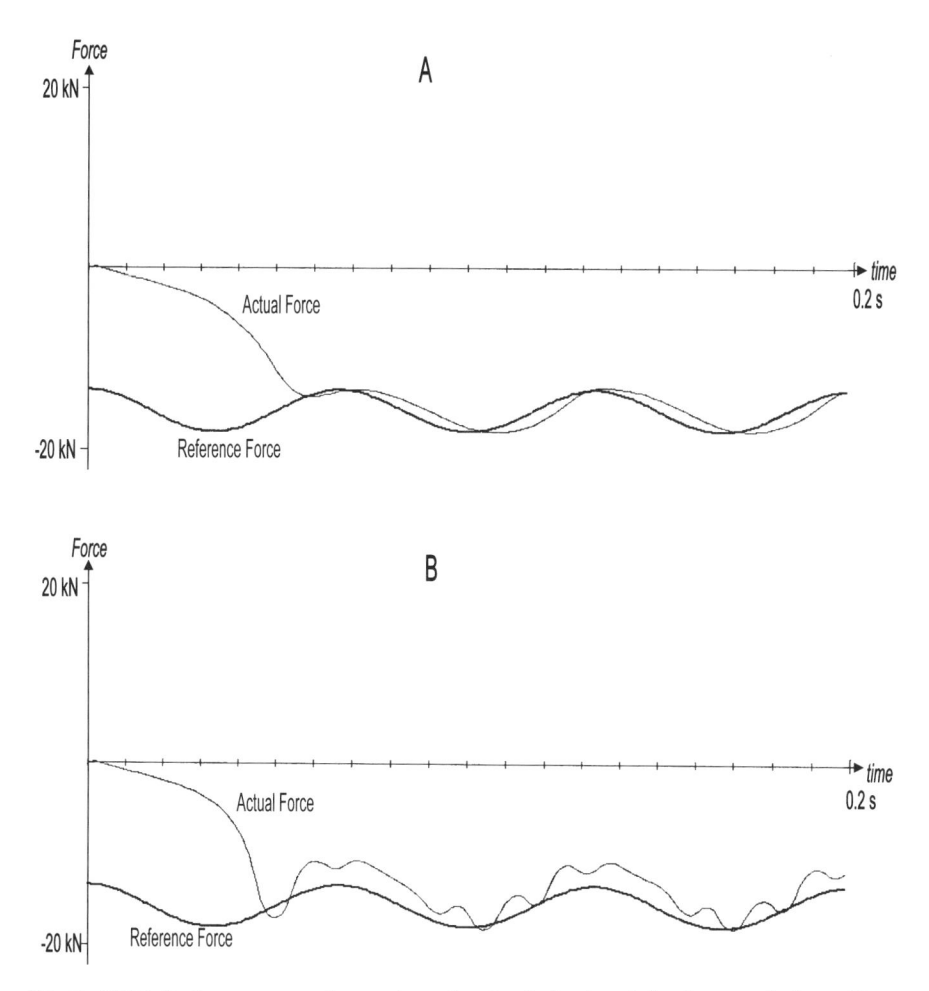

Figure 12.16 Performance of a reduced-order trajectory-following controller with a defined equilibrium point without robustness training.

trajectory with the help of feedback is faster than the changes of the reference trajectory. By contrast, the full-order feedforward component (Figure 12.20) does almost all the work.

12.4.4 Performance of Neurocontrollers on the Real Plant

The *real-time performance* has been tested only for the best neurocontroller structure, the trajectory-following controller with a defined equilibrium point trained for robustness. The performance has been approximately the same as in simulations, close to that shown in Figure 12.10*A* and *B*. (The combination of high frequency and high force whose simulation is given in Fig. 12.10*C* is not being run on the real plant in test operations. It has been included in the

304 ELASTOMERE TEST BENCH CONTROL

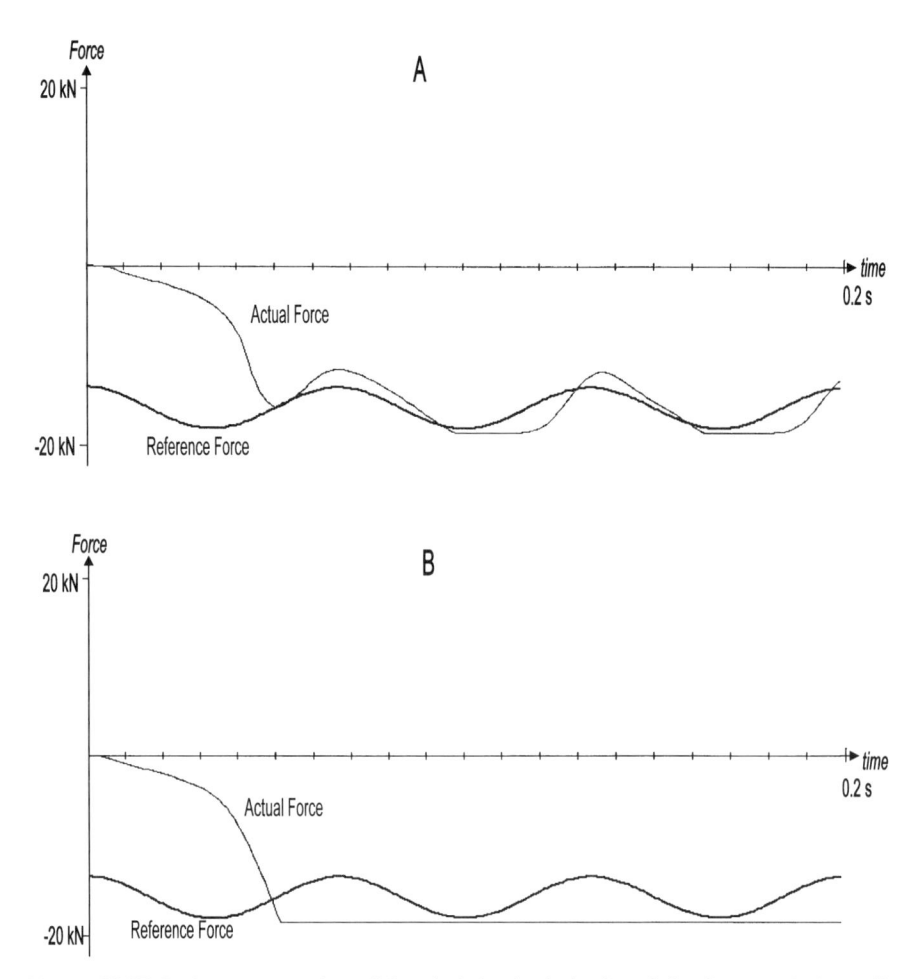

Figure 12.17 Performance of a (full-order) basic trajectory-following controller with robustness training.

training set to provide for a security margin.) The deviations have always been less than those shown in simulations with extreme gain variations in Figure 12.18—real gains obviously do not vary in such a broad range.

The largest control errors have occurred on the very margin of the state space range defined by sampled measurements (i.e., forces close to 1 kN and to −15 kN). Obviously, it is vital for sampled data used for identification to cover all regions of the state space that may be reached in a real-time operation. It is, in fact, desirable, to go *beyond* the range of routinely occurring states. Only then, close correspondence between model and plant is guaranteed.

Anyway, the deviations from reference trajectories have been, except for the transition between different working points, always under 9 percent, and mostly under 4 percent of the scheduled force amplitude. This is a substantial

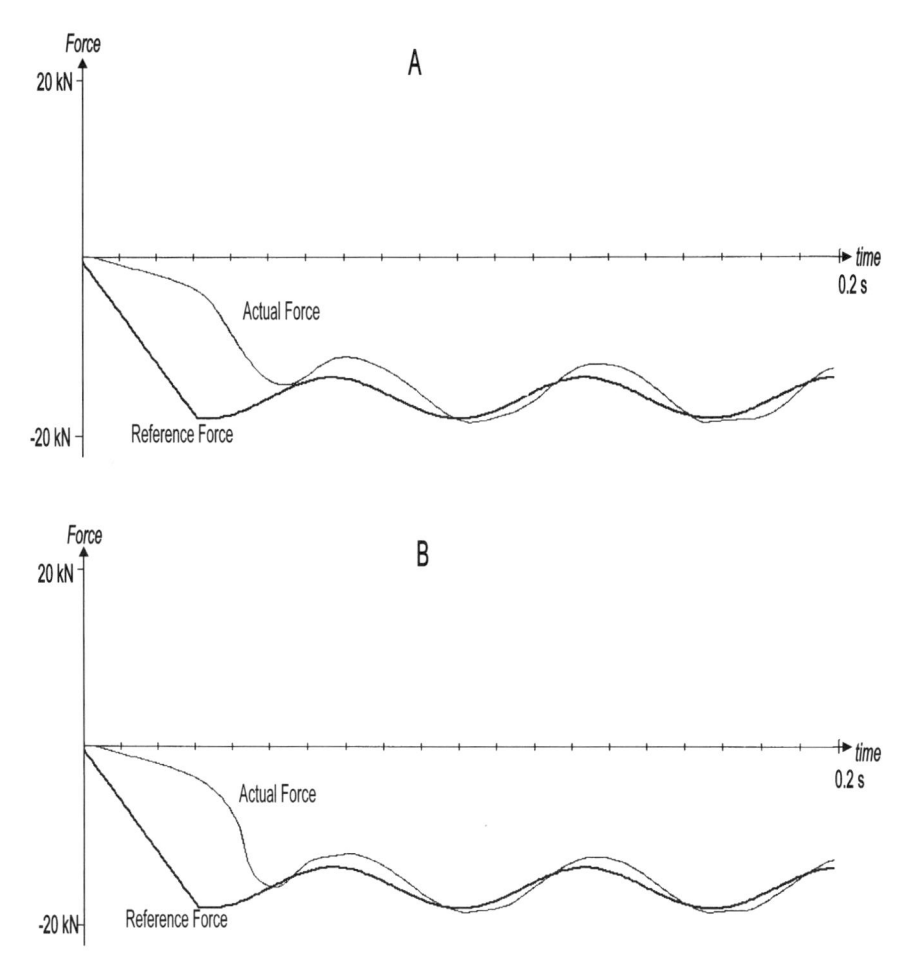

Figure 12.18 Performance of a full-order trajectory-following controller with a defined equilibrium point with robustness training.

improvement if compared with the original gain-scheduling controller with over 20 percent deviation. The transitions between different reference sinusoids have been stable.

Besides the control quality, the most important point has been that, in contrast to the classical design (and also fuzzy control design experimented with later), no long series of experiments on the plant had to be performed. The first trial consisting of sampling data, plant identification, controller training, and real-time test has shown that the range of sampled data is too limited. Already, in the second trial, controller performance has been completely satisfactory.

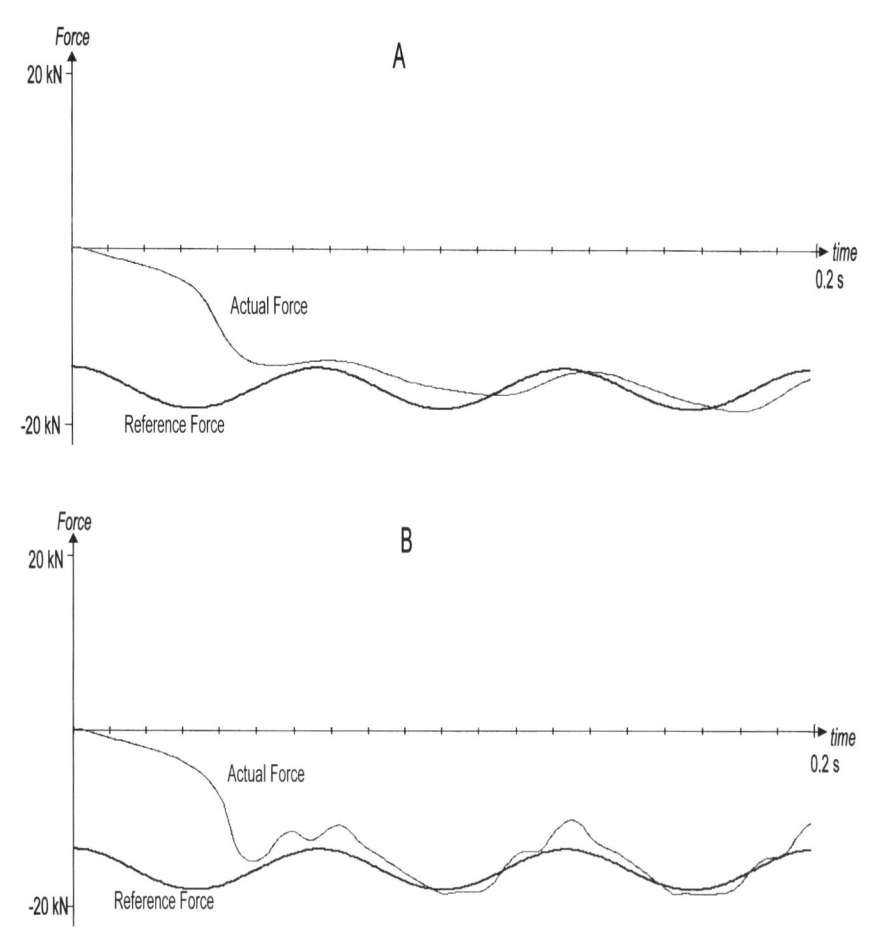

Figure 12.19 Performance of a reduced-order trajectory-following controller with a defined equilibrium point with robustness training.

Table 12.3 Cost function minima with plant gain variations

Controller Structure	Cost Function Minimum
Basic trajectory-following controller	1.04
Full-order controller with a defined equilibrium point	1.03
Reduced-order controller with a defined equilibrium point	0.59
Full-order controller—feedforward component only	0.60
Reduced-order controller—feedforward component only	5.35

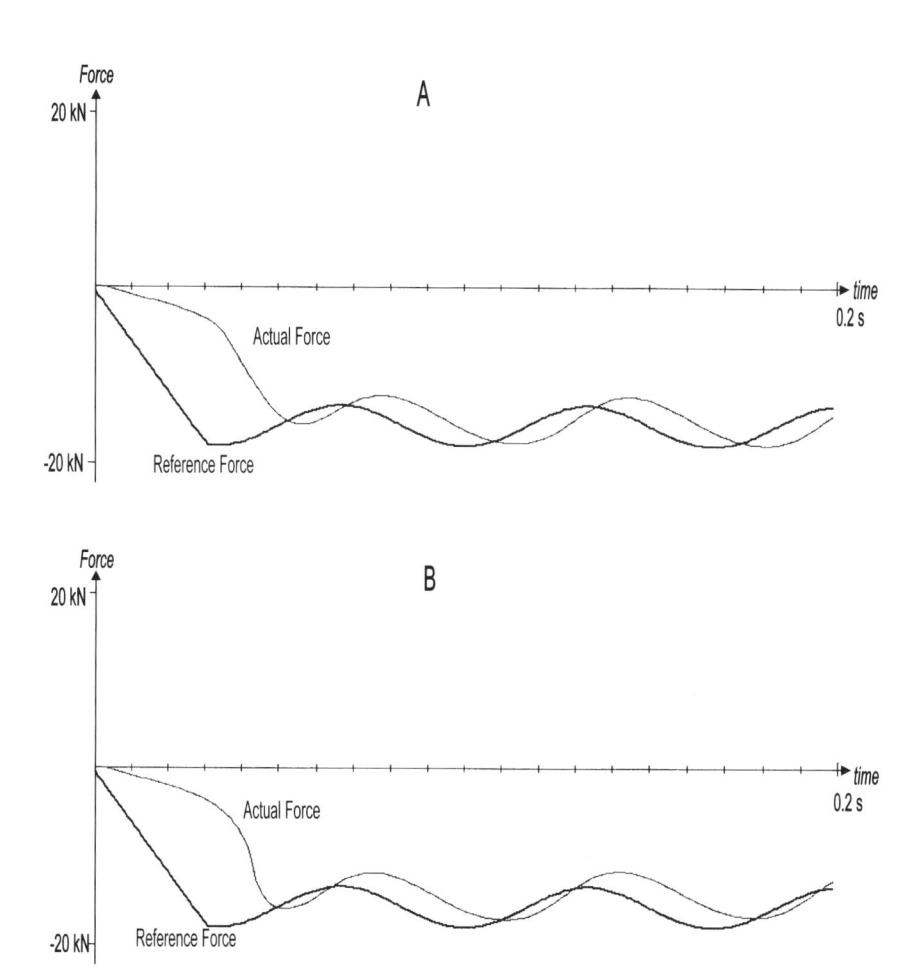

Figure 12.20 Performance of the feedforward component of a full-order controller with robustness training.

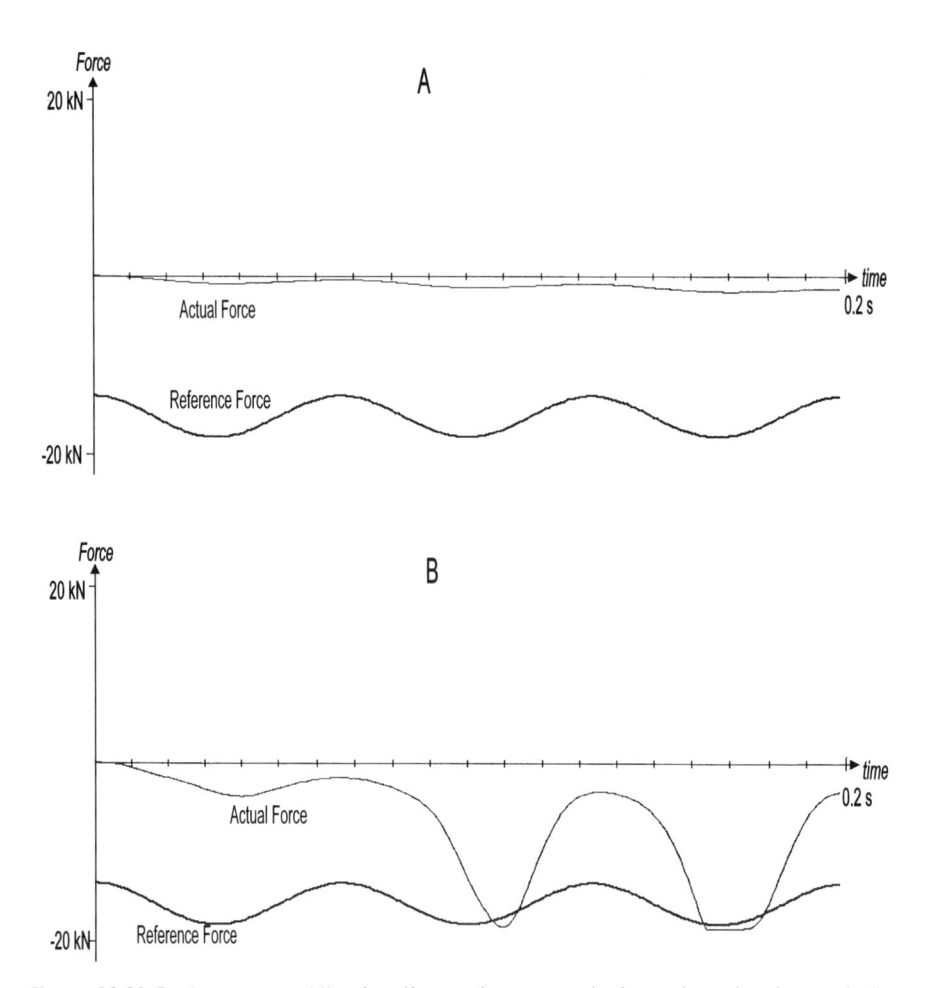

Figure 12.21 Performance of the feedforward componet of a reduced-order controller with robustness training.

13

DRIVE TRAIN TEST BENCH CONTROL

The second case study is also concerned with a test bench. The drive train test bench whose control is presented here is a large device on which a complete car drive train (combustible engine, automatic transmission box, torque transducer, and rear axle) are fixed (Fig. 13.1). (It has also been introduced briefly and commented on in Section 5.3.3). Two direct-current (dc) engines are connected to both sides of the rear axle to produce a prescribed load.

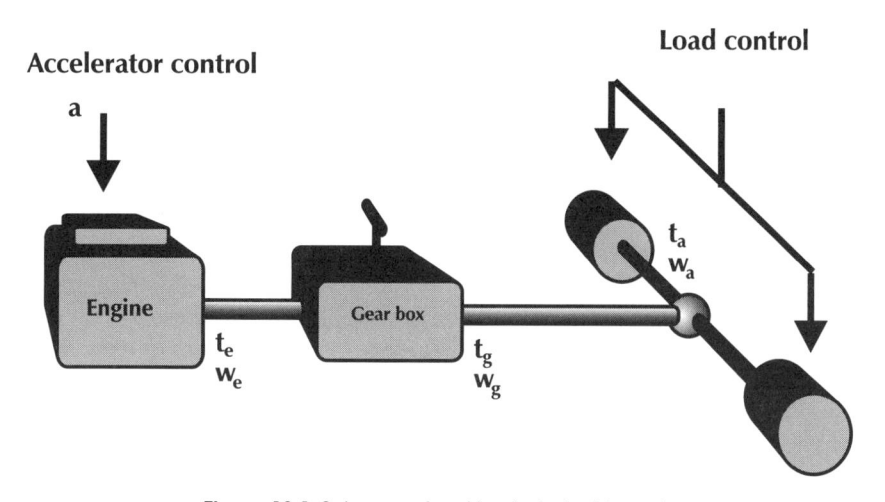

Figure 13.1 Scheme of a drive train test bench.

A drive train test bench is used to simulate realistic load conditions and to measure the behavior of all drive train components (in particular, of the engine) under these conditions.

The control task is of the trajectory-following type. Two variables, the torque and the angular velocity of the drive train, measured on the rear axle, are to follow schedules sampled in various operating conditions of a real car. The emphasis is on reproducing the dynamics of the schedule, that is, fast changes of torque and angular velocity.

Particular difficulties occurring with the dynamic drive train control result from the following properties of individual drive train components:

1. The torque of a combustible engine is nonlinearly dependent on the accelerator position and on the angular velocity of the engine. The latter dependence is nonmonotonic. The gains vary over a range of 1 to 30. The inherent reaction delays are approximately 0.2 s, that is, the same order of magnitude as that of time constants and control actions involved. (A delay of size approaching the time constants of the plant represents a formidable problem in control theory and practice. In commonsense terms, it means that the control action comes too late to effectively influence the state of the plant.)
2. The torque transducer gain depends on the input–output angular velocity ratio in a nonlinear way. The ratio is not directly measurable on the test bench concerned.
3. The transmission box changes the gears according to a complex rule not completely disclosed to the test bench engineer.

The complexity of a complete drive train makes it hardly tractable by analytic control design methods. Although nonlinear classical controllers of considerable sophistication have been used, the necessity of frequent fine-tuning with high manpower costs (5 days for each engine type) motivated the use of neurocontrol.

13.1 PLANT MODEL IDENTIFICATION

A particular feature of this application has been the use, as an alternative to a plant identification with the help of neural networks, of an analytic nonlinear model with free parameters instantiated by sampled measurements.

The drive train test has two action variables: The accelerator controls the car engine and voltage controls dc engines. The test bench can be run in two modes. In the first mode, torque is controlled by the accelerator and angular velocity by voltage. In the second mode, the roles of both actions are interchanged. The configuration considered here is being run in the first configuration.

PLANT MODEL IDENTIFICATION

The angular velocity control by dc engines is uncomparably easier than the torque control by accelerator, and it works with a negligible control error. This is why the angular velocity is assumed to be equal to the reference angular velocity and the corresponding controller is omitted from this case study presentation.

The analytic model represents the following aspects of the plant components:

Execution of Accelerator Action The actuator together with the accelerator lever is modeled as a second-order system with a dead time and a hysteresis.

Engine Torque Development The engine torque is modeled as a nonlinear function of accelerator position and of current angular velocity. The action is delayed by a dead time and a time constant.

Torque Transducer A very rough model of torque transducer assumes a common input–output ratio for both torque and angular velocity. This ratio depends on input torque and output angular velocity.

Automatic Transmission Box The transmission box model consists of a gear schedule with a hysteresis and a ramp-formed gear ratio dynamics.

The constant axle transmission ratio is neglected. The plant inputs are

```
AccRef     \\ Reference accelerator position
WAxleRef   \\ Axle angular velocity
           \\ (Assumption: reference value = real value)
```

State variables are

/***** **Accelerator** *****/

```
Acc0             \\ Delayed reference accelerator position
dAcc0            \\ Time derivative of Acc0
Acc              \\ Accelerator input into the engine
TorqEng0         \\ Engine torque before delay
TorqEng          \\ Engine torque at engine output
TorqAxle         \\ Axle torque
WEng             \\ Angular velocity at engine output
WTT              \\ Angular velocity at transmission box output
TransRatio       \\ Transmission ratio
                 \\    of torque transducer
GearRatio0       \\ Reference transmission ratio
                 \\    of transmission box
GearRatio        \\ Transmission ratio at the transmission
                 \\ box output
```

Reference accelerator position is the position as set from outside, for example, by a controller. This value is subject to delay. The reference transmission ratio denotes the theoretical transmission ratio corresponding to the current angular velocity. This ratio is not reached immediately but on a ramp-formed dynamic trajectory.

Only the variables Acc, WEng, and TorqAxle are directly measurable on the test bench concerned. The model equations are the following:

```
/***** Accelerator *****/

n_dAcc0      = dAcc0
               + DT * (dead(AccRef,TdAcc) - Acc0
               - 2*TcAcc*DcAcc*dAcc0)
               / TcAcc^2;
n_Acc0       = Acc0 + DT * dAcc0;
n_Acc        = hyst(Acc,n_Acc0,HystWi);

/***** Engine Torque *****/

n_TorqEng0   = engine(Acc,WEng,
               Flat1,TorqMin1,TorqMax1,Pow1,
               Flat2,TorqMin2,TorqMax2,Pow2,
               Flat3,TorqMin3,TorqMax3,Pow3);
n_TorqEng    = TorqEng
               + DT * (dead(TorqEng0,TdEng)
                 - TorqEng) / TcEng;

/***** Gear Ratios *****/

n_TransRatio = torq_trans(WTT,TorqEng);
n_GearRatio0 = gear_box(GearRatio0 ,
               WAxleRef - (Acc > 0.95?120:0));
n_GearRatio  = GearRatio
               + DT * RampGB * sgn(GearRatio0
                 - GearRatio);

/***** Torque *****/

n_TorqAxle   = n_TorqEng*n_TransRatio*n_GearRatio;

/***** Angular Velocities *****/

n_WTT        = n_WAxleRef/n_GearRatio;
n_WEng       = n_WTT    /n_TransRatio;
```

The function dead(Var,TT) returns the value of variable Var delayed by TT. Function hyst(Output,Input,Width) is modeling a hysteresis.

sgn(X) is a usual sign function returning 1, −1, or 0 in dependence on the sign of X.

The equations for AccO and its time derivative dAccO represent numeric integration of the second-order equation

$$T^2 \ddot{a} = a_{\text{ref}} - a - 2T d\dot{a} \qquad (13.1)$$

with delayed a_{ref}.

There are three application-specific functions. The function gear_box(Ratio,W) returns the transmission ratio for axle angular velocity W. It contains a hysteresis and thus requires the last value of transmission ratio as an additional argument. The angular velocity is corrected in kick-down situations with accelerator position over 0.95.

The transmission ratio of the torque transducer is roughly approximated by the function torq_trans(W,T).

The engine torque depends on accelerator position and the angular velocity of the engine. The engine torque function is parametrized by $3 \times 4 = 12$ parameters corresponding to low, middle, and high angular velocities. The torque for each angular velocity range is modeled as a function

$$T_{\text{eng}} = T_{\text{max}} - (T_{\text{max}} - T_{\text{min}}) \left(\frac{1 - \min(a, a_{\text{flag}})}{1 - a_{\text{flat}}} \right)^P \qquad (13.2)$$

with parameters a_{flat} (the range of low accelerator positions for which the torque practically does not increase), T_{min} (minimum torque), T_{max} (maximum torque), and P (the power of approximating curve).

This plant model has been first parametrized with theoretical values of individual constants. Figure 13.2 shows the forecasts for two state variables essential for the success: the accelerator position Acc (Chart *A*) and the rear axle torque TorqAxle (Chart *B*). The smooth curves are the forecasts. The nonsmoothness of sampled measurements is partially due to oscillations of measurement circuits (at approximately 11 Hz, which, in turn, may be a discrete-sampling alias of a higher frequency). The correspondence between the forecasts of this fixed analytic model and sampled measurements has not been satisfactory. It is not only delays and scaling that are inaccurate. It has not been possible to change the individual constants so that the results would be substantially improved for all measurement series.

To improve this model, a subset of model constants has been considered as free parameters and identified using an optimization method. The training data set consisted of six measurement series with 700 samples each.

The identified model constant are the accelerator dead time TdAcc, the engine time constant, and dead time TcEng and TdEng, as well as the engine torque function parameters Flat1, TorqMin1, TorqMax2, Pow1, Flat2, TorqMin2, TorqMax2, Pow2, Flat3, TorqMin3, TorqMax3 and Pow3. The forecasts of this model are shown in Figure 13.3. A substantial improvement over the fixed model has obviously been

314 DRIVE TRAIN TEST BENCH CONTROL

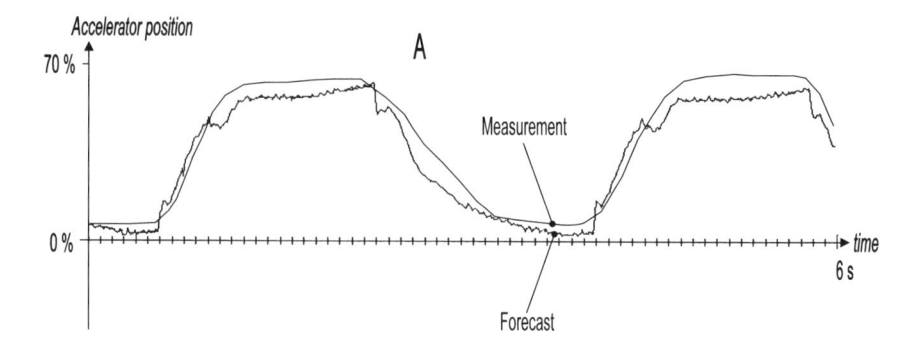

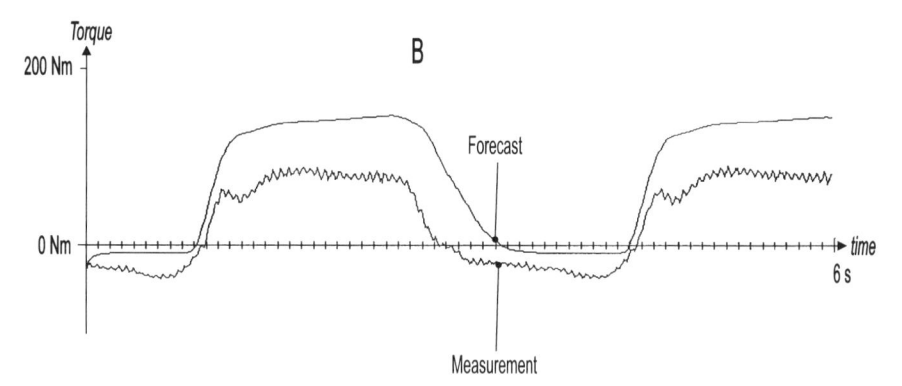

Figure 13.2 Fixed analytical model: Forecasts versus measurements.

achieved. The high-frequency oscillations are smoothed out. This is only partially desirable because these oscillations are due not only to measurement disturbances but also to oscillating input. However, the high-frequency dynamics is not important for control, except for its disturbing effects.

Furthermore, a neural-network-based plant model has been identified. The structure of the drive train make clear that the accelerator constitutes a separate block that influences the remaining parts of the plant only via Acc (i.e., the influence of AccRef can be omitted.

The plant model specification is then:

```
output
  Acc        eq_nr: 1
  TorqAxle   eq_nr: 2
  WEng       eq_nr: 3
input
  AccRef
  WAxleRef
```

PLANT MODEL IDENTIFICATION   315

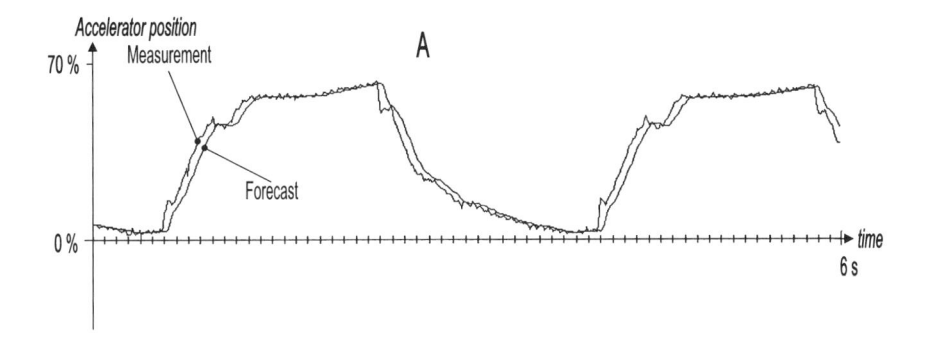

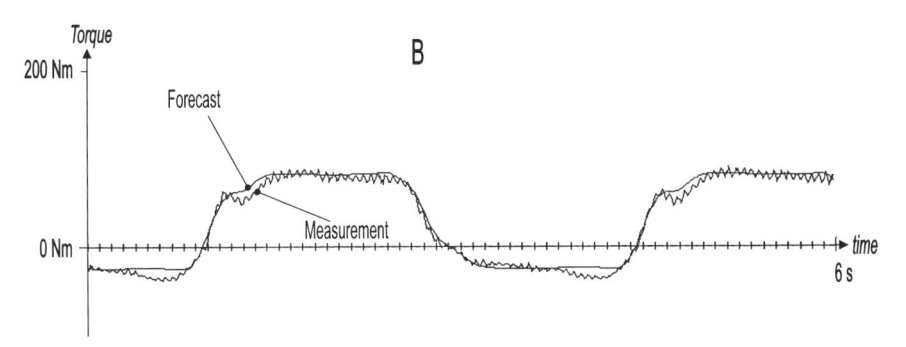

Figure 13.3 Parametrized analytical model: Forecasts versus measurements.

```
dependences
   DD Acc         depends_on:      Acc,D Acc,AccRef,D AccRef
                  gain_depends_on: Acc,D Acc,AccRef,D AccRef
   DD TorqAxle    depends_on:      TorqAxle,D TorqAxle,WEng,D WEng,
                                   Acc,D Acc,WAxleRef,D WAxleRef
                  gain_depends_on: TorqAxle,D TorqAxle,WEng,D WEng,
                                   Acc,D Acc,WAxleRef,D WAxleRef
   DD WEng        depends_on:      WEng,D WEng,TorqAxle,D TorqAxle,
                                   Acc,D Acc,WAxleRef,D WAxleRef
                  gain_depends_on: WEng,D WEng,TorqAxle,D TorqAxle,
                                   Acc,D Acc,WAxleRef,D WAxleRef
```

The neural-network representation of this model is by three separate networks of structure depicted in Figures 13.4 to 13.6, each corresponding to one output variable of the model (for notational rules, see Note 12.1). The overall structure connecting the three partial model networks is that of Figure 13.7.

The forecasts of this model are shown in Figure 13.8. It can be observed that this time, also, the high-frequency component has been captured to a consider-

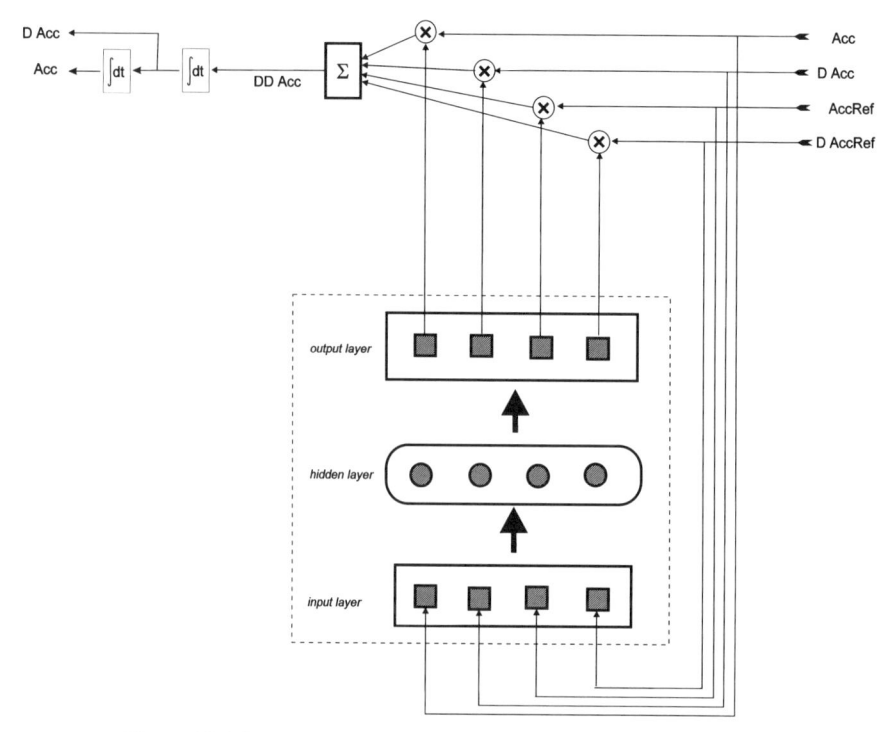

Figure 13.4 Neural network structure for modeling variable Acc.

able extent. An obvious explanation for why this is not the case using the analytic model with free parameters is the following: The analytic model contains no structural components that can be instantiated to such fast oscillating blocks. A prestructured analytic model identifies only those aspects of the behavior that it is deliberately designed for. In other words, the functional class that can be identified is constrained by the model structure. However, the neural-network-based plant model had 260 free parameters to be instantiated, in contrast to 15 parameters of the analytic model. The computing time necessary for identification has been correspondingly higher. The number of samples used had to be increased to ten to reach sufficient generalization.

The substance of this experience can be formulated in the form of the following rule:

A prestructured analytic model with free parameters will efficiently identify those aspects of behavior that are explicitly modeled and correspondingly parametrized. By contrast, a neural-network-based model of sufficient order is free to identify any behavior that dominates in the sampled measurements. The cost for this universality is a (frequently considerably) increased computational expense.

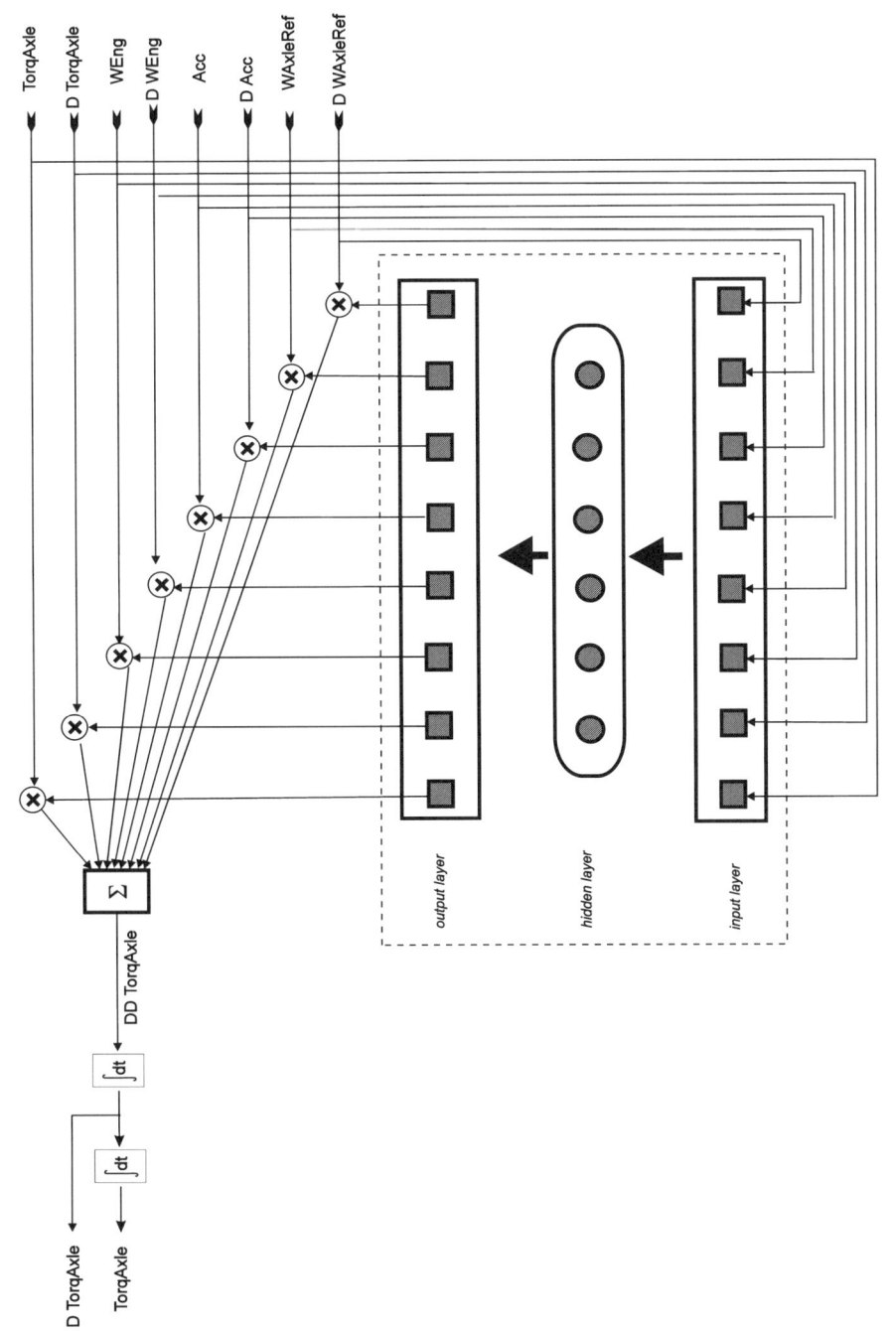

Figure 13.5 Neural network structure for modeling variable `TorqAxle`.

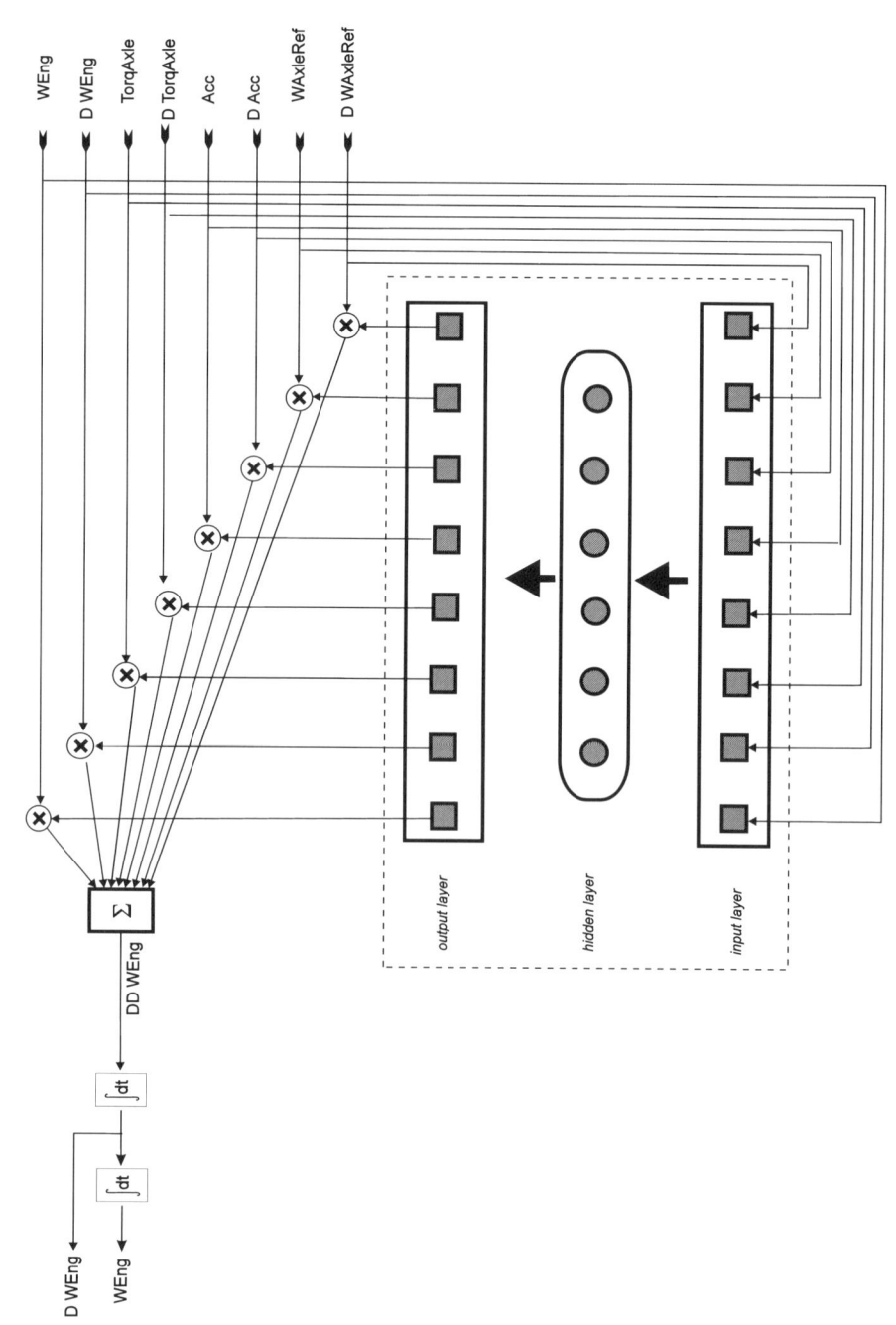

Figure 13.6 Neural network structure for modeling variable WEng.

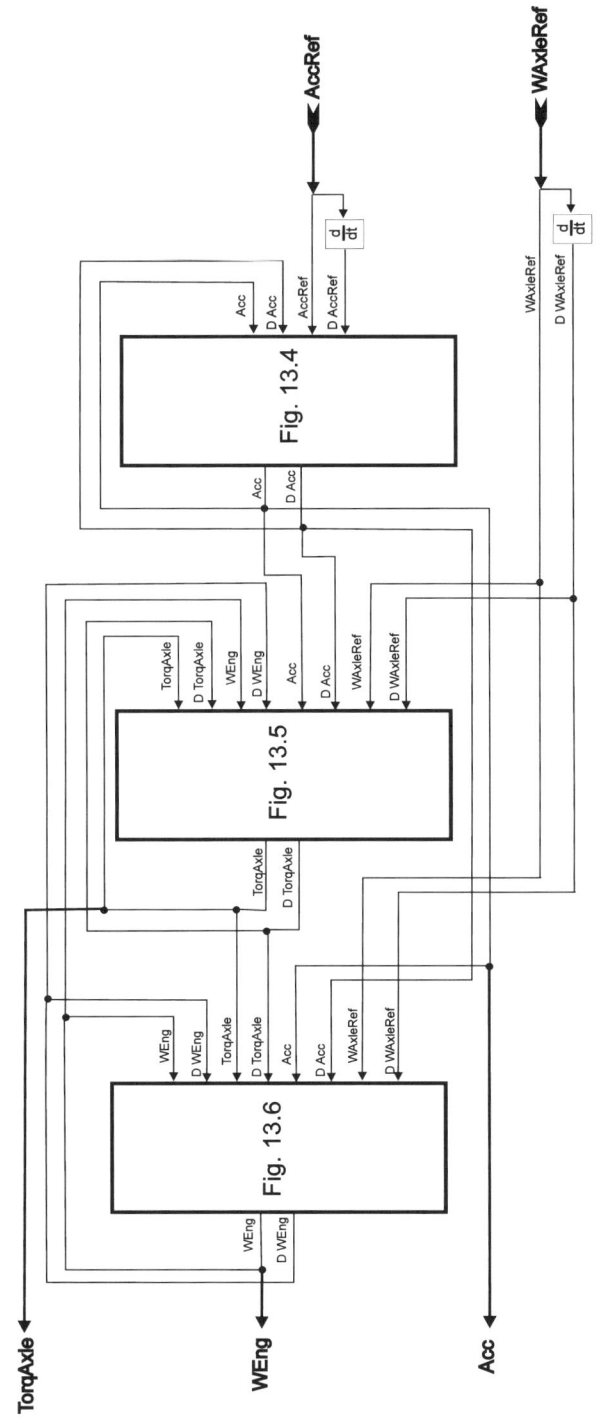

Figure 13.7 Overall neural network structure for drive train test bench plant identification.

320 DRIVE TRAIN TEST BENCH CONTROL

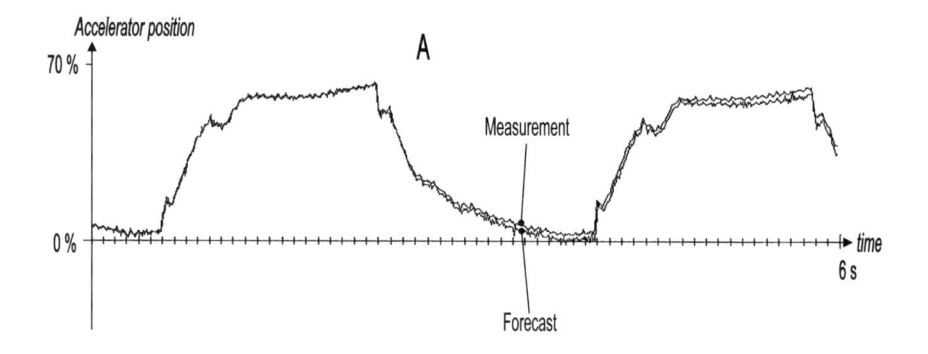

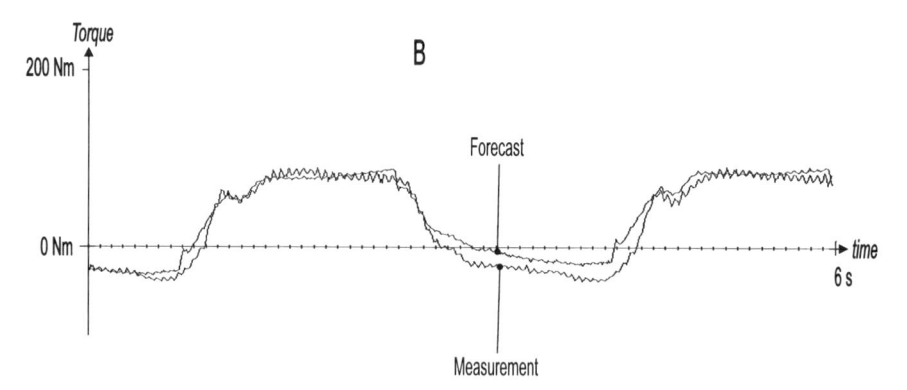

Figure 13.8 Neural-network-based model: Forecasts versus measurements.

13.2 CONTROLLER TRAINING

In face of the necessity of frequent reoptimization of the neurocontroller for new workpieces, a particular emphasis has been laid on acceptable computing times. One way to reach this is by reducing the number of inputs of the controller, and thus the number of free neural network parameters to be optimized.

The dynamics of engine torque is by far the most important aspect of the model. This is why it was decided to neglect the dynamics of angular velocities—their changes are substantially slower than those of torque and can be viewed as systematic disturbances. The behavior of the accelerator mechanism also has been neglected, for the opposite reason. Its dynamic behavior is too fast to be considered; in time units relevant for engine torque, the accelerator mechanism can be viewed as accurately following the input.

The engine model is first order in the analytic form, and second order in the form identified by neural networks. To account for effects of neglected variables, on the one hand, and nonmeasurable variables, on the other hand, the order (i.e., the assumed observability index) of the controller has been increased to three. The neglected measurable variables Acc and WEng have been treated as working point modifiers—they have been provided as additional gain-modifying controller inputs in the gain vector representation. Incomplete feedback, omitting higher derivatives of the action variable has turned out to be sufficient. The controller specification is as follows:

```
D AccRef depends_on    : ErrTorqAxle,D ErrTorqAxle,DD ErrTorqAxle
        gain_depends_on: ErrTorqAxle,D ErrTorqAxle,DD ErrTorqAxle,
                         Acc,WEng
```

which corresponds to the neural-network structure of Figure 13.9 (for notational rules, see Note 12.1).

In terms of the concepts presented in Chapter 6, this controller can be viewed as a trajectory following variant of the controller with action integration (6.91). The variable ErrTorqAxle is defined as

```
n_ErrTorqAxle = n_TorqAxleRef - n_TorqAxle;
```

The reasons for taking this structure have been that (1) the performance was satisfactory and (2) the theoretical foundations have not been elaborated to a sufficient degree at the time of the neurocontroller development. With our present knowledge, the controller with separate feedforward component modeled Eqs. (6.87) and (6.88) would probably be tried because of its good trajectory following properties. On the other hand, the relatively high number of free parameters of the controller with a separate feedforward component might prevent its use.

For neurocontroller training, external reference trajectories of rear axle torque and angular velocity must be defined. A simplified model of typical trajectories uses a step from Tr1 to Tr2 at time tTr1 for torque, and a ramp from Wr1 to Wr2 during the time interval from tWr1 to tWr2 for angular velocity:

```
n_TorqAxleRef = step(T,Tr1,Tr2,tTr1);
n_WAxleRef    = ramp(T,Wr1,Wr2,tWr1,tWr2);
```

The goal of control has been, in addition to exact trajectory following, the overshoot avoidance for steplike trajectory segments. A cost function reflecting this goal is

```
 TorqAxleRef - TorqAxle)^2 *
 ((TorqAxle - TorqAxleRef)*(Tr2 - Tr1) > 0?100:1);
```

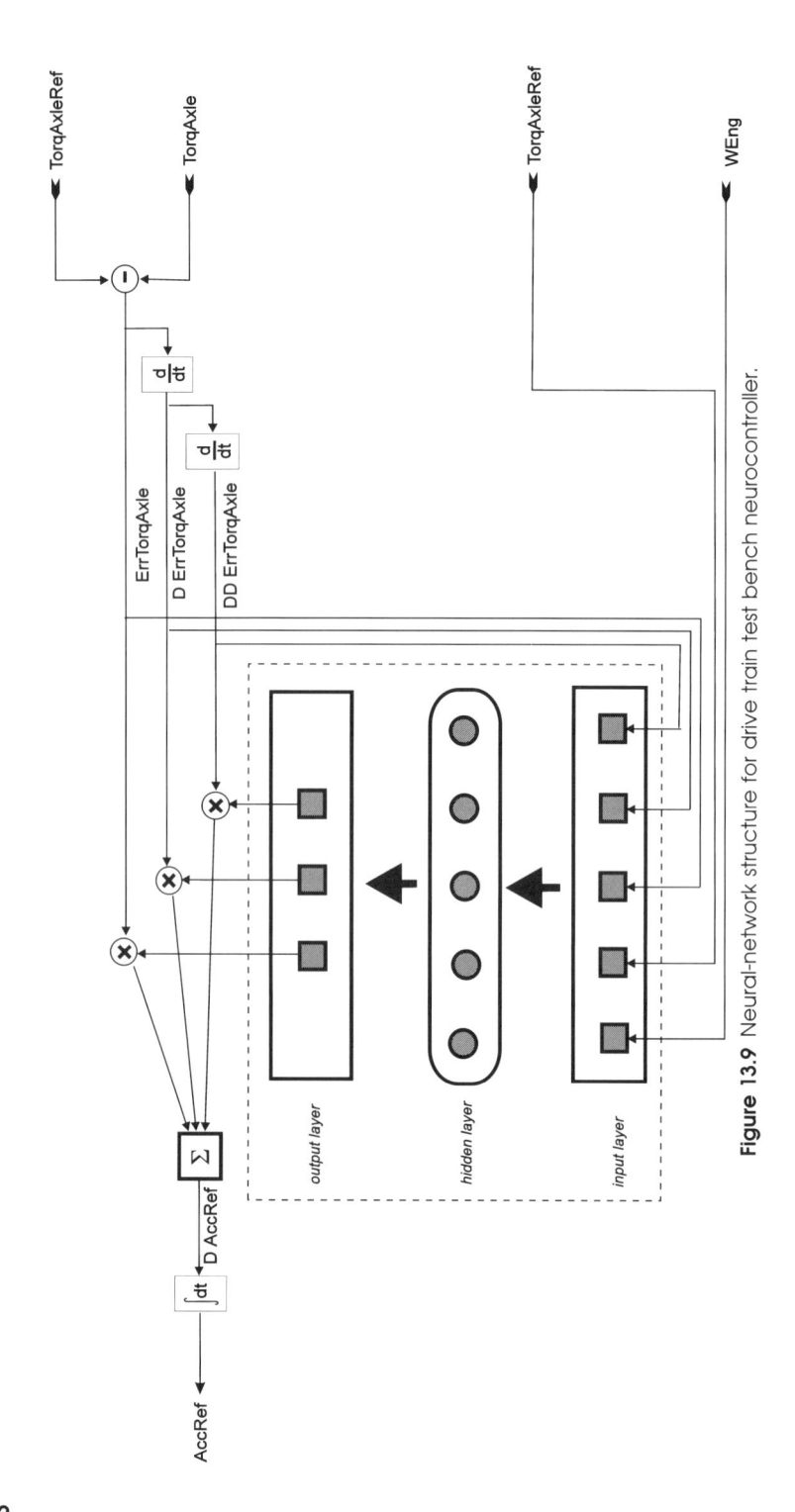

Figure 13.9 Neural-network structure for drive train test bench neurocontroller.

The first factor is the usual square error. The second factor decides whether the square error will be weighted by 100 or by one. It is a conditional expression delivering 100 if the sign of TorqAxle - TorqAxleRef is the same as the sign of Tr2 - Tr1, that is, if there is an overshoot relative to the direction of the training step.

Because the oscillating measurement errors mentioned in Section 13.1 have not been able to be removed, dedicated training for this type of disturbance became unavoidable. This could be done by a simple modification of the definition of ErrTorqAxle:

```
n_ErrTorqAxle = n_TorqAxleRef - n_TorqAxle + 10*cos(6.28*11*F);
```

With such a measurement error, an overshoot avoidance term had to be adapted so as not to penalize small "overshoots" caused by perpetual excitation of the controller by measurement oscillations:

```
(TorqAxleRef - TorqAxle)^2 *
((TorqAxle - TorqAxleRef)*(Tr2 - Tr1) > 1?100:1);
```

(The modification consists in changing >0 to >1.)

The simulation performance of a neurocontroller trained with an identified analytic model is shown in Figure 13.10. Two reference trajectory steps at time 0.1 s, a positive (*A*) and a negative one (*B*), have been used.

The controller trained with a neural-network-based plant model exhibited almost identical performance. The decision in favor of the controller using an analytic model has been justified by substantially lower computational expense for plant identification.

13.2.1 Field Test

In the field test, the neurocontroller has satisfied all specifications concerning the speed of response in a way slightly superior to that of sophisticated classical controller based on inversion of plant nonlinearities and gain scheduling. No charts from the field test are available, but the step response behavior has been very close to that of Figure 13.10; no substantial differences between the performance in simulation and on the real plant have occurred. The neurocontroller has been clearly superior in stability throughout all working points and under the given persistent disturbance. It is not the ability to cope with this particular disturbance but to be trained for an *arbitrary* disturbance that is of high value for practice. Another feature in which the neurocontroller has been clearly superior has been overshoot avoidance.

The most important achievement has been that the manpower expense for controller tuning for a new workpiece type is reduced from previously five work-days to virtually zero (except for a computational expense spent by the training algorithms).

324 DRIVE TRAIN TEST BENCH CONTROL

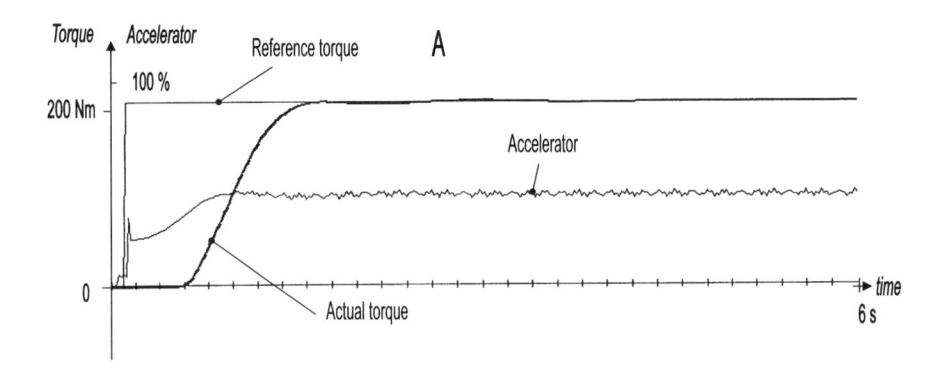

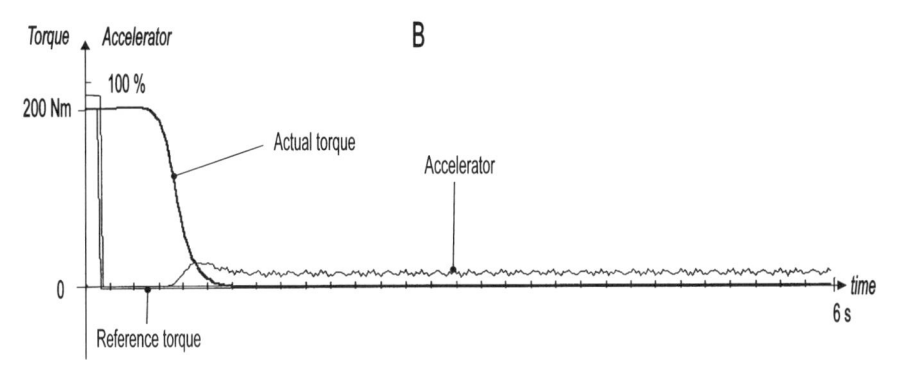

Figure 13.10 Neurocontroller behavior.

These positive results lead to the decision to implement neurocontrol in commercial drive train test benches in near future [63].

14

LATERAL CONTROL OF AN AUTONOMOUS VEHICLE

Some control applications seem to be trivial at first glance. The plant may be known to be nearly linear, its parameters easily measurable, dimension relatively low, sampling precise, and real-time requirements uncritical. Then, it is tempting to immediately exclude neurocontrol from consideration—why perform an expensive numeric optimization if the controller can be designed instantly using some classical method based on a couple of matrix operations?

However, there are cases with these characteristics for which there is still a reason to use neurocontrol, perhaps because some classical assumption is violated. Bounded control action and nonquadratic cost function are the most frequent assumption violations. Then, it is worth considering the use of neurocontrol because of the virtually unlimited freedom in formulating control goals and constraints. One such example is the lateral control of an autonomous road vehicle, which is presented in this chapter.

14.1 AN AUTONOMOUS ROAD VEHICLE

The autonomous road vehicle considered here possesses a camera-based vision system for model-based prediction and search of road markings (with algorithms described by Franke and Mehring [34] and Franke [35]). The vision system output is processed to determine aggregate values such as road width, curvature, lateral vehicle offset, camera yaw angle, and tilt angle. These values are used by various control and surveillance subsystems.

Lateral control (see Mecklenburg et al. [94]) uses the measurements of road curvature, vehicle offset, and vehicle yaw angle. (For additional neurocontrol applications in an autonomous vehicle not addressed here, see Winter [159] and Wagner [141].) The goal of lateral control is to keep the vehicle in the middle of the lane, avoiding fast lateral acceleration (for comfort reasons) as long as the lateral offset of the vehicle is not critical.

14.2 MODELING LATERAL CONTROL

A simplified model of lateral position of a vehicle on a straight road is

$$\dot{\beta} = \left(-2\frac{k}{mv}\right)\beta + \left(\frac{v}{a} - \frac{k}{mv}\right)\lambda$$
$$\dot{y} = v(\psi + \beta) \qquad (14.1)$$
$$\dot{\psi} = \frac{v}{a}\lambda$$

with state variables being the offset from the lane center (y), the yaw angle (ψ), and the drift angle (β). The inputs to the plant defined by these are the steering angle (λ) and vehicle velocity (v). The former input is simultaneously the control action. The latter input is, formally, a measurable disturbance. The constants used in the model are: m (vehicle mass), a (wheel base), and k (lateral friction coefficient).

In TREG formalism, the model is the following:

```
n_beta   = beta + DT * (-2*k/(m*vel)*beta
                +(vel/a-k/(m*vel))*lambda);
n_yps    = yps  + DT * vel*(psi+beta);
n_psi    = psi  + DT * lambda*vel/a;
```

This model is nonlinear only in the vehicle velocity `vel`. This velocity is changing slowly in comparison with other variables involved. Therefore, the plant model can be viewed as a linear model parametrized by the velocity.

The values of constants m and a are known with high reliability, and a good estimate of k can be done. Thus, the analytic plant model can be considered sufficiently accurate for the given control application. This model is used directly, without measurement-based model identification.

14.3 CONTROLLER TRAINING

The approximately linear plant model presented in Section 14.2 would not itself motivate the use of neurocontrol sufficiently—it suggests that some standard linear control design method might be successfully applied. Serious arguments in favor of a neurocontrol solution result from the particular control goal. The primary control objective is to keep the car in the lane center. However, it is further required that large offsets (over about 40 cm) are left as quickly as possible (safety requirement), and smaller offsets (below approximately 40 cm) are tolerated in favor of keeping steering movements slow (comfort requirement).

In addition, for a given offset, the lateral acceleration should remain approximately constant in a wide speed range—the control should not become nonsmooth at higher speeds. The nonlinearity of the last requirement makes the task a nonlinear control task, but the fact that the only nonlinear model variable is the slowly changing vehicle velocity would suggest the use of velocity-dependent gain scheduling.

It is the conflict between the first two requirements that constitutes the main difficulty of the lateral control problem. The attempts to treat both situations as different state contexts with different controllers have lead to nonsmooth transitions between both contexts. Smoothing the solution would certainly be feasible, but the analytic expense for this makes the task surpass the analytic complexity boundary beyond which neurocontrol is superior.

Two of three state variables, the offset yps and the yaw angle psi, are delivered by the vision system. This high proportion of measurable variables suggests that output feedback might be sufficient. Additionally, the open loop itself is nearly stable—a nonsteered vehicle does not tend to amplify lateral disturbances. So, the argument in favor of complete state feedback, the stabilization of an unstable plant, is not particularly relevant here.

On the other hand, it is important to provide for robustness against slowly changing systematic disturbances such as lateral wind or transversal slope of the road. The simplest means to reach such robustness use the integrated control error equations (6.89) and (6.90) or integrating the control action (6.91). Integrating the control action is preferable because of the necessity of resetting the integral with error integration (see Section 6.1).

```
D lambda depends_on     : lambda, yps, psi
        gain_depends_on: lambda, yps, psi, vel
```

The nonlinear gains of the controller are modified by the vehicle velocity vel.

Because the closed-loop equilibrium point is always at the zero state vector, there is no reason to consider the use of a separate feedforward component for equilibrium point definition. The neural-network structure of this neurocontroller is shown in Figure 14.1 (for notational rules, see Note 12.1).

The cost function consists of two terms: a term minimizing the offset yps (the security component) and a term minimizing the lateral acceleration DD yps (the comfort component). Lateral acceleration is proportional to (D LAMBDA)*vel; at a high speed, the effect of the same steering action is larger than at a low speed. The offset minimization should apply overproportionally to large offsets in comparison with small ones. This is accounted for by

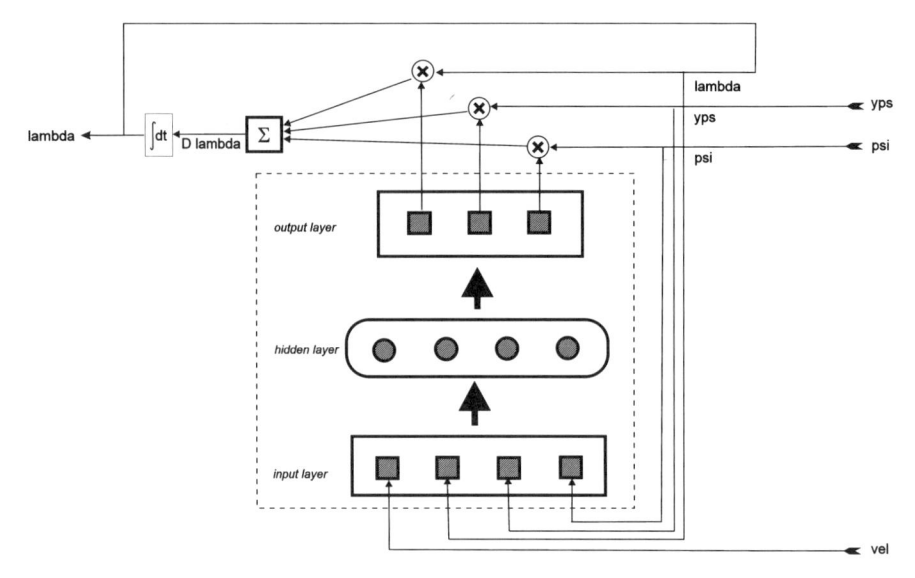

Figure 14.1 Neural-network structure for lateral control neurocontroller.

different powers of both terms. The priorities of both goals have been weighted in the following way:

```
ps^6 + 2.5*((D lambda)*vel)^2 + (yps*yps0 < 0.0?5000:0)*yps^2
```

The additional conditional term for yps such that its sign is opposite to the sign of the initial offset yps0 of each training example represents a constraint penalty for overshooting the lane center.

14.4 RESULTS

The neurocontroller has been trained by six training examples with initial offsets of 1 m and 0.3 m and velocities 10, 15, and 20 m/s. It has been optimized over a horizon of 200 time periods, 80 ms each. Simulation results with this controller and initial offsets of 1 m and 0.4 m are shown in Figure 14.2. The qualitative difference of both trajectories can be observed. For an initial value of 1 m, the offsets comes down to values near zero in a shorter time than for an initial value of 0.4 m. This behavior cannot be reached by classical linear controllers because their action is proportional to the state (Fig. 14.3).

14.4.1 Field Test

The controllers have been tested in the autonomous road vehicle on a public highway between Stuttgart and Ulm. An objective evaluation of controllers in

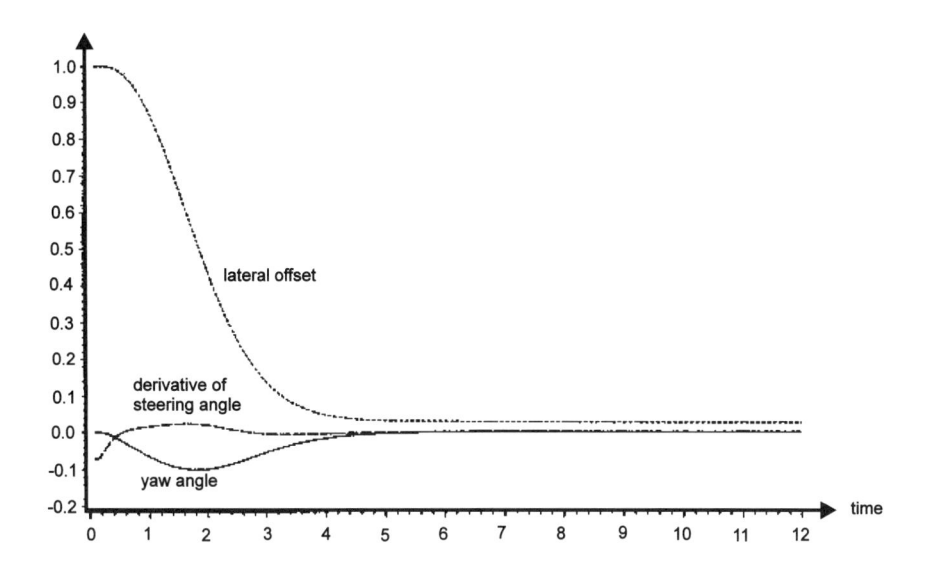

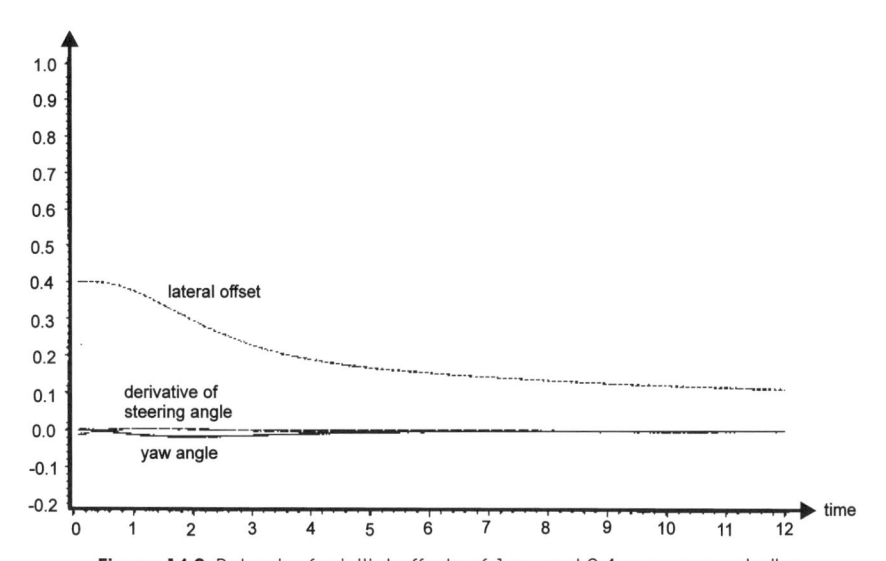

Figure 14.2 Behavior for initial offsets of 1-m and 0.4-m neurocontroller.

the field has been more difficult than in simulation experiments, because of lack of directly observable optimality criterion, and the impossibility of guaranteeing reproducible road and traffic conditions in comparisons. However, the conclusions made from the simulations could be confirmed qualitatively.

The experience with long-distance driving (up to 32 km without interruption by the operator) has been reported by Franke et al. [36].

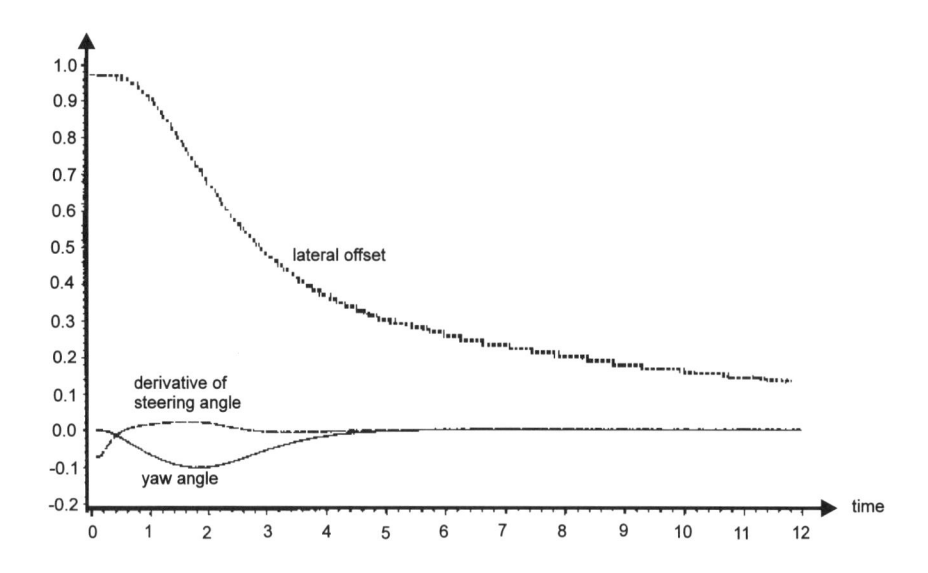

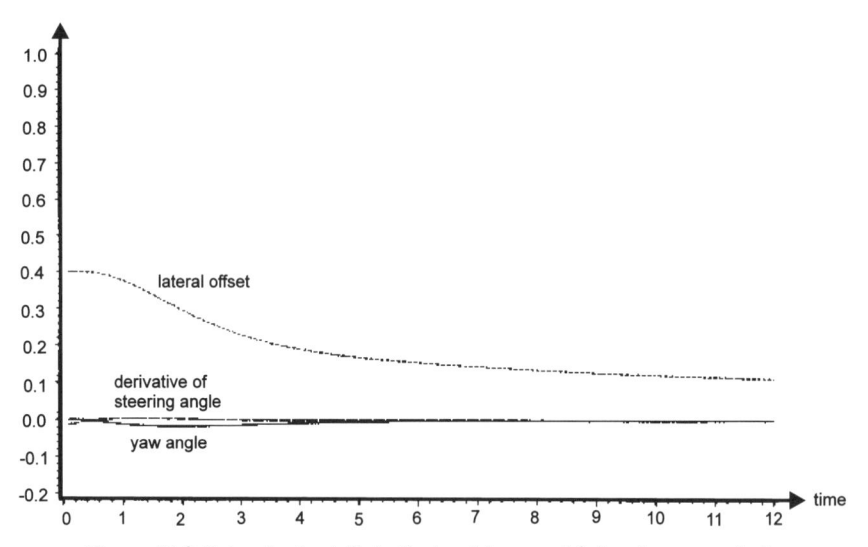

Figure 14.3 Behavior for initial offsets of 1-m and 0.4-m linear controller.

15

BIOLOGICAL WASTEWATER TREATMENT CONTROL

The most complex case study in this book is concerned with neurocontrol of a wastewater treatment plant. This application has been investigated in the joint research project "Innovative Automation Technology in the Waste Water Treatment" ("Innovative Automatisierungstechnik in der Abwasserwirtschaft"—IAA) of AEG Automation Department, AEG-AAT, ISA—Institut für Siedlungs- und Abwasserwirtschaft, Aachen, Ruhr Verband, Essen, and Daimler-Benz Research. The neurocontrol application has been developed by Daimler-Benz Research. Substantial contributions to the neurocontrol solution have been made by Fetzer [30].

Wastewater treatment control is a task with many difficult aspects. The most important difficulties involved are the following:

1. The underlying processes are of a biochemical nature. Although a (computationally very expensive) structural mathematical model has been formulated, its numerous parameters have values about which there is still no complete agreement by domain specialists.
2. The model is too complex to provide a fair chance for analytic control design to succeed.
3. The model is computationally too expensive to be used in a numeric controller optimization—a day of plant operation (approximately 700 sampling periods) takes about an hour of simulation, if numeric stability of computation is to be retained.

332 BIOLOGICAL WASTEWATER TREATMENT CONTROL

4. The control goal consists of (a) satisfying legal constraints for wastewater pollution and (b) cost terms. It cannot be expressed in simple terms such as plant stabilization or following reference trajectories.

Note 15.1 Using neural networks in wastewater treatment control has also been reported by Werbos, McAvoy, and Su [150]. The approach applied there is that of classical predictive control with a neural-network-based plant model.

15.1 BIOLOGICAL WASTEWATER TREATMENT PROCESS

Wastewater treatment is a process of elimination of a broad spectrum of contaminants from wastewater. The contaminants are of various types: soluble and solid, organic and inorganic. Solid contaminants can frequently be extracted mechanically by filtering or sedimentation. Soluble contaminants must be transformed to solid ones. This can be done by adding chemicals that react with particular contaminants to build a solid sediment. Under relatively simple conditions concerning physical parameters such as minimum temperature or chemical parameters such as bounds on pH value of the treated wastewater, this process can be described with sufficient accuracy by reaction equations of the type

$$\dot{c}_o = k \prod_i c_i \qquad (15.1)$$

stating that the concentration of the reaction output compound changes proportionally to the product of concentrations of reaction input compounds.

Other soluble contaminants can be made solid by a biologic process. Certain types of microorganisms feed on certain contaminants, thus absorbing the contaminant into their bodies. Their bodies constitute a solid matter, biologic sludge, that can be removed by sedimentation. A trivial analogy to chemical treatment would be to add the mass of living microorganisms in some appropriate proportion to the contaminant concentration, but this is neither practically feasible nor economical. In practice, the natural growth of sludge is used, and this sludge is kept in the process by feeding it physically from the process output back to the process input. This procedure is called an *activated sludge process* [e.g., 122].

This procedure is applied primarily to three types of contaminants:

1. Organic carbon compounds called *substrate*
2. Nitrites and nitrates
3. Ammonia

Two groups of bacteria are involved:

1. Heterotrophic bacteria consuming organic matter
2. Autotrophic bacteria consuming anorganic matter

The first of the biochemical processes related to the contaminants listed above is the substrate consumption by oxidation dissolved in the wastewater. The concrete reaction equations involved depend on the type of carbon compound, and there is a vast number of various organic carbon compounds that can be contained in wastewater. A very schematic equation is

$$C + O_2 \rightarrow CO_2 \tag{15.2}$$

Under the condition of free oxygen absence, some heterotrophic bacteria are capable of using nitrate oxygen for oxidation. The reaction involved is

$$4NO_3^- + 4H^+ + 5C \rightarrow 2N_2 + 2H_2O + 5CO_2 \tag{15.3}$$

This process is called *denitrification*.

The process involving autotrophic bacteria consists of two chained subprocesses. In the first subprocess, ammonia is transformed into nitrite. In the subsequent one, nitrite is transformed into nitrate. Both subprocesses require free oxygen. The complete process transforming ammonia to nitrate is called *nitrification* and is described by the equation

$$NH_4^+ + 2O_2 \rightarrow NO_3^- + H_2O + 2H^+ \tag{15.4}$$

The rates of these biochemical processes are governed by rules formally analogous to the rules valid for pure chemical processes. However, reaction rates of individual biochemical subprocesses are nonlinearly dependent on the plant state and environmental variables such as temperature. Frequently, the dependence is nonmonotonic, some processes take place only in a narrow band of concentrations.

Bacteria kept in the process are mortal: the die off, and their bodies are decomposed by hydrolysis, among other components, into carbon substrate and ammonia. This makes obvious that the overall biochemical process including (1) carbon substrate transformation with oxidation alternatively by free oxygen and nitrate oxygen, (2) bacterial mass hydrolysis, and (3) ammonia transformation into nitrate is obviously cyclic (Fig. 15.1). As ammonia is transformed into nitrate, the nitrate can be eliminated into atmospheric nitrogen with the help of a sufficient concentration of the carbon substrate, and the carbon substrate can be eliminated by a large mass of bacteria, which raises the ammonia.

Obviously, activated sludge process consists of a set of simultaneous and cyclic subprocesses. It uses stream feedback, called *recirculation*, organized so that the concentrations of individual components decrease toward the plant exit. However, the many different working points with different rates of elimination of individual contaminants make it difficult to find an optimum for a given pollution structure.

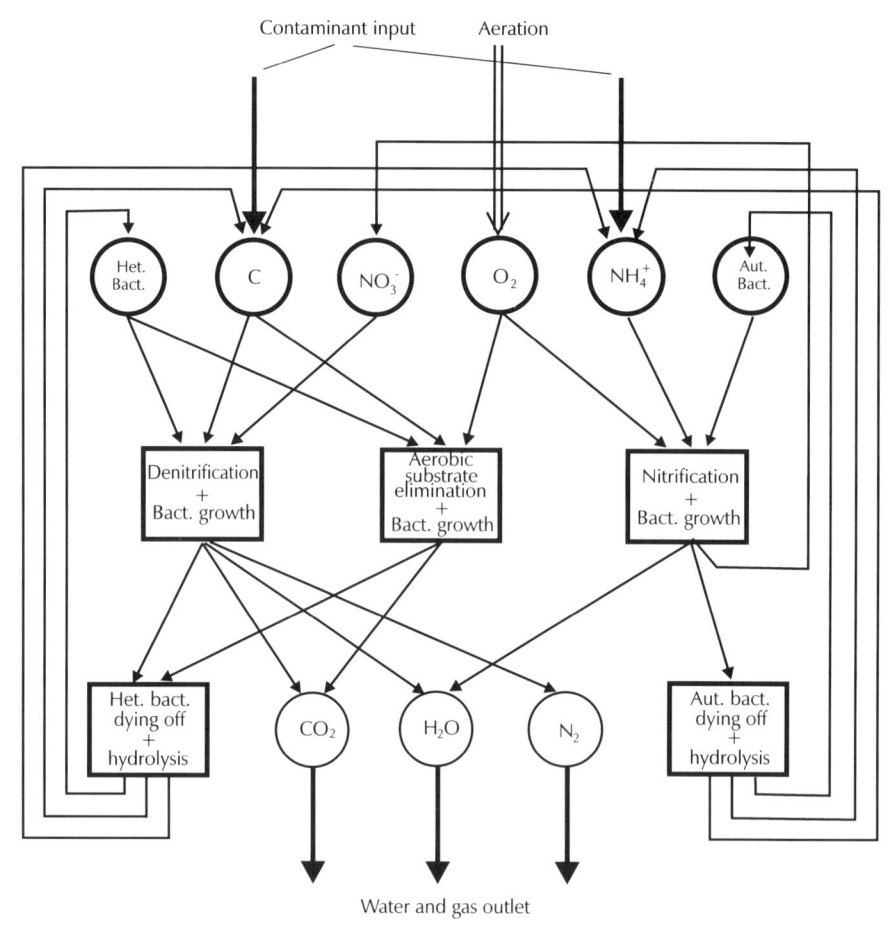

Figure 15.1 Scheme of activated sludge process.

There is a limited number of action variables. The three most important are the following:

1. Rate of air supply into the activated sludge basins by aeration devices.
2. Recirculation stream, that is, the rate at which wastewater is pumped from the outlet of the activated sludge basin (or chain thereof) back to the inlet.
3. Rate of sludge removal from the process.

The sludge increase and decrease rates are too low for the sludge concentration to be effectively scheduled with a daily periodicity to account for a periodically changing contaminant concentration. This is why the last action variable represents, in fact, no genuine degree of freedom. In the long run, the sludge concentration cannot but be kept constant; the varying removal rate can only be a consequence of varying the working point of the overall activated sludge process.

Equally modest is the number of available online measurements. Only the measurement of concentrations of simple anorganic components is relatively simple. This includes the measurements of these concentrations:

Free oxygen concentration
Nitrate concentration
Ammonia concentration

The direct substrate concentration is difficult to measure for two reasons. First, there are very different substrate components. This problem can be solved by defining a normed magnitude, the *biological oxygen demand*, and measuring it by experiment, that is, by letting a defined quantity of microorganisms consume it. It is clear that this measurement method is not easy to use under real-time conditions. There are also indirect, less precise, or more expensive methods (e.g., based on gas outlet measurements [135]). The second problem is even more serious. Once mixed with activated sludge, the substrate is difficult to discern from the sludge itself. This is why, at present, it is realistic to measure the substrate concentration only at the wastewater treatment plant inlet.

Even more difficult is the measurement of the active sludge concentration. Without laboratory tests, it is impossible to separate the living part of sludge from the dead part. For routine processing in current wastewater treatment plants, it can be assumed that the quantity of active sludge is not measurable.

15.2 STRUCTURE OF A WASTEWATER TREATMENT PLANT

This case study is concerned with a wastewater treatment plant involving biochemical processes (i.e., neglecting rough filtering devices at the inlet) as given in Figure 15.2. Important components are the four aeration basins for the activated sludge procedure and the sludge sedimentation basin.

There are two feedback streams:

1. Sludge feedback from the sludge sedimentation basin to the inlet of the first aeration basin
2. Recirculation feedback from the outlet of the first aeration basin

All four aeration basins are identical. Each receives an independent aeration input. The first basin is used as a dedicated anaerobic basin for the denitrification process. This is reached by setting its aeration to zero. The remaining three basins can be used in various modes. In particular, the second is used alternately in aerobic and anaerobic modes depending on the individual concentrations of ammonia and nitrate; the basin is used for elimination of both contaminants whose concentration is more critical.

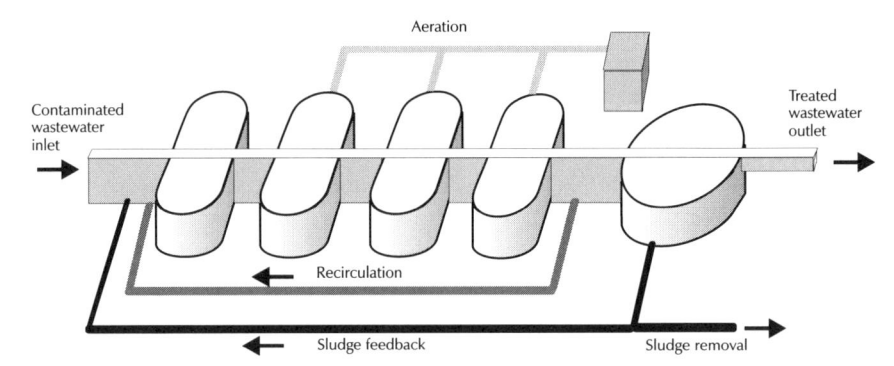

Figure 15.2 Structure of the wastewater treatment plant used in the case study.

15.2.1 Control Goals

The construction of this wastewater treatment plant is such that the elimination of organic carbon contaminants is assumed to be always sufficient. By contrast, nitrogen components, that is, ammonia and nitrate, are viewed as critical. There are two legal constraints. The maximum ammonia concentration at the fourth aeration basin outlet is 6 g/m^3. The maximum of the combined ammonia and nitrate concentration is set to 18 g/m^3. The key goal of aeration control is that both constraints are always satisfied.

Another goal concerns financial expenses. Various types of expenses can be assigned to control actions but to different extents. In the version presented here, only the costs directly attributable to the aeration have been considered. The linear dependence of the costs on the aeration volume is a good approximation.

15.2.2 Indirect Neurocontrol Design

Wastewater treatment is a process with critical reliability requirements. There are legal sanctions for every deliberate act leading to the deterioration of the contaminant output. Under such conditions, even testing new controllers is a risky enterprise; every new controller may potentially lead to such deterioration in some working point.

This leads to the requirement that the neurocontroller performance be tested by extensive simulations. The simulation model used is that of International Association for Water Quality [see 49], referred to here as the IAWQ-model.

The sensitivity of wastewater treatment has one more consequence for neurocontroller design. In Section 5.6, we found that, with nonlinear plants, the sampled data must be representative for the complete state space. Theoretically such sampling could be reached by deliberate excitation of the plant. However, deliberate excitation of the wastewater treatment plant is a criminal act (see above), since the generation of high contaminant concentrations is the result. There are two alternatives:

1. To wait for such data to arise by spontaneous variations in the contaminant inlet (and also deficiencies of current control systems).
2. To substitute such critical data by their approximations with the help of a model.

To make neurocontroller design of wastewater treatment plants acceptable in industrial setting, only the second alternative is sufficiently flexible.

There is another reason for using a simulation model for generating data. The measurement equipment in most wastewater treatment plants is limited. As stated in Section 15.1, it is possible to measure chemical concentrations such as the oxygen or nitrate concentration routinely, but it usually is not done in all aeration basins. Other relevant quantities, such as the substrate concentration or current quantity of microorganisms, cannot be measured at all.

To assess the loss of information about plant dynamics in simplified models omitting certain state variables, *all* data, in the first project phase, have been generated by the model for potential use for plant identification. After this assessment, it will be decided to which extent the complete data (which may not necessarily correspond with the reality) generated by the model can be substituted by the incomplete (but real) measured data.

If an analytic model is available, why not use it for neurocontroller training directly? There are two reasons: First, the model used is very complex. As stated at the beginning of this chapter, execution time for a simulation of one day on the plant takes, at best, several minutes. For global optimization, about a million of such simulations are necessary. This is why optimization using models with execution times over a second is definitely infeasible. Second, although the IAWQ-model captures all relevant processes in a qualitatively satisfactory manner, its quantitative correspondence with the data of a particular plant is still not satisfactory. It has to be extensively calibrated. It is desirable that in the future, real measurements are taken at least for usual, uncritical working points of the plant. This is why the following procedure has been adopted:

1. Generation of measurement series by the IAWQ model.
2. Identification of a neural-network-based plant model using these data.
3. Neurocontroller training.
4. Embedding the neurocontroller in the IAWQ model to perform extensive tests.
5. Implementing the neurocontroller on the real plant.

Note 15.2 This indirect identification procedure can be used for the reasons given above. It makes it possible for all state variables contained in the IAWQ model to be treated as measurable. In the general case, the use of such an indirect procedure might not be justified. Then, the model has to be identified by real data, and contain either only strictly measurable variables or hidden, nonmeasurable states. This problem is addressed in Section 15.4.

15.3 PLANT IDENTIFICATION

The available nonredundant action variables of Section 15.1, that is, the rate of air supply and the recirculation stream, suggest that the most important processes that can be influenced by neurocontrollers are those taking place in the aeration basins (together with the anaerobic basin of identical structure). It is also the aeration basins in which the most complex biochemical processes take place. This is why plant identification has been limited to the aeration basins.

The IAWQ model works with 13 components. After a brief analysis of the process structure, it has been assumed that a large part of dynamics can be captured by considering six of these components:

Heterotrophic bacteria concentration (XBH)

Autotrophic bacteria concentration (XBA)

Biologic oxygen demand (BOD)—a sum of dissolved and solid substrate normalized to the oxygen concentration necessary to its biologic transformation

Free oxygen concentration (SO)

Ammonia concentration (SNH)

Nitrate and nitrite concentration (SNO)

It has been assumed that these (measurable, with help of the IAWQ model; see Note 15.2) variables characterize a complete state of a particular basin. That is, in terms of input–output models, an observability index of one has been postulated.

The state variables listed appear on each of four aeration basins as well as in the input into the identified part of the plant. Additional inputs into the plants are the flow through the plant i_Q and the air volumes l1_SO, l2_SO, l3_SO, and l4_SO pumped by aeration devices into the individual basins.

The plant model specification is the following:

```
output
o1_BOD     eq_nr:   1
o1_XBH     eq_nr:   2
o1_XBA     eq_nr:   3
o1_SO      eq_nr:   4
o1_SNO     eq_nr:   5
o1_SNH     eq_nr:   6
   .
   .
   .
o4_BOD     eq_nr:   1
o4_XBH     eq_nr:   2
o4_XBA     eq_nr:   3
o4_SO      eq_nr:   4
```

```
o4_SNO      eq_nr:   5
o4_SNH      eq_nr:   6

input
i_BOD
i_XBH
i_XBA
i_SO
i_SNO
i_SNH

i_Q
l1_SO
l2_SO
l3_SO
l4_SO

dependences

D o1_BOD depends_on      : o1_BOD,o1_XBH,o1_XBA,o1_SO,o1_SNO
        gain_depends_on  : o1_BOD,o1_XBH,o1_XBA,o1_SO,o1_SNO
D o1_XBH depends_on      : o1_BOD,o1_XBH,o1_SO,o1_SNO
        gain_depends_on  : o1_BOD,o1_XBH,o1_SO,o1_SNO
D o1_XBA depends_on      : o1_XBA,o1_SO,o1_SNH
        gain_depends_on  : o1_XBA,o1_SO,o1_SNH
D o1_SO  depends_on      : o1_BOD,o1_XBH,o1_XBA,o1_SO,o1_SNH,l1_SO
        gain_depends_on  : o1_BOD,o1_XBH,o1_XBA,o1_SO,o1_SNH,l1_SO
D o1_SNO depends_on      : o1_BOD,o1_XBH,o1_XBA,o1_SO,o1_SNO,o1_SNH
        gain_depends_on  : o1_BOD,o1_XBH,o1_XBA,o1_SO,o1_SNO,o1_SNH
D o1_SNH depends_on      : o1_XBH,o1_XBA,o1_SO,o1_SNH
        gain_depends_on  : o1_XBH,o1_XBA,o1_SO,o1_SNH

   .
   .
   .

D o4_BOD depends_on      : o4_BOD,o4_XBH,o4_XBA,o4_SO,o4_SNO
        gain_depends_on  : o4_BOD,o4_XBH,o4_XBA,o4_SO,o4_SNO
D o4_XBH depends_on      : o4_BOD,o4_XBH,o4_SO,o4_SNO
        gain_depends_on  : o4_BOD,o4_XBH,o4_SO,o4_SNO
D o4_XBA depends_on      : o4_XBA,o4_SO,o4_SNH
        gain_depends_on  : o4_XBA,o4_SO,o4_SNH
D o4_SO  depends_on      : o4_BOD,o4_XBH,o4_XBA,o4_SO,o4_SNH,l4_SO
        gain_depends_on  : o4_BOD,o4_XBH,o4_XBA,o4_SO,o4_SNH,l4_SO
D o4_SNO depends_on      : o4_BOD,o4_XBH,o4_XBA,o4_SO,o4_SNO,o4_SNH
        gain_depends_on  : o4_BOD,o4_XBH,o4_XBA,o4_SO,o4_SNO,o4_SNH
D o4_SNH depends_on      : o4_XBH,o4_XBA,o4_SO,o4_SNH
        gain_depends_on  : o4_XBH,o4_XBA,o4_SO,o4_SNH
```

340 BIOLOGICAL WASTEWATER TREATMENT CONTROL

To reduce the number of model parameters, the dependencies have been pruned, taking into account the assumed structure of individual processes. This pruning is modest—unfortunately, most state variables are directly influenced by most of the remaining ones. It can be assumed that the results would not be very different if a complete model were taken.

However, another property of the plant has been exploited to reduce the number of plant model parameters. Since all four basins considered are identical, they are represented by identical networks. This is expressed by the attributes eq_nr of output variables. The state equations of output variables with identical values of this attribute are represented by the same network.

One particular process has been represented explicitly. It is the process of concentration change by the flow through the chain of basins. Under the assumption of homogeneous concentrations in every basin, this dependence is completely determined by the conservation of flow. If concentration c in a basin of volume V changes during time Δ_t by Δ_c through biochemical processes, and the inlet flow of magnitude q exhibits concentration d_i, the concentration dynamics can be described by

$$c_{t+1} = c_t + \Delta_c + \Delta_t \frac{q}{V} d_i - \Delta_t \frac{q}{V} c_t = c_t + \Delta_c - \Delta_t \frac{q}{V}(c_t - d_i) \qquad (15.5)$$

Suppose that the concentration value taking into account the change by biochemical processes, $c_t + \Delta_c$, is delivered by the neural network based model. This value is to be diminished by $\Delta_t(q/V)(c_t - d_t)$. For all relevant variables, this is represented in the following listing:

```
o1_BOD = o1_BOD - (o1_BOD - i_BOD) * i_Q * DT/BB_VOLUME;
o1_XBH = o1_XBH - (o1_XBH - i_XBH) * i_Q * DT/BB_VOLUME;
o1_XBA = o1_XBA - (o1_XBA - i_XBA) * i_Q * DT/BB_VOLUME;
o1_SO  = o1_SO  - (o1_SO  - i_SO ) * i_Q * DT/BB_VOLUME;
o1_SNO = o1_SNO - (o1_SNO - i_SNO) * i_Q * DT/BB_VOLUME;
o1_SNH = o1_SNH - (o1_SNH - i_SNH) * i_Q * DT/BB_VOLUME;

o2_BOD = o2_BOD - (o2_BOD -o1_BOD) * i_Q * DT/BB_VOLUME;
o2_XBH = o2_XBH - (o2_XBH -o1_XBH) * i_Q * DT/BB_VOLUME;
o2_XBA = o2_XBA - (o2_XBA -o1_XBA) * i_Q * DT/BB_VOLUME;
o2_SO  = o2_SO  - (o2_SO  -o1_SO ) * i_Q * DT/BB_VOLUME;
o2_SNO = o2_SNO - (o2_SNO -o1_SNO) * i_Q * DT/BB_VOLUME;
o2_SNH = o2_SNH - (o2_SNH -o1_SNH) * i_Q * DT/BB_VOLUME;
  .
  .
  .
```

The basin volume BB_VOLUME is a constant. As is obvious from this listing, chaining order of the basins is taken into account. The neural network structure of the plant model is that of Figures 15.3 to 15.5 (for notational rules, see Note 12.1). The scheme of Figure 15.3 represents a model for variable o1_BOD

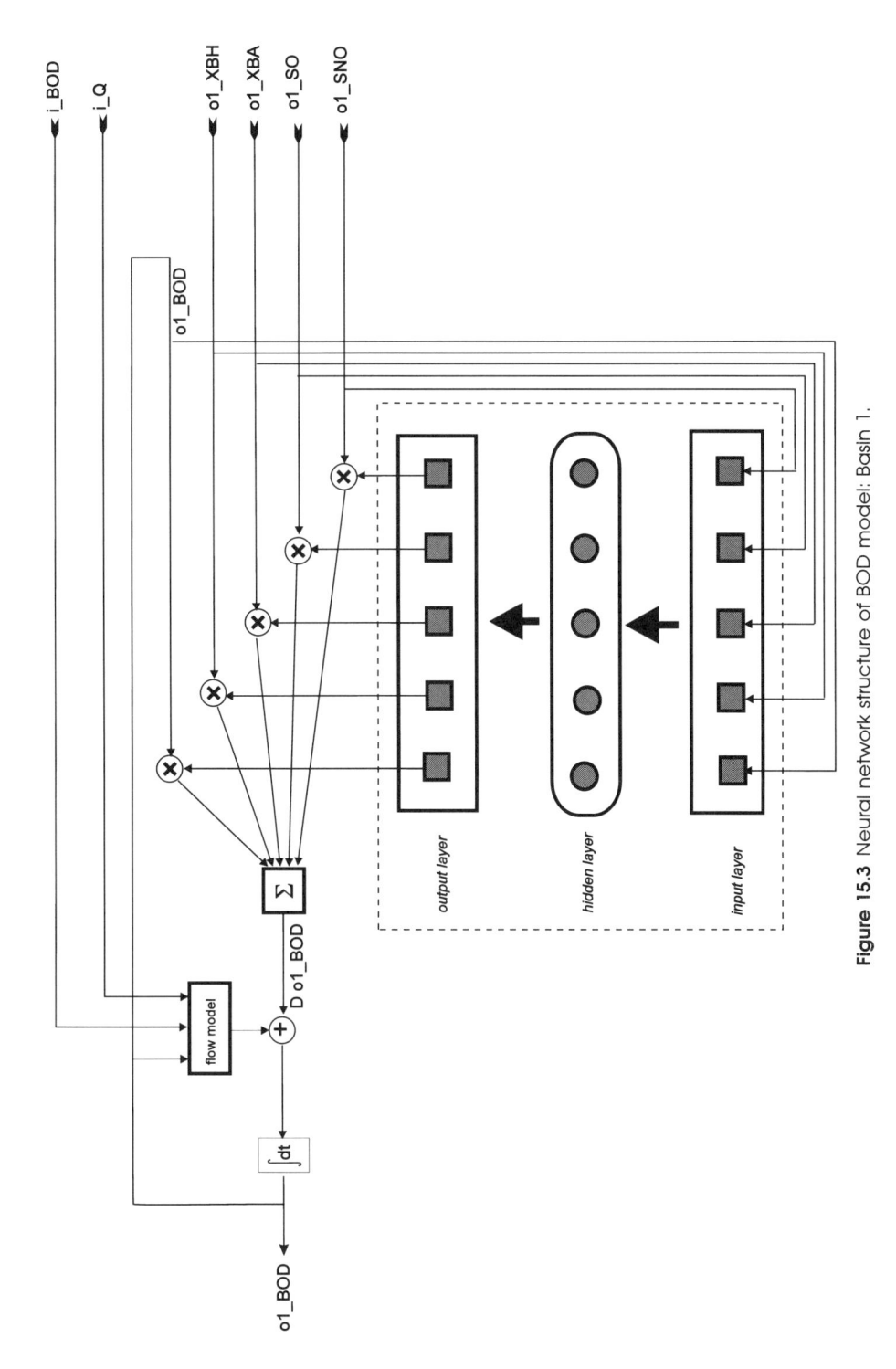

Figure 15.3 Neural network structure of BOD model: Basin 1.

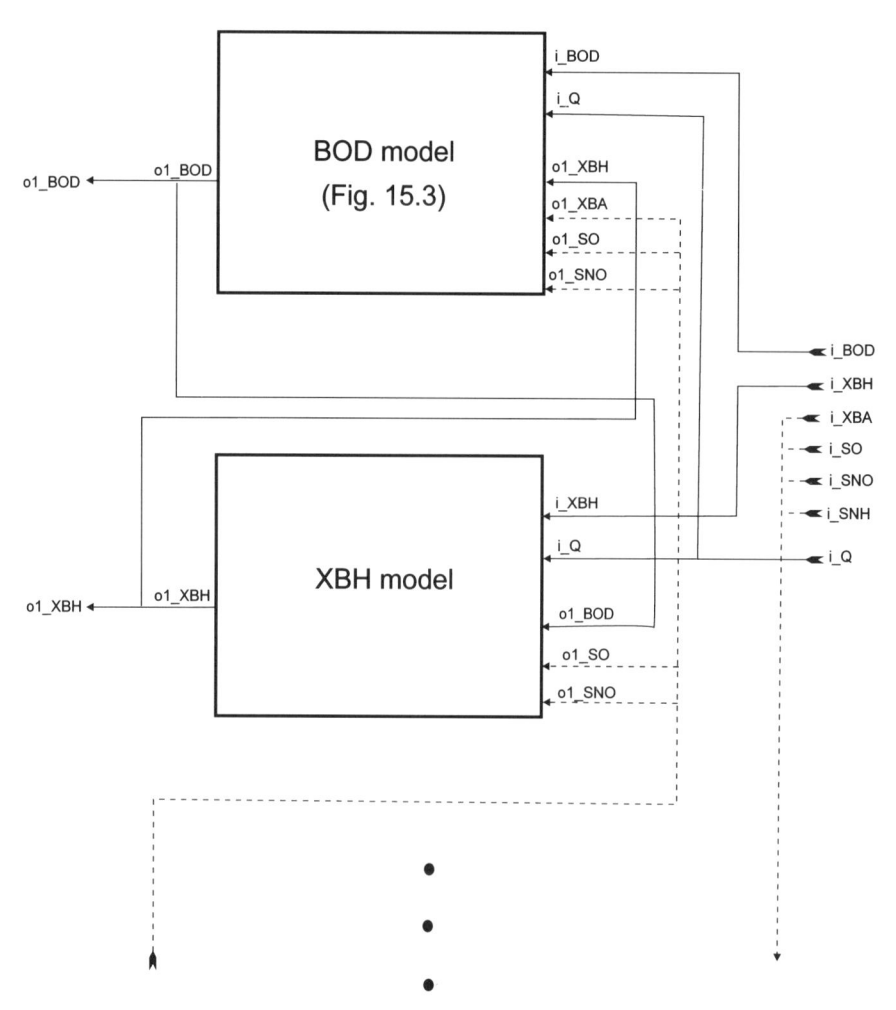

Figure 15.4 Neural-network structure for the model of Basin 1.

of Basin 1. The remaining five variables are modeled in an analogous way. The flow model block refers to the last term of Eq. (15.5), that is, $-(q/V)(c_t - d_i)$. For Basin 1 and variable `o1_BOD`, this term amounts to `-(Volume/i_Q)*(o1_BOD-i_BOD)`.

The individual neural network models are connected by mutual feedbacks (Fig. 15.4) to represent a model of one aeration basin.

The overall scheme of the plant model is presented in Figure 15.5. The inputs into the first basin are identical with the external input into the aeration basin chain. The inputs into the ith basin ($i > 1$) are the outputs from the ith basin (the inputs into the model of Basin 2 are `o1_BOD`, `o1_XBH`, etc.). The total wastewater flow is conserved throughout all aeration basins and thus

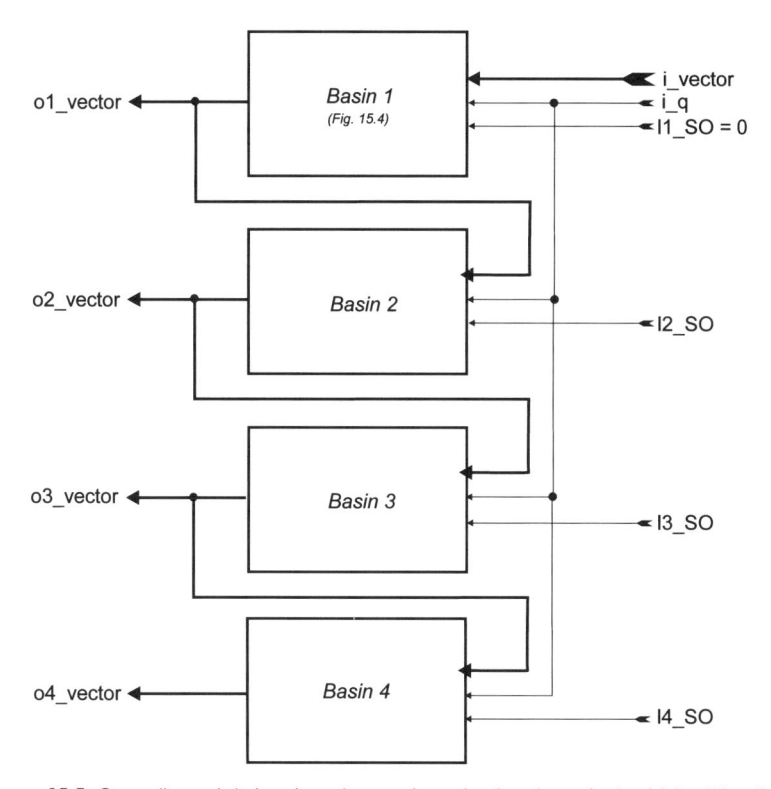

Figure 15.5 Overall model structure for wastewater treatment plant identification.

variable i_Q enters the models of all basins (instead of specific flows o1_Q, o2_Q, etc.). For the models of all four aeration basins to remain isomorphic (they are represented by a single set of neural networks with a single set of coefficients), they must all have an aeration input. Basin 1 has been assigned a fixed anaerobic regime, and so its aeration input is set constantly to zero.

Three series of sampled measurements have been collected. Each series consists of one day's measurements, that is, 720, two-minute sampling periods. One series corresponds to a typical observed daily schedule of contaminant concentrations at the inlet, run under an existing classical control system. In the remaining two series, perturbations of the aeration input and inlet flow have been introduced.

The results of identification for one of the perturbed time series are given in Figures 15.6 through 15.11. Throughout this chapter, the units are in milligrams per liter if not specified otherwise. Because the substrate is heterogeneous, it is normed to an equivalent measure expressed by the term *Biological Oxygen Demand (BOD)*. The charts show the trajectories of measured and forecast concentrations in the last aeration basin. The only exception is free oxygen, which has been almost constant in the fourth basin and is illustrated by the trajectory of the second basin. Bold lines represent

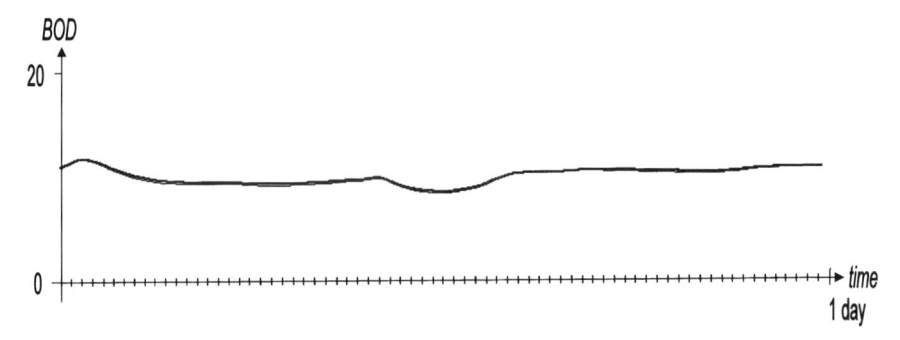

Figure 15.6 Trajectories of substrate concentration (BOD).

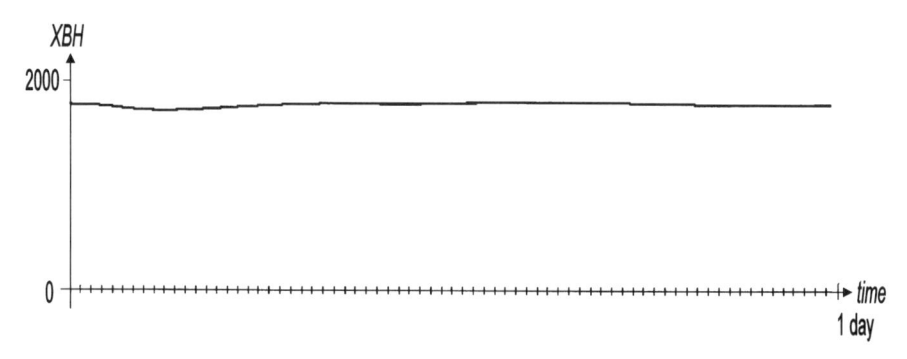

Figure 15.7 Trajectories of heterotrophic bacteria concentration (XBH).

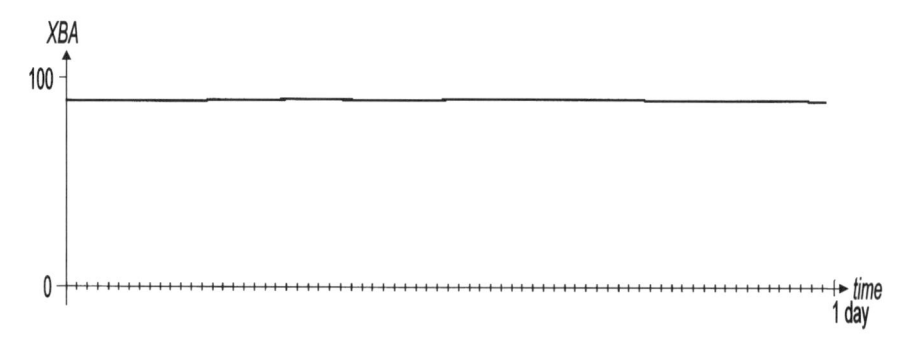

Figure 15.8 Trajectories of autotrophic bacteria concentration (XBA).

measured concentrations, thin lines the concentrations forecast by the neural network based model. Good forecasts in the last basin are the most important because the upper bounds of contaminants apply to the last basin. They are also the most difficult to reach because forecast errors are accumulated throughout the chain of aeration basins.

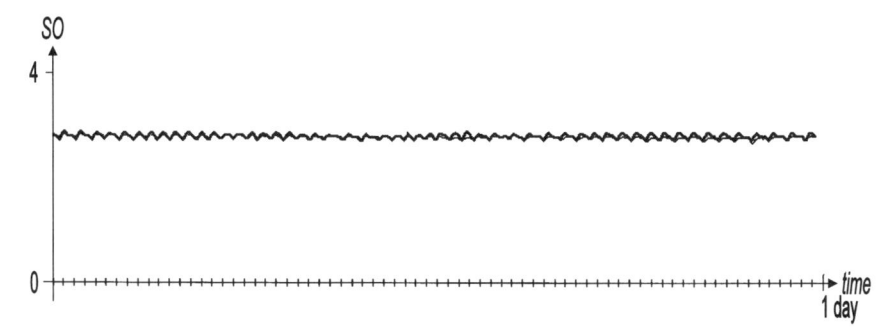

figure 15.9 Trajectories of free oxygen concentration (SO).

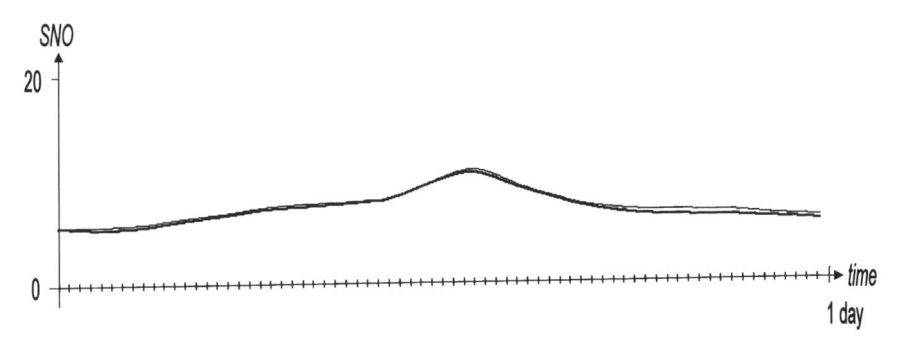

Figure 15.10 Trajectories of nitrate concentration (SNO).

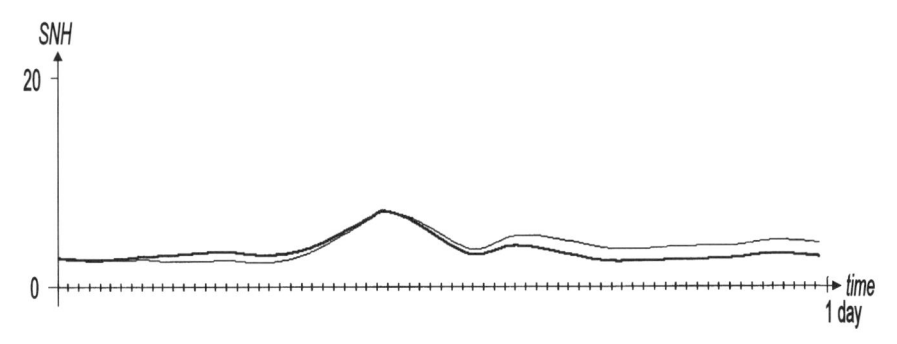

Figure 15.11 Trajectories of ammonia concentration (SNH).

The mean multistep squared forecast error, normed to the variances of individual variables, is 0.118. According to past experience with various applications, this value indicates sufficient correspondence between the model and the plant for neurocontroller training.

346 BIOLOGICAL WASTEWATER TREATMENT CONTROL

15.4 IDENTIFICATION OF HIDDEN STATES

It has been stated in Note 15.2 that the indirect identification approach of Section 15.2, relying upon data generated by the IAWQ model, is not general. It depends on the assumption that the IAWQ model is a sufficiently accurate approximation of reality. Should this assumption prove to be false, the usual approach using measured data would have to be applied. Then, it would need to account for the fact that not all six state variables are measurable in real time.

As shown in Section 15.1, real-time measurement of sludge variables XBA and XBH is virtually impossible. The substrate variable BOD measurements are theoretically conceivable but practically impossible with the equipment currently existing in wastewater plants. So, it would be desirable to omit the sludge variables XBA and XBH altogether and to test the possibility of omitting BOD.

It can be observed in Figures 15.7 and 15.8 that the bacteria concentrations are almost constant. This can be explained by the fact that the bacteria growth and dying processes are very slow, with a typical time frame of days to weeks. Then, the influence of aeration control on them is limited as long as certain conditions are satisfied. These conditions concern a sufficient substrate concentration in the inlet and sufficient sludge recirculation. The first condition is mostly satisfied; it is the motivation for the existence of the wastewater treatment plant itself. The second condition is satisfied by a deliberate sludge recirculation strategy that operates at substantially slower rates than aeration control does. This justifies the hypothesis that the variables describing bacteria concentrations can be omitted without large information loss.

By contrast, it is obvious from Figure 15.6 that the same conclusion cannot be made about substrate concentration BOD. If this variable cannot be measured, two alternative assumptions can be made, and subsequently tested: (1) If the substrate concentration is observable in the sense of Chapter 5, it can be accounted for by increasing the order of the plant model (i.e., increasing the observability index assumed), and (2) if the substrate concentration is a hidden state in the sense of Section 3.3, it can be identified even if the corresponding measurements are not available.

In accordance with these principles, the plant model can be identified by three additional versions. In the first version, four state variables BOD, SO, SNO, and SNH are considered. All of them are assumed to be measurable. The neural network structure of the plant model is analogous to that of Figures 15.3 to 15.5, omitting variables XBH and XBA and the corresponding model blocks.

In the second version, only three measurable output variables SO, SNO, and SHN are considered but with observability index two; that is, the first derivatives of these variables are included on the right-hand side of input–output equations.

In the last version, BOD is viewed as a nonmeasurable recurrent hidden state. This is reached simply by excluding this variable from the error function—its error is not compared with the measurement, and the plant model, consequently, is optimized with regard to this variable. In other words, the neural-network structure of the plant model is analogous to that of Figures 15.3 to 15.5; the only difference consists in omitting the error of the hidden state BOD (which has no counterpart in measurements) from the error function. The initial state of BOD for the 300-step forecast is set to a reasonable value; the remaining trajectory of this variable is free.

The first attempt to identify a plant model of this structure has led to the nontrivial discovery of the BOD state within the working range of the first and the second aeration basin (Fig. 15.2), whereas, for the third and the fourth basin, this state diverges strongly (Fig. 15.13). It has been necessary to enclose this state between reasonable lower and upper bounds, that is, to extend the error function by a penalty term for BOD being outside these bounds. This has led to the stable hidden state BOD. Its trajectory is given in Figure 15.13. Its comparison with a measured (i.e., simulated by the IAWQ model) BOD trajectory cannot be interpreted literally; the model has not been trained for reaching

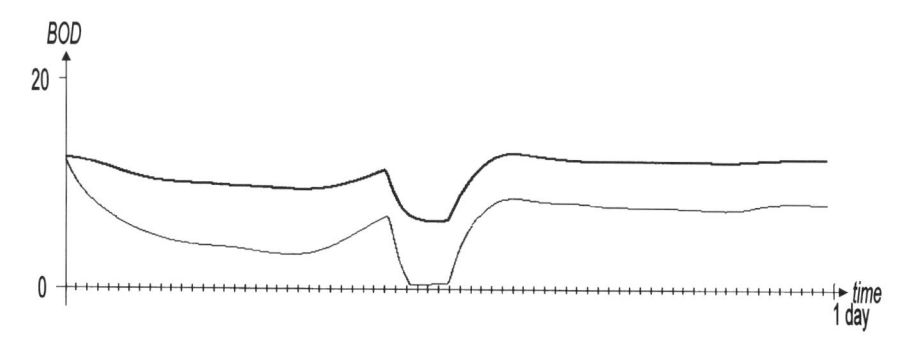

Figure 15.12 Trajectories of substrate concentration (BOD) viewed as a hidden state, the second basin.

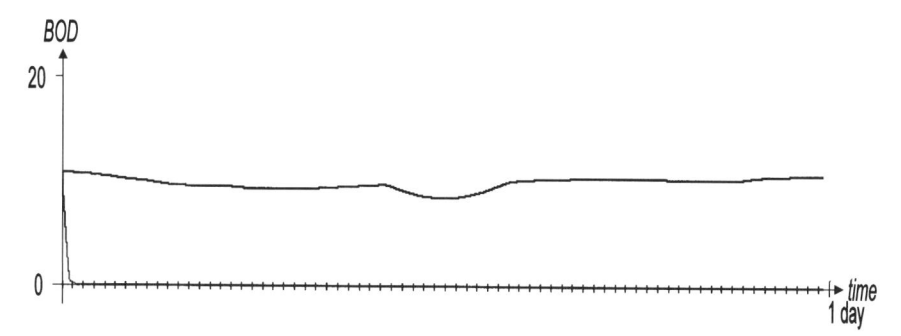

Figure 15.13 Trajectories of substrate concentration (BOD) viewed as a hidden state, the fourth basin.

Table 15.1 Mean squared errors of multistep forecasts

State Vector	Observability Index	Error
XBH,XBA,BOD,SO,SNO,SNH	1	0.118
BOD,SO,SNO,SNH	1	0.141
hidden BOD,SO,SNO,SNH	1	0.147
SO,SNO,SNH	2	0.147

absolute correspondence. It is sufficient that this hidden state exhibits, to some extent, similar dynamics to the measured state.

Nevertheless, the ultimate criterion for assessing the utility of including the hidden BOD state can only be its contribution to modeling measurable output variables SO, SNO, and SHN. It is these variables that can be used directly as neurocontroller inputs, and capturing their dynamics in the plant model has a decisive impact on the performance of the neurocontroller. The mean squared errors of multistep forecasts of SO, SNO, and SHN for all three versions of the model, as well as for the original model including bacteria concentrations, are listed in Table 15.1. The trajectories of the most critical measured variable, the ammonia concentration in the fourth aeration basin, are graphed in Figures 15.15, 15.16, and 15.17.

It can be observed that some information has been lost (the error is growing from 0.118 to 0.141) in leaving off the bacteria concentrations, and some more in making substrate concentration nonmeasurable (error growing from 0.141 to 0.147). An even larger error has been exhibited by the original hidden-state model without bounds: 0.156. This indicates that a meaningful (boundedness being one of the possible criteria of meaningfulness) hidden state may contribute to the overall information. However, increasing the observability index to two makes it possible to reach the same forecast quality (error 0.147)

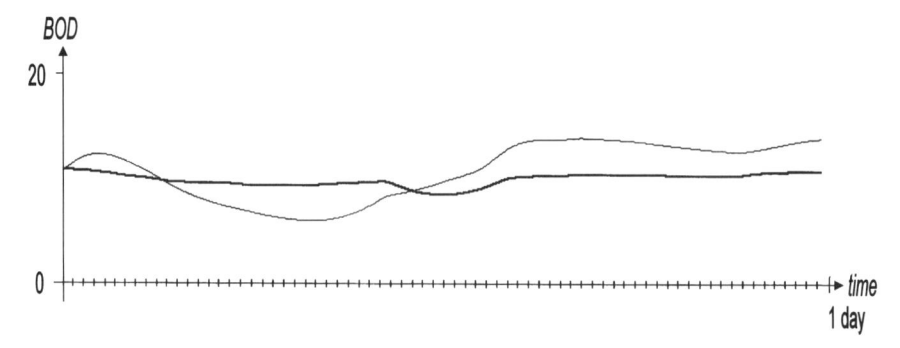

Figure 15.14 Trajectories of substrate concentration (BOD) viewed as a bounded hidden state, the fourth basin.

IDENTIFICATION OF HIDDEN STATES **349**

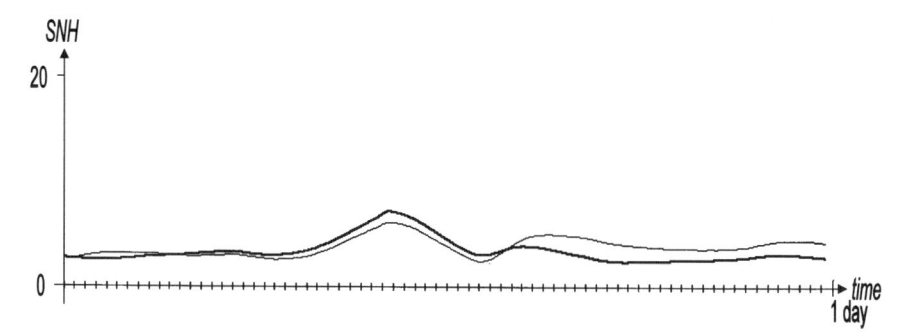

Figure 15.15 Trajectories of ammonia concentration (SNH), four-state model.

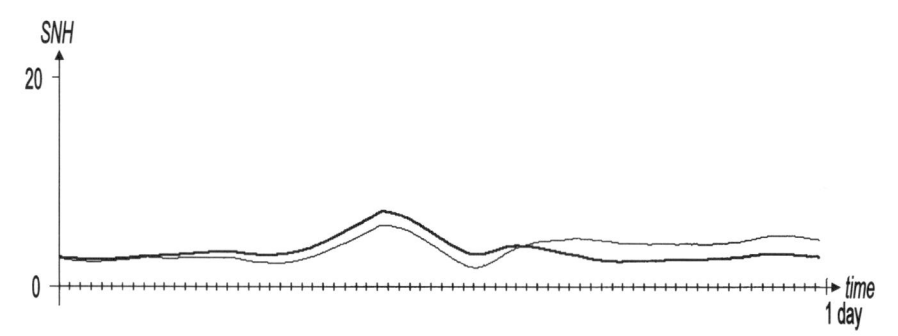

Figure 15.16 Trajectories of ammonia concentration (SNH), four-state model with hidden state.

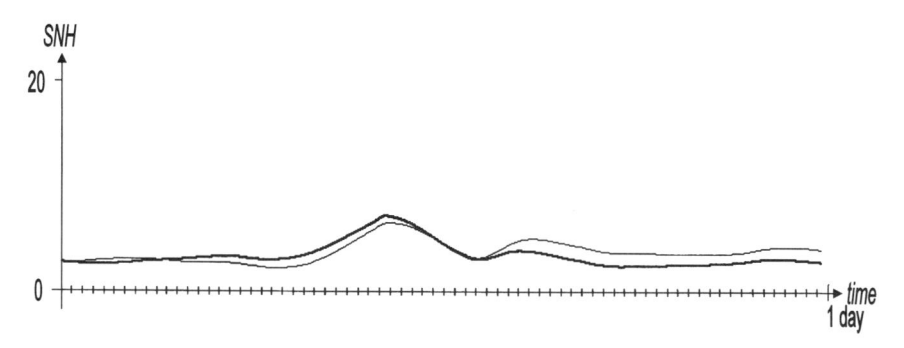

Figure 15.17 Trajectories of ammonia concentration (SNH), three-state model with observability index two.

with measurable outputs only. To summarize, this application does not deliver an incontestable case in favor of the hidden state identification approach of Section 3.3: The approach based on classical observability cannot be outperformed.

15.5 CONTROLLER TRAINING

In this section, neurocontroller training specification is presented. For this presentation, the neural network based plant model with four state variables per aeration basin of Section 15.3 has been used. To train the neurocontroller, a complete closed loop of the wastewater treatment plant is to be modeled. This closed loop includes not only the control feedback but also physical feedback flows by recirculation and sludge feedback.

Besides the four aeration basins modeled with the help of neural networks, there is a sedimentation basin. This device remains to be modeled analytically. For controller training, a very simple model is sufficient. It is assumed that the sedimentation basin is capable of perfect separation of solid components from soluble ones. (In reality, the separation is not perfect but still very efficient if the flow is sufficiently slow.) From the four state variables, SO, SNO, and SHN represent completely dissolved components. The substrate BOD is supposed to consist of BOD_S*BOD solid substance and (1-BOD_S)*BOD dissolved substance.

These are the definitions of model variables:

```
o1_BOD       // substrate concentration   - basin 1
o1_SO        // free oxygen concentration - basin 1
o1_SNO       // nitrate concentration     - basin 1
o1_SNH       // ammonia concentration     - basin 1
   .
   .
   .
o4_BOD       // substrate concentration   - basin 4
o4_SO        // free oxygen concentration - basin 4
o4_SNO       // nitrate concentration     - basin 4
o4_SNH       // ammonia concentration     - basin 4

i_BOD        // substrate concentration   - input to basin 1
i_SO         // free oxygen concentration - input tobasin 1
i_SNO        // nitrate concentration     - input to basin 1
i_SNH        // ammonia concentration     - input to basin 1
i_Q          // total flow                - input to basin 1

l1_SO        // aeration rate             - basin 1
l2_SO        // aeration rate             - basin 2
l3_SO        // aeration rate             - basin 3
l4_SO        // aeration rate             - basin 4

in_BOD       // substrate concentration   - plant inlet
in_SO        // free oxygen concentration - plant inlet
in_SNO       // nitrate concentration     - plant inlet
```

```
in_SNH        // ammonia concentration      - plant inlet
in_Q          // total flow                 - plant inlet

NK_BOD        // substrate quantity         - sedimentation basin
RS_BOD        // substrate concentration in sludge feedback

E_o4_SNH      // ammonia  concentration excess over 80 % of
              //    upper bound at basin 4 outlet
E_o4_SN       // nitrogen concentration excess over 80 % of
              //    upper bound at basin 4 outlet

o2_SO_ref     // reference free oxygen concentration - basin 2
o3_SO_ref     // reference free oxygen concentration - basin 3
o4_SO_ref     // reference free oxygen concentration - basin 4

E_o2_SO       // control error of oxygen concentration - basin 2
E_O3_SO       // control error of oxygen concentration - basin 3
E_O4_SO       // control error of oxygen concentration - basin 4
```

Besides the variables used in plant modeling, there are the following auxiliary variables:

Concentrations and total flow at plant inlet (i.e., before joining the recirculation and sludge feedback flows)

Quantity and concentration of substrate in the sedimentation basin

Excess of ammonia and total nitrogen components concentrations over the "emergency margin" of 80% of legal upper bounds

Reference values for free oxygen concentration in aeration basins 2, 3, and 4 (basin 1 being anaerobic)

Control errors of free oxygen concentration

The following symbolic constants have been used

```
MAX_SNH   6        // legal upper bound of ammonia  concentration
                   //    at basin 4 outlet
MAX_SN    18       // legal upper bound of nitrogen concentration
                   //    at basin 4 outlet
RZ        40000    // recirculation  flow
RS        50000    // sludge feedback flow
US        625      // sludge removal  flow
BOD_S     0.95     // proportion of solids in substrate
                   //    at basin 4 outlet
NK_Q      10880    // sedimentation basin volume
```

The neurocontroller is to be trained for various daily schedules of inlet concentrations and flows. These correspond to the measurable plant inputs θ in

Eq. (6.116). Each concentration and the total flow are characterized by (1) its mean value and (2) the amplitude of its fluctuations. The parameter set of each training example consists just of these mean values and amplitudes:

```
is_BOD    // mean of plant-inlet         substrate concentration
ia_BOD    // fluctuation amplitude of substrate concentration
is_SO     // mean of plant-inlet         free oxygen concentration
ia_SO     // fluctuation amplitude of free oxygen concentration
is_SNO    // mean of plant-inlet         nitrate concentration
ia_SNO    // fluctuation amplitude of nitrate concentration
is_SNH    // mean of plant-inlet         ammonia concentration
ia_SNH    // fluctuation amplitude of ammonia concentration
is_Q      // mean of plant-inlet         flow
ia_Q      // fluctuation amplitude of flow
```

With these variable definitions, the closed loop for neurocontrol training is specified in the following way:

```
/* Inlet computation */

n_in_BOD = is_BOD * (1 + ia_BOD*sin(6.28*T));
n_in_SO  = is_SO  * (1 + ia_SO *sin(6.28*T));
n_in_SNO = is_SNO * (1 + ia_SNO*sin(6.28*T));
n_in_SNH = is_SNH * (1 + ia_SNH*sin(6.28*T));
n_in_Q   = is_Q / (1 + sqrt(ia_Q )*sin(6.28*T))^2;

/* Inlet + Sludge Feedback + Recirculation -> Aeration Basin 1 */

n_i_Q   = n_in_Q + n_RZ + (RS-US);
n_i_BOD = (n_in_BOD*n_in_Q + o4_BOD*n_RZ + RS_BOD)n_i_Q;
n_i_SO  = (n_in_SO *n_in_Q + o4_SO *n_RZ              )/n_i_Q;
n_i_SNO = (n_in_SNO*n_in_Q + o4_SNO*n_RZ              )/n_i_Q;
n_i_SNH = (n_in_SNH*n_in_Q + o4_SNH*n_RZ              )/n_i_Q;

/* Aeration Basin 1 */

n_l1_SO  = 0;
n_o1_BOD = nn(1);
n_o1_SO  = nn(2);
n_o1_SNO = nn(3);
n_o1_SNH = nn(4);

    .
    .
    .

/* Aeration Basin 4 */

n_o4_BOD = nn(13);
```

```
n_o4_SO  = nn(14);
n_o4_SNO = nn(15);
n_o4_SNH = nn(16);

/* Sedimentation Basin */

n_NK_BOD = NK_BOD + DT * (  o4_BOD*BOD_S * (i_Q-RZ)
                          - (NK_BOD/NK_Q) *   RS          );
n_RS_BOD = n_NK_BOD/NK_Q * (RS-US);

/* Errors */

n_E_o4_SNH = positive( n_o4_SNH             - 0.8*MAX_SNH );
n_E_o4_SN  = positive( n_o4_SNH + n_o4_SNO  - 0.8*MAX_SN  );
n_E_o2_SO  = n_o2_SO_ref - n_o2_SO;
n_E_O3_SO  = n_o3_SO_ref - n_o3_SO;
n_E_O4_SO  = n_o4_SO_ref - n_o4_SO;
```

As can be seen, the equations of closed-loop specification comprise the following groups:

Computation of Instantaneous Plant Inlet Concentrations and Flow The daily schedule is supposed to be sinusoidal. Since the fluctuations in flow have sharp peaks, they are represented by a square of the sum of the sinusoid and offset.

Computation of Input into the First Aeration Basin This input is a sum of plant inlet, recirculation flow, and sludge feedback flow. Both the recirculation flow and the sludge feedback flow are assumed to be constant, although they vary in practice in a certain range.

Neural-Network-Based Part of Plant Model The identified part of the plant model represented by neural networks consists of the four aeration basins. The first basin being anaerobic, its aeration input is set to zero.

Substrate Quantity and Concentration in the Sedimentation Basin The quantity of substrate in the bottom part of the sedimentation basin is computed by integration of the solid substrate flow at the basin inlet and of the loss of substrate by sludge feedback flow.

Error Terms The error terms are (1) the excesses of contaminants in the fourth basin outlet over the emergency margins, and (2) deviations of free-oxygen concentrations from their reference values.

Different rates of the fast aeration process, on one hand, and relatively slow biochemical processes, on the other hand, have led to the implementation of two-stage control. In the first stage, optimal reference values for oxygen concentrations are determined. In the second stage, an aeration rate necessary for keeping this concentration is determined.

The goals for the first stage consist in preventing the violation of the upper bound for ammonia, preventing the violation of the upper bound for nitrogen components (i.e., for the sum of ammonia and nitrate), and minimizing the cost caused (and evaluated) by the total aeration volume.

Second-stage goals are maintaining the real oxygen concentration close to the reference value determined by the first stage, preventing oscillations of the second-stage controller by penalizing the second derivative of the control error (the best results have been reached with the fourth power of the second derivative, allowing small fluctuations but preventing large ones), and minimizing the cost caused (and estimated) by the total aeration volume.

The complete cost function with weights taking into account the typical magnitudes of individual terms is:

```
E_o4_SNH                                          * 1e5    // > first
+ E_o4_SN                                         * 1e6    // > stage
+ I2_SO           + I3_SO         + I4_SO                  // > goals
(E_o2_SO^2        + E_O3_SO^2     + E_O4_SO^2)    * 1e4    // > second
+ (DD E_o2_SO^4 + DD E_O3_SO^4 + DD E_O4_SO^4)  * 1e-5;  // > stage goals
```

To show the flexibility of the neurocontroller approach, one can consider an alternative neurocontroller using a cost function with an additional term:

```
o4_SNH              /  6 * 1e6
(o4_SNH+o4_SNO)     / 18 * 1e6
```

This term leads to minimizing the contaminant outflow even if it is below the legal limits.

The first-stage control problem is a nonstandard one. It is neither a stabilization nor a trajectory-following task. There are no genuine reference values, except for (1) the equivalence class of states that do not violate upper bounds for ammonia and total nitrogen, (2) the clearly unrealistic reference values of zero ammonia and nitrogen, or (3) the equally unrealistic zero total aeration rate corresponding to zero costs.

In addition, a concrete plant selected for real-time implementation of the neurocontrollers does not provide facilities for measuring ammonia concentrations in all aeration basins. The available sensors are for free oxygen in basins 2, 3, and 4 (in basin 1, oxygen concentration is assumed to be zero), nitrate in basins 2 and 4, and ammonia in basin 4. The observed variance of measured values suggests that computation of the first numeric derivative is possible but with very limited precision. These effects have led to the formulation of a neurocontroller structure whose output are the first derivatives of reference oxygen concentrations, with inputs consisting of reference oxygen concentrations, current aeration rates, measured ammonia and nitrate concentrations, and excesses of ammonia and nitrogen concentrations over the emergency limits.

Conceptually, the first-stage controller works with a partially inverted plant, extended by including dynamics of the second-stage controllers. Its input is the reference oxygen concentration, with implicit assumption that the second-stage controllers are capable of guaranteeing equality between the reference and real oxygen concentration. The outputs of this inverted plant are the measurable concentrations of ammonia and nitrate, and the current aeration volumes:

```
D o2_SO_ref,D o3_SO_ref,D o4_SO_ref
    depends_on      : o2_SO_ref,o3_SO_ref,o4_SO_ref,
                      12_SO,13_SO,14_SO,o2_SNO,o4_SNH,
                      E_o4_SNH,E_o4_SN
    gain_depends_on : o2_SO_ref,o3_SO_ref,o4_SO_ref,
                      12_SO,13_SO,14_SO,o2_SNO,o4_SNH,
                      E_o4_SNH,E_o4_SN
```

The role of reference values is played by the excess-over-bounds variables, whereas the reference value for costs is zero aeration volume.

The neural-network structure of the first-stage controller is sketched in Figure 15.18 (for notational rules, see Note 12.1). Shaded triangles represent multiplexers and demultiplexers, compressing and decompressing ten lines corresponding to o2_SO_ref, o3_SO_ref, o4_SO_ref, 12_SO, 13_SO, 14_SO, o2_SNO, o4_SNH, E_o4_SNH, and E_o4_SN to and from one vector line.

The equilibrium point of the closed loop can be viewed as either undefined or trivially defined by means of the unrealistic values mentioned above. To classify this controller according to Chapter 6, it can be viewed as a controller with integration of action (6.110) with the assumption of an observability index equal to one. Alternatively, it can also be viewed as an incomplete stabilization controller of type (6.104).

The second-stage controllers (Fig. 15.19) possess clear reference values (o2_SO_ref, etc., produced by the first-stage controller) and are instances of action-integrating structure (6.103) with observability index two. Their definition is:

```
D 12_SO   depends_on      : E_o2_SO,D E_o2_SO
          gain_depends_on : E_o2_SO,D E_o2_SO,o2_SO_ref

D 13_SO   depends_on      : E_o2_SO,D E_o2_SO
          gain_depends_on : E_o2_SO,D E_o2_SO,o2_SO_ref

D 14_SO   depends_on      : E_O4_SO,D E_O4_SO
          gain_depends_on : E_O4_SO,D E_O4_SO,o4_SO_ref
```

The overall structure of the two-stage neurocontroller is given in Figure 15.20.

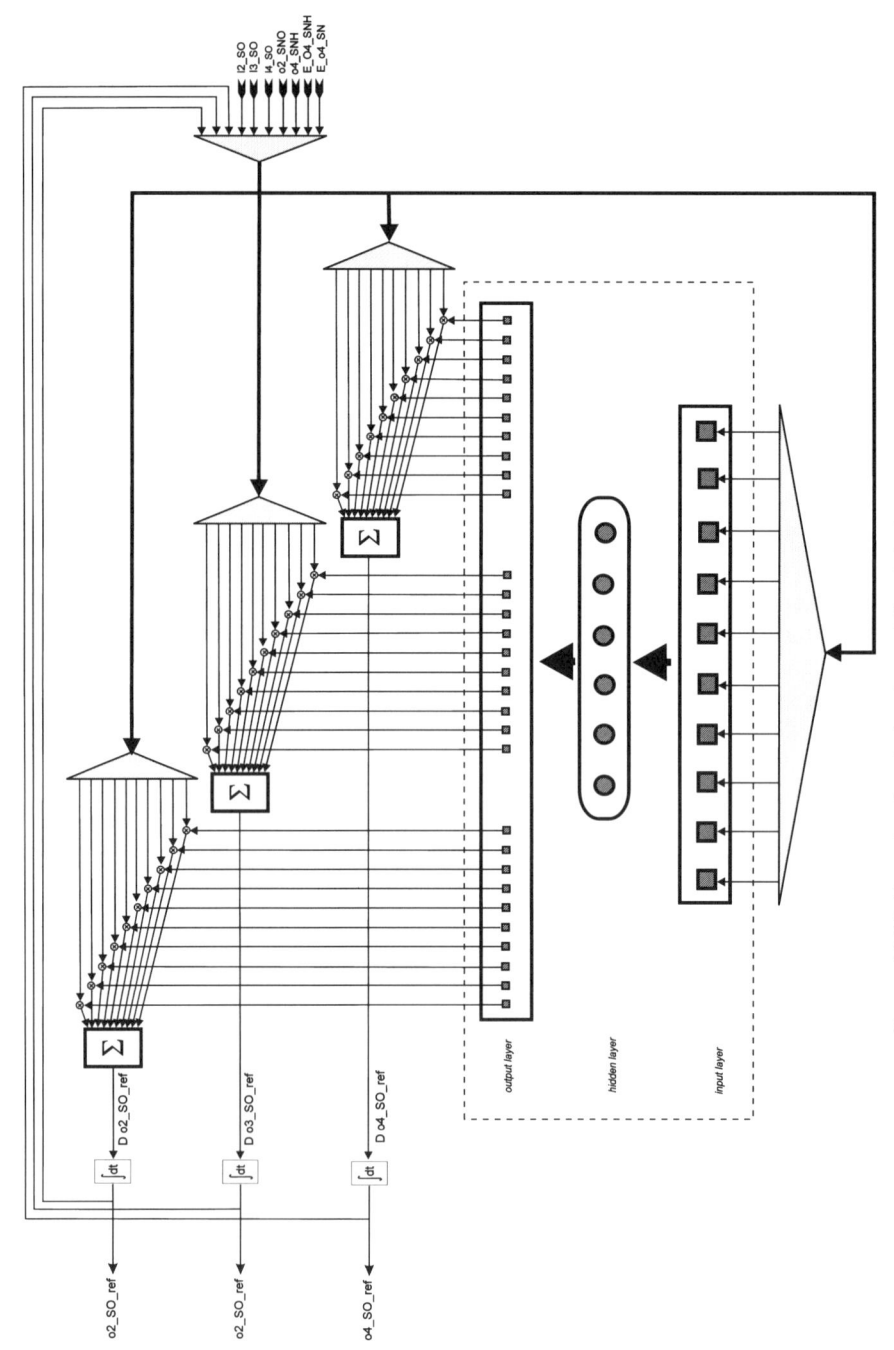

Figure 15.18 Neural network structure of the first-stage neurocontroller.

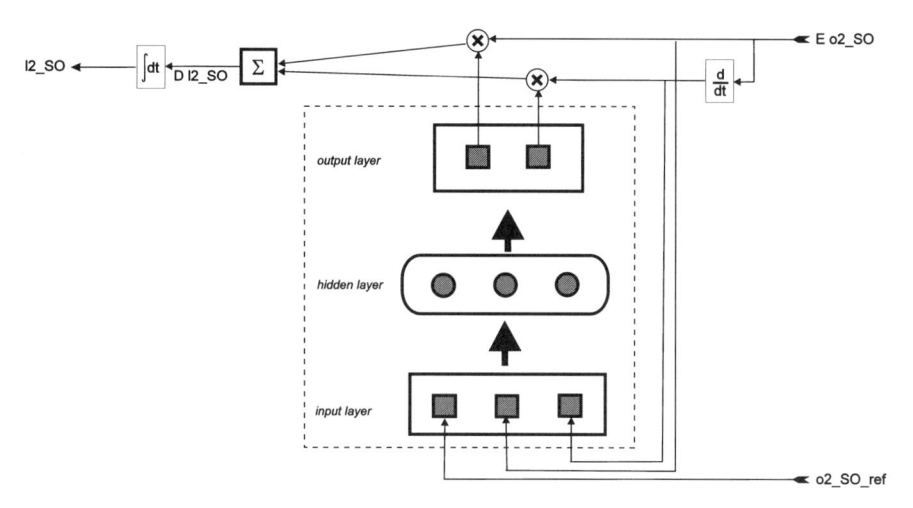

Figure 15.19 Neural network structure of the second-stage neurcontroller.

15.6 RESULTS

The behavior of both neurocontrollers for a relatively high contaminant load is shown in Figures 15.21 through 15.23 (aeration-volume minimizing neurocontroller) and Figures 15.24 through 15.26 (neurocontroller simultaneously minimizing the nitrogen outflow). These results are based on simulations with help of the IAWQ model. The field test results are briefly addressed at the end of this section.

For comparison, corresponding charts are presented (Figs. 15.27 through 15.29) for currently implemented classical control (direct reference value scheduling for first stage, gain scheduling + PID controllers for the second stage).

The most important characteristics are summarized in Table 15.2. The figures in the table are taken from the second day of the charted time range in order to eliminate the influence of the initial state. The contaminant concentrations of all control systems are within the legal limits. In neither case, the upper bounds of 18 g/m^3 for total nitrogen and 6 g/m^3 for ammonia have been violated.

Both neurocontrollers exhibit substantial reduction of total nitrogen concentration, both in the average and in the peaks. The aeration-minimizing neurocontroller allows higher peaks (but still below 80 percent of the limit) for ammonia, but its average ammonia concentration is better than that of the classical controller. Better total nitrogen values, but higher ammonia peaks of this neurocontroller can be explained by the fact that nitrate elimination is aeration-saving, whereas ammonia elimination is aeration-intensive. The nitrogen-minimizing neurocontroller is better in both average and maximum ammonia concentrations.

Note that both neurocontrollers follow an aeration strategy different from that of the classical controller (Figs. 15.23 and 15.29). While the classical

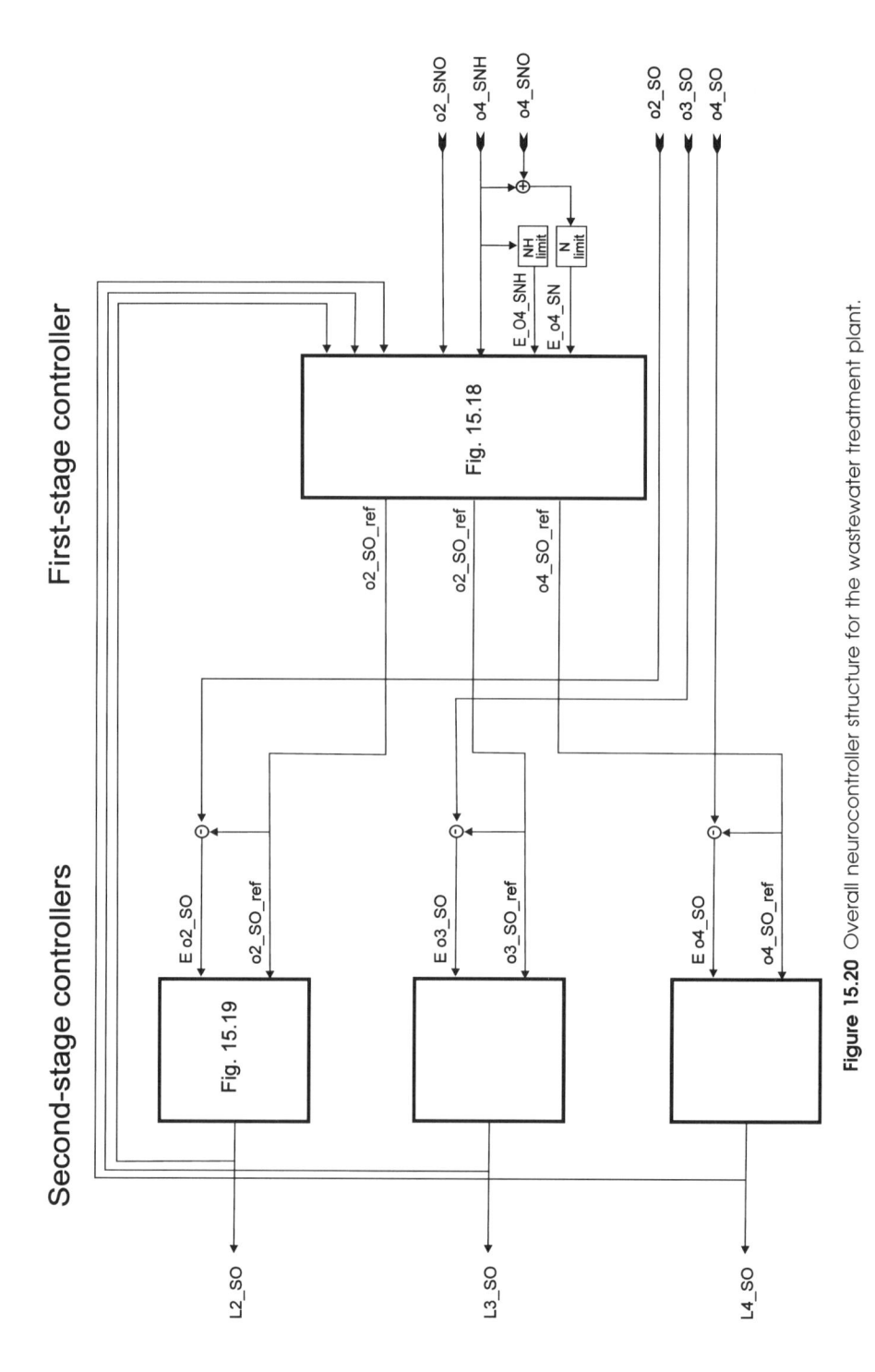

Figure 15.20 Overall neurocontroller structure for the wastewater treatment plant.

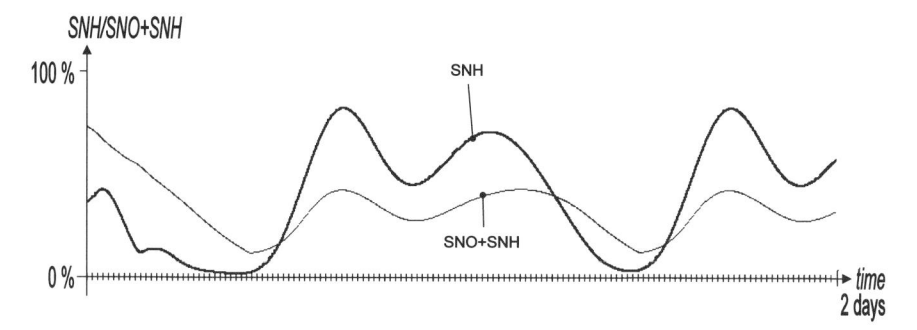

Figure 15.21 Trajectories of the ammonia concentration (SNH, bold line) and total nitrogen (SNH + SNO) in a percentage of the upper bounds with an aeration-volume minimizing neurocontroller.

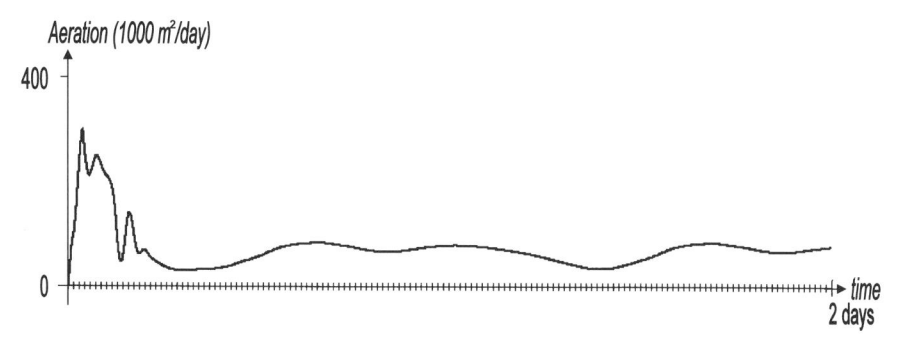

Figure 15.22 Trajectories of the total aeration rate with the aeration-volume minimizing neurocontroller.

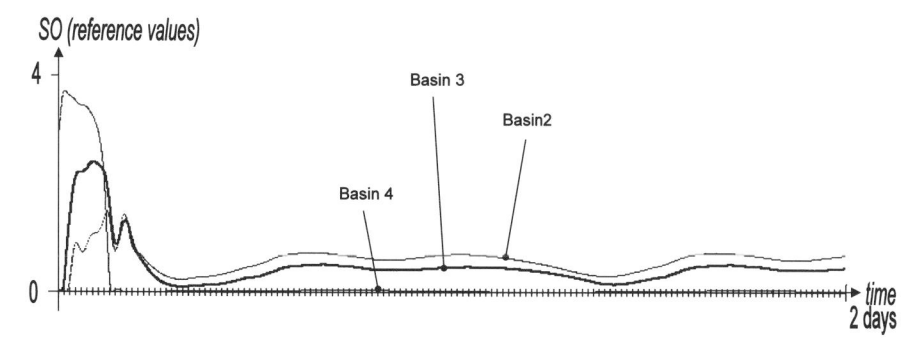

Figure 15.23 Trajectories of the reference oxygen concentrations with the aeration-volume minimizing neurocontroller.

controller keeps aeration on in basins 3 and 4 and alternates the aerobic and anaerobic regime in basin 2, neurocontrollers tend to switch off the aeration of basin 4. The latter strategy represents reverting denitrification and nitrification phases along the aeration basin chain. That both solutions are possible has to

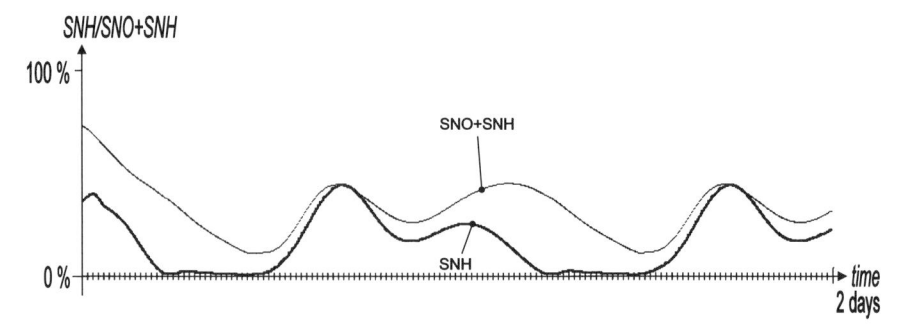

Figure 15.24 Trajectories of the ammonia concentration (SNH, bold line) and total nitrogen (SNH + SNO) as a percentage of the upper bounds with a neurocontroller minimizing nitrogen outflow.

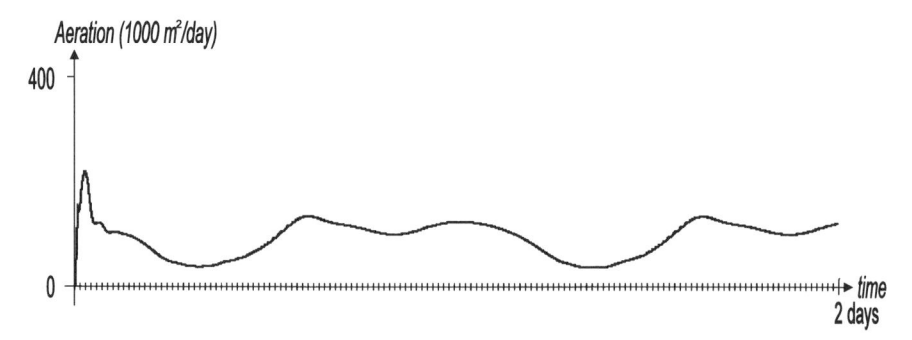

Figure 15.25 Trajectories of the total aeration rate with a neurocontroller minimizing nitrogen outflow.

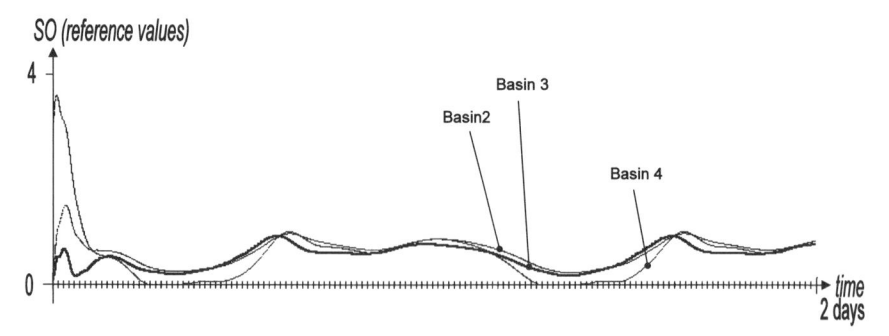

Figure 15.26 Trajectories of the reference oxygen concentrations with a neurocontroller minimizing nitrogen outflow.

do with the cyclic nature of the nitrification/denitrification process mentioned in Section 15.1. For a human expert, it is difficult to determine which of the two quantitatively different solutions is optimal for a given working point; numeric optimization used in neurocontroller training is obviously superior.

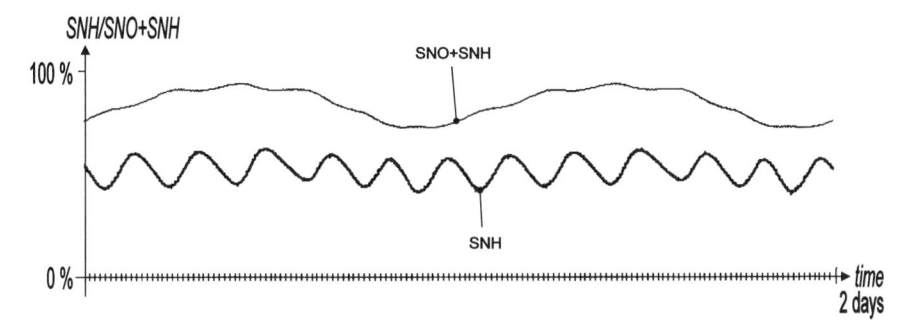

Figure 15.27 Trajectories of ammonia concentration (SNH, bold line) and total nitrogen (SNH + SNO) as a percentage of the upper bounds with classical control.

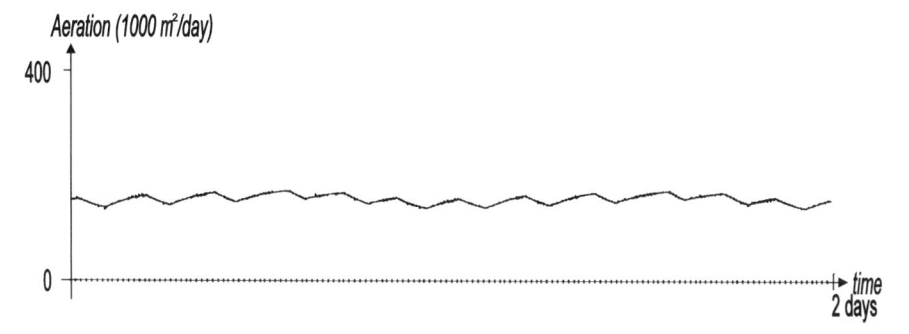

Figure 15.28 Trajectories of the total aeration rate with classical control.

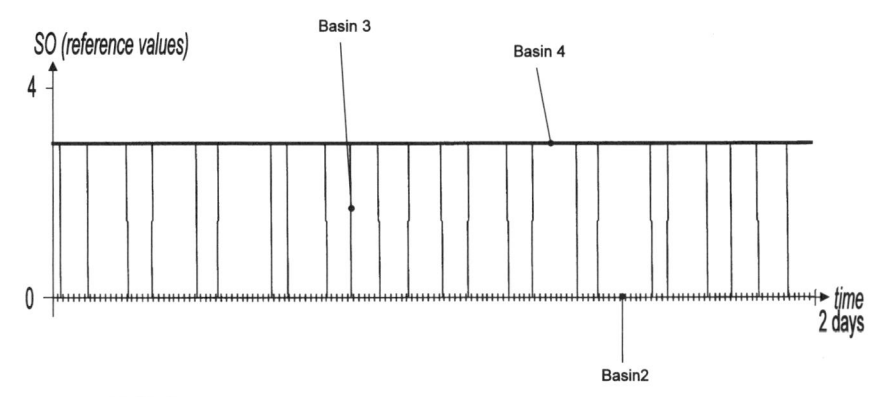

Figure 15.29 Trajectories of reference oxygen concentrations with classical control.

The most important result is that total aeration volume with an aeration-minimizing neurocontroller is as low as merely 36.7 percent of that of the classical controller. This can be directly projected into the cost savings. Similar results have been reached with other contaminant concentration schedules. The aeration savings have been varying between 15 and 70 percent. For

Table 15.2 Important quality measures of neurocontrollers and classical controllers

	Classical Controller	Aeration-Minimizing Neurocontroller	Nitrogen-Minimizing Neurocontroller
Maximum of total nitrogen concentration	16.82	7.82	8.21
Average of total nitrogen concentration	16.03	5.47	5.86
Maximum of ammonia concentration	3.78	4.95	2.73
Average of ammonia concentration	3.65	2.86	0.91
Total aeration volume	165,463	60,707	91,325

a very high load, the saving potential is low because full aeration capacity is required. In low-load phases, large savings are possible with the help of intelligent control strategies.

Even the nitrogen-minimizing neurocontroller needs only approximately 55 percent of the aeration of the classical controller. Taking into account that its contaminant values are clearly superior to those of classical control, it becomes obvious that optimizing the control allows the user not only to weight priorities in a flexible manner but also to reach improvements in all disciplines simultaneously.

To this date, the complete neurocontroller has not been tested on the real wastewater treatment plant. Instead, the control strategy pursued by the neurocontroller has been observed with the help of simulations, analyzed by wastewater treatment experts, and incorporated into a nonlinear controller mimicking the neurocontroller. This unconventional approach is necessary for legal justification of implementing the neurocontroller.

Already with this intermediary control system, excellent results have been reached: energy savings between 15 and 30 percent, without increased contaminant concentrations. Therefore, the field test of the neurocontroller is planned in a short time.

16

CONCLUSION: APPLICATION POTENTIAL OF NEUROCONTROL

16.1 MAIN ASSET OF NEUROCONTROL: GENERALITY

The case studies presented, together with tens or even hundreds of other published industrial applications, show an important asset of neurocontrol. It is the *generality* of the approach. There are really few (if any) comparably general methods in nonlinear control. It is almost revolutionary to use the same algorithm for applications focusing on fast trajectory following throughout all working points of a highly nonlinear plant (e.g., dynamic test bench control problems), on the one hand, and applications that require determining a complete *strategy* to reach nonstandard control goals with a loose relationship to the process variables (e.g., wastewater treatment plant control), on the other hand. It contrasts to the line of typical control theory textbooks: "Use method A for problems of type X, method B for problems of type Y, and so on. . . ." There are sophisticated and elegant theories behind these numerous approaches. But the industrial view is to look for a method that, once developed, is able to solve whole series of applications routinely for a return on the investment. This makes the industrial application potential of neurocontrol really impressive.

The principal practical benefits from having a general algorithmic method for nonlinear controller design can be summarized in the following way:

1. Design times are substantially reduced. It is no longer necessary to perform a detailed analysis of the application and to select the most appropriate solution method, or even to develop a dedicated method. It is sufficient to define the general framework, typically consisting of the following:

- List of measurable variables
- Assessment of observability index
- List of important and representative training cases
- Specification of control goal

The remaining work is done by the numerical algorithm.
2. The design is automatically reproducible if the task has changed within the framework formulated. Typical tasks are control redesigns after plant modifications or test bench control after workpiece changes.
3. The flexibility in formulation of control goals is increased. The control goal need not be forced to some prescribed scheme such as trajectory following, stabilization, or optimization of quadratic terms.

As far as the quality of control solutions is concerned, the experience of this author is that neurocontrollers are usually superior to existing controllers. The reason for this is certainly not that classical nonlinear design methods are not sophisticated enough but rather that they are difficult to use, specific to particular problem classes, and not sufficiently focused on omnipresent practical problems such as making the plant model consistent with measured data or making the controller robust against plant variations and disturbances of various types.

16.2 NEUROCONTROL IS A NUMERIC METHOD

To see neurocontrol in the right relationship to other methods, it is important to realize that neurocontrol is a numeric method. There are two reasons for using numeric methods instead of analytic ones. First, numeric methods are substantially easier and cheaper to implement on a computer. It is cheaper to compute for a week than to let a human expert analyze the problem for a couple of hours.

The second reason is that numerical methods are more universal (although their statements are frequently less informative). Think of the number of tricks there are to solve some narrow class of nonlinear differential equations. Every engineer has experienced this, delight for some, torture for others. Most of these, and almost all other differential equations, can be solved (i.e., simulated) by a single method, for example, the Runge-Kutta method. Although simulation does not allow general conclusions, it is informative enough to solve practical problems. With the growing complexity of systems considered, the proportion of problems analytically tractable shrinks continually.

Are then analytic methods obsolete? Definitely not. They provide general insights instead of algorithms for individual solutions. Neurocontrol is a good example. The theory determines the structures of plant models and neurocontrollers appropriate for the task. Within such a well-founded framework, the numeric algorithm does the dirty work.

This also makes clear that the quality of numerical algorithms and of their implementation is the key to the success of neurocontrol. There is a large body of knowledge about, and also a lot of practical experience with, numeric optimization. This knowledge and experience should be drawn upon as much as possible.

16.3 NEUROCONTROL PRACTICE

It is well known that the numeric way to solve neurocontrol problems is by lengthy computations. Frequently, the easiest (and sometimes the only) way to make a bad solution good is to compute more. If optimizing from scratch and using a personal computer as platform, computing times of hours to days must be expected. For fine-tuning after small changes of the control problem framework, several minutes to a few hours may be sufficient if local optimization and an efficient algorithm is used.

In particular, to avoid too many optimization runs and to get a good solution, the problem formulation must be carefully prepared. With general and automatic methods, it is tempting to make sloppy formulations, put some data "into the pot," and expect the algorithm to make the best of it. However, it is too easy to make the task so nonsensical that no optimization method in the world will be able to produce something that can be called a controller.

The following list suggests how to avoid formulating nonsense:

1. The sampled measurements for plant identification should be consistent with some unique model. They must not contain drastic measurement outliers (a couple of them may be enough for a breakdown), and all measurements should be taken under similar conditions so that possible changes (of the plant or of the measurement equipment between individual measurements) are negligible.
2. The range of values for which the plant model has been identified must not be exceeded during the controller training. It is important to keep in mind that the plant model virtually *does not exist beyond this range*, and may produce arbitrary values or even numeric overflows.
3. The cost function together with training examples used is to be properly formulated up to the last numeric consequence. It happens easily that goals are formulated that are infeasible together with a particular training example. The best way to figure out inconsistencies is by simulating the examples and the cost function.

16.4 METHOD SELECTION FOR INDUSTRIAL APPLICATIONS

In selecting appropriate methods from a neurocontrol repository, an industrial developer will apply other criteria than an academic researcher. A method becomes industrial not because it is exciting but because it works. On the other hand, it is exciting enough to find and use methods that really work.

Neurobiology has been, and still is, a key source of inspiration for neurocontrol. This does not necessarily mean that a similarity to the way neurobiological systems operate (or to what it is currently assumed to be) is the best criterion for industrial method selection.

A good example of this is the decision between batch and incremental operation, frequently discussed in this book. Using the incremental approach seems more "biological." It really seems to be the case that neurobiological systems, at least at the lowest processing level, are using this approach. In the architecture of neurobiological systems, the systems are frequently genuinely committed to local interactions between neurons. By contrast, current computer-based systems only "play the game of local interactions"—in reality, they can equally well do the opposite whenever needed. In industrial applications, it is the result that counts. It makes no sense to make a task that is already difficult enough yet more difficult.

There are many arguments in favor of batch schemes. Some of them have been presented and justified earlier in this book. They concern the spectrum and quality of optimization algorithms that can be used with batch schemes, speed, robustness and accuracy of their convergence, and stability. So it is right to use a batch scheme whenever there are no serious arguments against it.

Similar arguments apply to the choice between online adaptiveness and offline optimization. Definitely, there are applications for which online adaptiveness is a crucial part of the task. However, for most applications, adaptiveness is either relatively unimportant or simply infeasible. Frequently, the decision in favor of online adaptation is made with the assumption that it is not much more difficult than offline optimization. This assumption is completely wrong. For online adaptation, either efficient algorithms have to be given up in favor of much less efficient ones (for incremental adaptive schemes), or the computing power available must be sufficient to do within seconds or even milliseconds what otherwise takes hours. This makes very small the number of applications for which an online solution can be of comparable quality to that of an offline solution.

16.5 RELATIONSHIP TO THE FOUNDATIONS OF NEURAL NETWORKS

A neural fundamentalist may object that an approach to neurocontrol based on batch and offline computations drifts away so far from the neural-network origins that it is no longer neurocontrol but simply control. However, there are still a sufficient number of fundamental features that pragmatic neurocontrol, when viewed as a numeric algorithm, shares with neurobiological systems.

The most important of them is the principle of improving the performance of a system with (and only with) the help of measurements received from the system's environment. There have been important achievements concerning this objective in the field of classical control and in other fields elaborated before neurocontrol, and it is only natural that an interdisciplinary view is required for such a basic question.

Another feature is the use of network structures for functional approximation. In contrast to other functional approximation approaches, this approach uses a number of nonlinear elements of the same type. This is an important achievement of the science of neural networks and is conceptually analogous to the way neurobiological systems are constructed. It makes a difference whether it has to be decided which elements are to be taken (as in the most classical functional approximation methods) or only how many. The fact that relatively small numbers of neurons are used in neurocontrol is due to both the limitation of computing resources and the efficiency of numeric algorithms used in artificial systems. It is at least a possibility that numeric algorithms are so efficient in comparison with the natural ones that such small networks are sufficient.

Note 16.1 Nature tends to substitute quality for quantity. This may or may not be right for artificial systems. To answer this question, more about the behavior of *very large* neural networks would have to be known and corresponding algorithms be proposed and evaluated.

Finally, the process of optimization is a very natural one. Whichever optimization method is adopted, it will always consist in evaluating the performance and selecting the better-performing combination. All numeric approaches have some counterparts in natural algorithms. The rules of a Hebbian type use a rudimentary form of the analytic gradient. Reinforcement learning rules are related to the numeric gradient (see Note 2.4). Various types of genetic algorithms are a prototype for global optimization based on combinatorial search under limited resources.

16.6 WHAT THE FUTURE MAY HOLD

This book has focused on methods that are sufficiently mature to be routinely applicable to problems of an industrial scale and complexity. This resulted in a certain commitment to the closed-loop optimization and plant inversion approaches.

However, the critic-based approach may become very productive in the future. It allows a less specific description of control goals—their instantiation may, to a certain degree, be a part of the solution. This is certainly an attractive property for a broad class of problems, in particular, for large, interacting, and hierarchic systems. Whether and when the critic-based approach will be applicable to typical industrial problems depends, among other factors, on how far one will succeed in formulating the task so that efficient and robust numeric methods can be applied to it.

Another change of the current neurocontrol landscape might result from future developments concerning large neural networks. It is not only the hardware support for large networks but also dedicated parallel algorithms that are sufficiently efficient to make profitable the substitution of algorithmic quality by network quantity as in natural systems.

REFERENCES

1. Ackermann, J. *Sampled Data Control Systems.* Springer, New York, 1988.
2. Ackley, D.H., Hinton, G.E., and Sejnowski, T.J. A learning algorithm for Boltzmann machines. *Cogn Sci* 1985; **9**: 147–169.
3. Albus, J.S. A new approach to manipulator control: the Cerebellar Model Articulation Controller (CMAC). *Trans ASME J Dynam Syst Meas Contr* 1975; **97**: 220–227.
4. Albus, J.S. *Brain, Behavior, and Robotics.* BYTE Books, 1981.
5. Alippi, C., Casamatta, F., and Furlan, L. Control of underexcitation mode of synchronous machines in power plants. *Proceedings of the 1995 IEEE International Conference on Neural Networks.* Perth, 1995.
6. Anderson, C.W. Learning to control an inverted pendulum using neural networks, *IEEE Contr Syst Magazine* 1989; **9**(3): 31–37.
7. Archetti, F. A sampling technique for global optimization. In: Dixon, L.C.W. and Szegö, G.P. (Eds.), *Towards Global Optimization.* Amsterdam, 1975.
8. Åström, K.J., and Wittenmark, B. *Adaptive Control.* Addison-Wesley, Reading, MA, 1989.
9. Åström, K.J., and McAvoy, T.J. Intelligent control: an overview and evaluation. In: White, D.A., and Sofge, D.A. (Eds.), *Handbook of Intelligent Control.* Van Nostrand Reihhold, New York, 1992; 3–34.
10. Baker, W.L., and Farrell, J.A. An introduction to connectionist learning control systems. In: White, D.A., and Sofge, D.A. (Eds.), *Handbook of Intelligent Control.* Van Nostrand Reihhold, New York, 1992; 35–64.
11. Barnes, C. et al. Applications of neural networks to process control and modeling, *International Conference on Artificial Neural Networks '91.* Espoo, 1991; 321–326.
12. Barto, A.G., Sutton, R.S., and Anderson, C.W. Neuronlike adaptive elements that can solve difficult learning control problems. *IEEE Trans Syst Man Cybernet* 1983: **13**: 834–846.
13. Bellman, R.E. *Dynamic Programming.* Princeton University Press, Princeton, NJ, 1957.
14. Bengio, Y., Simard, P., and Frasconi, P. Learning long-term dependencies with gradient descent is difficult. *IEEE Trans Neural Networks* 1994; **5**(2): 157–166.
15. Bernstein, I.H. *Applied Multivariate Analysis.* Springer-Verlag, New York, 1988.
16. Boender, C.G.E., Rinnooy Kan, A.H.G. Bayesian stopping rules for multistart global optimization methods. *Math Program* 1987; **37**(1): 59–80.

17. Boender, C.G.E., Rinnooy Kan, A.H.G., Timmer, G.T., and Stougie, L. A stochastic methods for global optimization. *Math Program* 1982; **22**(2): 125–140.
18. Brent, R.P. *Algorithms for Minimization without Derivatives*. Prentice Hall, Englewood Cliffs, NJ, 1973.
19. Brodt, M. *Entwicklung eines trainierbaren Controllers für ein Fahrzeug mit aktiver hydropneumatischer Federung* (The development of a trainable controller for a car with an active hydropneumatic suspension). Diploma Thesis, Esslingen Technical College, 1993; in German.
20. Brogan, W.L. *Modern Control Theory*. Prentice Hall, Englewood Cliffs, NJ, 1991.
21. Carrol, S.M., and Dickinson, B.W. Construction of neural nets using the Radon transform. *Proceedings of the International Joint Conference on Neural Networks*. Washington, DC, 1989; **1**: 607–611.
22. Chichinadze, V.K. The ψ-transform for solving linear and nonlinear programming techniques. *Automatica* 1969; **5**: 347–355.
23. Cybenko, G. Approximation by superposition of sigmoidal function. *Math Contr Syst Signals*, 1989.
24. Dennis, J.F., and Schnabel, R.B. *Numerical Methods for Unconstrained Optimization and Nonlinear Equations*. Prentice-Hall, Englewood Cliffs, NJ, 1983.
25. Dvoretzky, A. On stochastic approximation. In: *Proceedings of the Third Berkeley Symposium on Mathematical Statistics and Probability*, Berkeley, 1956; **1**: 39–55.
26. Elman, J.L. Finding structure in time. *Cogn Sci* 1990; **14**: 179–211.
27. Falb, P.L., and Wolowich, W.A. Decoupling in the design and synthesis of multivariable control systems. *IEEE Trans Auto Contr* 1967; **12**: 651–659.
28. Feldkamp, L.A., Puskorius, G.V., Davis, L.I., and Yuan, F. Decoupled Kalman Training of neural and fuzzy controllers for automotive systems. In: *Proceedings of the IEEE VTS Workshop on Fuzzy and Neural Systems, and Vehicle Applications*, Tokyo, Nov. 1991.
29. Feldkamp, L.A., and Puskorius, G.V. Training controllers for robustness: multistream DEKF. *Proceedings of IEEE International Conference on Neural Networks*. Orlando, Fl, 1994; 2377–2382.
30. Fetzer, K., Regelung einer Kläranlage mit Hilfe Neuronaler Netze (Control of a wastewater treatment plant with help of neural nets), Ulm University, Dept. for Measurement Technology, Control and Microtechnology, 1995; in German.
31. Föllinger, O. *Regelungstechnik*. Hüthig, Heidelberg, 1994; in German.
32. Föllinger, O. *Nichtlineare Regelungen*. Oldenbourg, Munich, 1993; in German.
33. Fogel, D.B. *Evolutionary Computation: Toward a New Philosophy of Machine Intelligence*. IEEE Press, Piscataway, NJ, 1995.
34. Franke, U., and Mehring, S. A transputer based image processing system for real time vehicle guidance. PROMETHEUS Workshop, Grenoble, Dec. 1990.
35. Franke, U. Real time 3D road modeling for autonomous vehicle guidance. *Proceedings of the 7th Scandinavian Conference on Image Analysis*. Aalborg, 1991; 316–323.
36. Franke, U., Fritz, H., and Mehring, S. Long distance driving with the Daimler-Benz autonomous vehicle VITA. PROMETHEUS Workshop, Grenoble, December 1991.

37. Freund, E. The structure of decoupled nonlinear systems. *Inter J Contr* 1973; **21**: 443–450.
38. Fu, J.-H., and Abed, E.H. Families of Lyapunov functions for nonlinear systems in critical cases. *IEEE Trans Auto Contr* 1993; **38**(1).
39. Funahashi, K. On the approximate realization of continuous mappings by neural networks. *Neural Networks* 1989; **2**: 183–192.
40. Ge., R.P. A filled function method for finding global minimizer of a function of several variables. *Math Program* 1990; **46**(2): 191–204.
41. Geman, S., and Geman, D. Stochastic relaxation, Gibbs distributions and the Bayesian restoration of images *IEEE Trans. Pattern Anal Mach Intell* 1984; **6**(6): 721–742.
42. Giles, C.L., and Maxwell, T. Learning, invariance, and generalization in high-order networks. *Appl Opt* 1987; **26**: 4972–4978.
43. Grossberg, S. *Studies in Mind and Brain*. Reidel, Dordrecht, 1982.
44. Grossberg, S. On the development of feature detectors in the visual cortex with applications to learning and reaction-diffusion systems. *Biol Cybernet* 1976; **21**: 145–159.
45. Grossberg, S. Adaptive pattern classification and universal recoding I: parallel development and coding of neural feature detectors. *Biol Cybernet* 1976; **23**: 121–134.
46. Grossberg, S. Competitive learning: from interactive activation to adaptive resonance. *Cogn Sci* 1987; **11**: 23–63.
47. Grossberg, S. (Ed.). *The Adaptive Brain*. North-Holland, Amsterdam, 1987.
48. Grossberg, S., and Kuperstein, M. *Neural Dynamics of Adaptive Sensory-Motor Control*. North-Holland, Amsterdam, 1986.
49. Henze, M., Grady, Jr., C.P.L., Gujer, W., et al. *Final Report—Activated Sludge Model No. 1*. International Association for Water Quality Scientific and Technical Reports, London, 1987.
50. Holland, J.H. *Adaptation in Natural and Artificial Systems*. University of Michigan Press, Ann Arbor, 1975.
51. Hopfield, J.J. Neural networks and physical systems with emergent collective computational abilities. *Proc Natl Acad Sci* 1982; **79**: 2554–2558.
52. Hrycej, T. Supporting supervised learning by self-organization. In: *Neurocomputing* 1992; **4**: 17–300.
53. Hrycej, T. A modular architecture for efficient learning. *Proceedings of the International Joint Conference on Neural Networks*. San Diego, 1990; **1**: 557–562.
54. Hrycej, T. Model-based training of neural controllers, case study: satellite attitude control. *Proceedings of the IEEE VTS Workshop on Fuzzy and Neural Systems, and Vehicle Applications*, Tokyo, Nov. 1991.
55. Hrycej, T. Neural-network-based car drive train control. *Proceedings of the IEEE Vehicular Technology Conference '92*, Denver, May 1992.
56. Hrycej, T. Model-based training method for neural controllers. In: *Proceedings of the International Conference on Artificial Neural Networks*, Brighton, 1992.
57. Hrycej, T. Plant identification using temporal sequences. *Proceedings of the International Conference on Artificial Neural Networks '94*, Sorrento, Italy, 1994.

58. Hrycej, T. Symmetric properties of neural networks for control applications. *Proceedings of the IEEE International Conference on Neural Networks*, Orlando, Fl, 1994.
59. Hrycej, T. *Modular Learning in Neural Networks*. Wiley-Interscience, New York, 1992.
60. Hrycej, T. Neural network control of test benches. *Proceedings ICARCV '94 — Third International Conference on Automation, Robotics and Computer Vision*, Singapore, 1994.
61. Hrycej, T. Identifying chaotic attractors with neural networks. *Proceedings of the 1995 IEEE International Conference on Neural Networks*, Perth, 1995.
62. Hrycej, T. Stability and equilibrium points in neurocontrol. *Proceedings of the 1995 IEEE International Conference on Neural Networks*, Perth, 1995.
63. Hrycej, T., and Selg W. Rear-axle test bench control with artificial neural networks. 28. Technisches Presse-Colloquium der AEG, Berlin, November 1993; in German.
64. Jameson, J. A neurocontroller based on model feedback and the adaptive heuristic critic. *Proceedings of the International Joint Conference on Neural Networks*, San Diego, 1990; **2**: 37–43.
65. Ji, C., Snapp, R.R., and Psaltis, D. Generalizing smoothness constraints from discrete samples. *Neural Computation* 1990; **2**: 188–197.
66. Johnson, C.D. Accommodation of external disturbances in linear regulator and servomechanism problems. *IEEE Trans on Auto Contr* 1971; **16**: 635–644.
67. Jordan, M.I. Serial Order: A Parallel Distributed Processing Approach. Technical Report No. 8604, Institute for Cognitive Science, University of California, San Diego, 1986.
68. Juen, G. Lageregelung elastischer Antriebssysteme, dargestellt an einem Radioteleskop. VDI-Fortschritt-Berichte, Reihe 8, Nr. 127, VDI-Verlag, 1977.
69. Kalman, R.E. Design of a self-optimizing control systems. *Trans ASME* 1958; **80**: 468–478.
70. Kalman, R.E. On the general theory of control systems. In: *Proceedings of the 1st International Congress on Automatic Control*, Moscow, 1960, 481–492.
71. Kawato, M., Furukawa, K., and Suzuki, R. A hierarchical neural-network model for control and learning of voluntary movement. *Biol Cybernet* 1987; **57**: 169–185.
72. Keller, H. Entwurf nichtlinearer Mehrgrößensysteme mit Hilfe der Ljapunow-Theorie und der Moore-Penroseschen Pseudoinversen. *Automatisierungstechnik* 1989; **37**: 386–391, 462–465.
73. Kiefer, J., and Wolfowitz, J. Stochastic estimation of the maximum of a regression function. *Ann Math Stat* 1952; **23**: 462–466.
74. Kirkpatrick, S., Gelatt., C.D., and Vecchi, M.P. *Science* 1983; **220**: 671.
75. Kirkpatrick, S. *J Stat Phys* **34**: 975–986.
76. Kohonen, T. *Self-Organization and Associative Memory*. Springer-Verlag, New York, 1984.
77. Kohonen, T., Barna, G., and Chrisley, R. Statistical pattern recognition with neural networks: benchmarking studies. In: *Proceedings of the IEEE International Conference on Neural Networks*, San Diego, 1988; 61–67.

78. Konigorski, U. Entwurf strukturbeschränkter Zustandsregelungen unter besonderer Berücksichtigung des Störverhaltens (Design of limited-structure state controllers with particular account of behavior under disturbances). *Automatisierungstechnik* 1987; **35**: 349–358 in German.
79. Kreisselmeier, G., and Birkhölzer, T. Numerical nonlinear regulator design. *IEEE Trans Auto Contr* 1994; **39**(1): 33–46.
80. Kushner, H.J., and Clark, D.S. *Stochastic Approximation Methods for Constrained and Unconstrained Systems.* Springer-Verlag, New York, 1978.
81. Kushner, H.J., and Gavin, T. A versatile method for the Monte Carlo optimization of stochastic systems. *Int J Contr* 1973; **18**: 963–975.
82. Kwakernak, H., and Siwan, R. *Linear Optimal Control Systems.* New York, Wiley, 1972.
83. Lambert, J.-M., and Hecht-Nielsen, R. Application of feedforward and recurrent neural networks to chemical plant predictive modeling. *Proceedings of the International Joint Conference on Neural Networks*, Seattle, 1991; **1**: 373–378.
84. La Salle, J.P., and Lefschetz, S. *Stability by Liapunov's Direct Method with Applications.* Prentice-Hall, Englewood Cliffs, NJ, 1963.
85. Lemaŕchal, C., Strodiot, J.-J., and Bihain, A. On a bundle algorithm for nonsmooth optimization. In: Mangasarian et al. (Eds.). *Nonlinear Programming,* Vol. 4 Academic Press, New York, 1981.
86. Levin, A.U., and Narendra, K.S. Control of nonlinear dynamical systems using neural networks: controllability and stabilization. *IEEE Trans on Neural Networks* 1993; **4**: 192–206.
87. Levy, A.V., and Montalvo, A. The tunneling algorithm for the global minimization of functions. *SIAM J Sci Stat Comp* 1985; **6**(1): 15–19.
88. Lyapunov, A.M. Probl'eme g'en'erale de la stabilit'e de mouvement. *Ann Fac Sci Toulouse* 1907; **9**: 203–474.
89. Lyapunov, A.M. On the general problem of stability of motion. *GITTL* 1950.
90. Ljung, L. *System Identification—Theory for the User.* Prentice Hall, Englewood Cliffs, NJ, 1987.
91. Ljung, L., and Söderström, T. *Theory and Practice of Recursive Identification.* MIT Press, Cambridge, MA, 1983.
92. Lowe, D. Adaptive radial basis function nonlinearities and the problem of generalization. *Proceedings of the 1st IEEE International Conference on Artificial Neural Networks*, London, 1989; 171–175.
93. Lukšan, L., Šiška, M., and Tuma, M. Interactive system for functional optimization (UFO). Technical Report No. 529, ICIS, Czech Academy of Sciences, Praha, 1992.
94. Mecklenburg, K., Hrycej, T., Franke, U., and Fritz, H. Neural control of autonomous vehicles. *Proceedings of the IEEE Vehicular Technology Conference '92*, Denver, May 1992.
95. Malinowski, A., Zurada, J.M., and Lilly, J.H. Inverse control of nonlinear systems using neural network observer and inverse mapping approach. In: *Proceedings of the 1995 IEEE International Conference on Neural Networks*, Perth, Australia, 1995.

96. Metropolis, N., Rosenbluth, A. Rosenbluth., M., Teller, A., and Teller., E. Equations of state calculations for fast computing machines. *J Chem Phys* 1953; **21**: 1087–1092.

97. Michie, D., and Chambers R. BOXES: an experiment in adaptive control. In: Dale, E., and Michie, D. (Eds.), *Machine Intelligence*, Vol. 2. Oliver and Boyd, 1968.

98. Müller, P.H., Nollau, V., and Polovinkin, A.I. *Stochastiche Suchverfahren* (Stochastic search procedures). VEB Fachbuchverlag, Leipzig, 1986; in German.

99. Murray-Smith, R., Neumerkel, D., and Sbarbaro-Hofer, D. Neural networks for modelling and control of a nonlinear dynamic system. *ISIC-92 IEEE International Symposium on Intelligent Control*, Glasgow, UK, 1992.

100. Munro, P. A dual backpropagation scheme for scalar reward learning. *Proceedings of the 9th Annual Conference of Cognitive Science Society*, 1987; 165–176.

101. Narendra, K.S. Adaptive control using neural networks. In Miller, W.T., Sutton, R.S., and Werbos, P.J. (Eds.), *Neural Networks for Control*. MIT Press, Cambridge, MA, 1991.

102. Narendra, K.S. Adaptive control of dynamical systems using neural networks. In White, D.A., and Sofge, D.A. (Eds.), *Handbook of Intelligent Control*. Van Nostrand Reihhold, New York, 1992.

103. Narendra, K.S., and Annaswamy, A.M. *Stable Adaptive Systems*. Prentice Hall, Englewood Cliffs, NJ, 1989.

104. Narendra, K.S., and Mukhopadhyay, S. Adaptive control of nonlinear multivariable systems using neural networks. *Neural Networks* 1994; **7**(5): 737–752.

105. Narendra, K.S., and Parthasarathy, K. Gradient methods for the optimization of dynamical systems containing neural networks. *IEEE Trans Neural Networks* 1991; **2**(2): 252–262.

106. Narendra, K.S., and Thathachar, M.A.L. *Learning Automata*. Prentice Hall, Englewood Cliffs, NJ, 1989.

107. Nelder, J.A., and Mead, R. *Computer J* 1965; **7**: 308–313.

108. Nerrand, O., et al. Training recurrent neural networks: why and how? An illustration in dynamical process modeling. *IEEE Trans Neural Networks* 1994; **5**(2): 178–184.

109. Nguyen, D., and Widrow, B. The truck backer-upper: an example of self-learning in neural networks. In *Proceedings of the International Joint Conference on Neural Networks*, Washington DC, 1989; **2**: 357–363.

110. Nicolis, G., and Prigogine, I. *Exploring Complexity*. W. H. Freeman, New York, 1989.

111. Pao, Y.-H., and Sobajic, D.J. Nonlinear process control with neural nets, *Neurocomputing* 1990; **2**: 51–59.

112. Parisini, T., and Zoppoli, R. Neural networks for feedback feedforward nonlinear control systems. *IEEE Trans Neural Networks* 1994; **5**(3): 436–449.

113. Parks, P.C. Lyapunov redesign of model reference adaptive control systems. *IEEE Trans Auto Contr* 1966; **11**: 362–367.

114. Pearl, J. *Probabilistic Reasoning in Intelligent Systems*. Morgan Kaufmann, London, 1988.

115. Poggio, T., and Girosi, F. A theory of networks for approximation and learning. MIT Artificial Intelligence Laboratory Memo No. 1140, July 1989.
116. Polak, E. *Computational Methods in Optimization.* Academic Press, New York, 1971.
117. Press, W.H., Teukolsky, S.A., Vetterling, W.T., and Flannery, B.P. *Numerical Recipes in C.* Cambridge University Press, Cambridge, 1992.
118. Prokhorov, D., Santiago, R., and Wunsch, D. Adaptive critic designs: a case study for neurocontrol. *Neural Networks* 1995; (5).
119. Psaltis, D., Sideris, A., and Yamamura, A.A. A multilayer neural network controller. *IEEE Contr Syst Mag* 1988; **8**: 17–21.
120. Puskorius, G.V., and Feldkamp, L.A. Decoupled extended Kalman filter training of feedforward layered networks. In *Proceedings of the Joint Conference on Neural Networks 91*, Seattle, 1991; 771–777.
121. Puskorius, G.V., and Feldkamp, L.A. Neurocontrol of nonlinear dynamical systems with Kalman filter trained recurrent networks. *IEEE Trans Neural Networks* 1994; **5**(2): 279–297.
122. Ramalho, R.S. *Introduction to Wastewater Treatment Processes.* Academic Press, New York, 1977.
123. Rinnooy Kan, A.H.G., and Timmer, G.T. Stochastic global optimization methods, part I: clustering methods, part II: multi-level methods. *Math Program* 1987; **39**(1): 26–78.
124. Robbins, H., and Monro, S. A stochastic approximation method. *Ann Math Stat* 1951; **22**: 400–407.
125. Rosenblatt, F. *Principles of Neurodynamics.* Spartan Books, Washington, DC, 1962.
126. Rumelhart, D.E., Hinton, G.E., and Williams, R.J. Learning internal representations by error propagation. In Rumelhart, D.E., and McClelland, J.L. (Eds.), *Parallel Distributed Processing*, Vol. 1. MIT Press, Cambridge, MA, 1986; Ch. 8.
127. Růžička, P. The design of neural net configurations by the methods of theory of tolerances (in Czech), Doctoral Dissertation, Czechoslovak Academy of Sciences, 1989.
128. Růžička, P., and Kober, R. NEUral network based CONtrollers, Project NEUCON—Final Report, Forschungsinstitut fur abwendungsorientierte Wissensverarbeitung, Ulm, Germany, 1992.
129. Sbarbaro-Hofer, D., Neumerkel, D., and Hunt, K. Neural control of a steel rolling mill. *ISIC-92 IEEE International Symposium on Intelligent Control*, Glasgow, UK, 1992.
130. Schuster, H. G. *Deterministic Chaos.* VCH Verlagsgesellschaft, Weinheim, Germany, 1988.
131. Schwefel, H.-P. *Evolution and Optimum Seeking.* John Wiley, New York, 1995.
132. Sing, L.B., and Wong, K.P. Transient stability assessment: an artificial neural network approach. *Proceedings of the 1995 IEEE International Conference on Neural Networks*, Perth, 1995.

133. Singhal, S., and Wu, L. Training multilayer perceptrons with the extended Kalman algorithm. In Touretzky, D.S. (Ed.), *Advances in Neural Information Processing Systems*, Vol. 1. Morgan Kaufmann, San Mateo, CA, 1989; 133–140.

134. Srinivasan, B., Prasad, U.R., and Rao, N.J. Back propagation through adjoints for the identification of nonlinear dynamic systems using recurrent neural models. *IEEE Trans Neural Networks* 1994; **5**(2): 178–184.

135. Tenno, R., and Uronen, P. Online state estimation and control for wastewater treatment processes. *Proceedings of the ICARCV '94—Third International Conference on Automation, Robotics and Computer Vision*, Singapore, 1994.

136. Thrun, S.B. The role of exploration in learning control. In White, D.A., and Sofge, D.A. (Eds.), *Handbook of Intelligent Control*. Van Nostrand Reihhold, New York, 1992.

137. Torn, A., and Žilinskas, A. *Global Optimization*. Lecture Notes in Computer Science. Springer-Verlag, Berlin, 1989.

138. Tsypkin, Y.Z. *Adaptation and Learning in Automatic Systems*. Academic Press, New York, 1971.

139. Tsypkin, Y.Z. *Foundations of the Theory of Learning Systems*. Nauka, Moskva, 1970; in Russian.

140. Unbehauen, H. *Regelungstechnik III—Identifikation, Adaption, Optimierung*. Vieweg-Verlag, Braunschweig, 1995; in German.

141. Wagner, J. Am Fahrzeugmodell trainierte neuronale Geschwindigkeitsregelung (Car speed control trained on a car model), Diploma Thesis, University Stuttgart, 1993; in German.

142. Watkins, C. Learning from delayed rewards, Ph.D.Thesis, Cambridge University, Cambridge.

143. Werbos, P. Beyond regression: new techniques for prediction and analysis in the behavioral sciences, Ph.D. Thesis, Harvard University, Cambridge, November 1974.

144. Werbos, P.J. Advanced forecasting methods for global crisis warning and models of intelligence. *Gen Syst Yearbook* 1977; **22**: 25–38.

145. Werbos, P.J. Generalization of back propagation with applications to a recurrent gas market model. *Neural Networks* 1988; **1**: 339–356.

146. Werbos, P.J. Backpropagation and neurocontrol: a review and prospectus. *Proceedings of the International Joint Conference on Neural Networks*, Washington DC, 1989; **1**: 209–216.

147. Werbos, P.J. Consistency of HDP applied to a simple reinforcement learning problem. *Neural Networks* 1990; **3**: 179–189.

148. Werbos, P.J. A menu of designs for reinforcement learning over time. In Miller, W.T., Sutton, R.S., and Werbos, P.J. (Eds.), *Neural Networks for Control*. MIT Press, Cambridge, MA, 1991.

149. Werbos, P.J. Neurocontrol and supervised learning: an overview and evaluation. In White, D.A., and Sofge, D.A. (Eds.), *Handbook of Intelligent Control*. Van Nostrand Reihhold, New York, 1992.

150. Werbos, P.J., McAvoy, T., and Su, T. Neural networks, system identification, and control in the chemical process industries. In White, D.A., and Sofge, D.A. (Eds.), *Handbook of Intelligent Control*. Van Nostrand Reinhold, New York, 1992.

151. White, D.A., and Jordan, M.I. Optimal control: a foundation for intelligent control. In White, D.A., and Sofge, D.A. (Eds.), *Handbook of Intelligent Control*. Van Nostrand Reinhold, New York, 1992.
152. White, D.A., and Sofge, D.A. (Eds.), *Handbook of Intelligent Control*. Van Nostrand Reinhold, New York, 1992.
153. White, D.A., Bowers, A., Iliff, K., et al. Flight, propulsion, and thermal control of advanced aircraft and hypersonic vehicles. In White, D.A., and Sofge, D.A. (Eds.), *Handbook of Intelligent Control*. Van Nostrand Reihhold, New York, 1992; 357–465.
154. Widrow, B., McCool, J. and Medoff, B. Adaptive control by inverse modeling. In *Proceedings of the Twelfth Asilomar Conference on Circuits, Systems, and Computers*, 1978.
155. Williams, R.J. On the use of backpropagation in associative reinforcement learning. In *Proceedings of the IEEE International Conference on Neural Networks*. San Diego, CA, 1988; **1**: 263–270.
156. Williams, R.J. Reinforcement learning in connectionist networks. ICS Report 8605, University of California, San Diego, 1986.
157. Williams, R.J. Adaptive state representation using recurrent networks. In Miller, W.T., Sutton, R.S., and Werbos, P.J. (Eds.), *Neural Networks for Control*. MIT Press, Cambridge, MA, 1991.
158. Williams, R.J., and Zipser, D. A learning algorithm for a continually running fully recurrent network. *Neural Computation* 1989; **1**: 39–40.
159. Winter, C. bstandsregelung von Fahrzeugen auf der Basis von neuronalen Netzen (Distance control of cars on the basis of neural networks), Diploma Thesis, Berufsakademie Stuttgart, 1992; in German.
160. Wunsch, D., and Prokhorov, D. Adaptive critic designs. In *Proceedings of the 1995 IEEE International Conference on Neural Networks*, Perth, Australia, 1995.
161. Yuan, F., Feldkamp, L., Puskorius, G., and Davis, L. A simple solution to bioreactor benchmark problem by application of Q-learning. In *Proceedings of the World Congress on Neural Networks*, Washington, DC, 1995.

INDEX

Activated sludge process, 332
Adaptive control, 10, 223
　batch schemes, 227
　incremental schemes, 227
Adaptive rules, globally stable, 44
Analytic control design with neural network models, 204
Analytic gradient, 71, 137, 201
Analytic models, neurocontrol with, 203, 320, 326
Annealing algorithms, 78
Asymmetric networks, 47

Backpropagation through time, 74
Batch learning schemes, 87, 135
　for adaptive control, 227
　for neurocontroller training, 199
　for plant identification, 134
Biological wastewater treatment control, 331

Canonical form of a recurrent network, 56
Chaotic systems, 126
Characteristic equation, 161, 167
Closed-loop optimization, 18
Clustering optimization methods, 82
Coarse coding, 46
Conjugate directions, 67
Conjugate gradient methods, 70
Constraints on control, 180
Continuous-time models, linear, 96
Controllability, 160, 171
Controller quality criterion:
　multistep, 191
　single-step, 191
Convergence speed, of plant identification, 143
Cost function:
　covering the domain of, 80
　for neurocontroller training, 189, 257
　for plant identification, 133, 255
Credit assignment problem, 12
Critic systems, 21

Decoupling controller, linear, 168
Decoupling model:
　linear, 95, 97
　nonlinear, 101, 112
Dependence:
　nontransitive, 117
　transitive, 117
Direct representation of mappings, 61, 118
Discrete-time models, linear, 93
Discretized continuous-time models, 102
Disturbances:
　measurable, 216
　random, 216, 220
　systematic, 215, 220
Drive train test bench control, 118, 309
Dynamic backpropagation, 74
Dynamic optimization, 8, 29

Elastomere test bench control, 277
Equilibrium points:
　and network symmetry, 49
　in robust control, 21
　of closed loop, 163, 176, 210
Error integration, 165, 174, 286
Error term for plant identification, 123
Error volume criterion, 127
Evolutionary algorithms, 82
Explicit plant models, 156, 310
External feedback, 55, 58

Feedback:
　external, 55, 58

379

Feedback *(continued)*
  internal, 55
Feedback networks, 54
Feedforward mapping, in neurocontroller, 182, 286
Feedforward networks, 54
First-order local optimization methods, 66
Frequency-domain specification of robustness, 221
Fundamental cost function, 12, 14, 19, 22, 24, 27

Gain scheduling, 181
Gain vector networks, 52, 260
Generalization and adaptiveness, 223
Globally stable adaptive rules, 44
Global optimization, 75, 88, 155, 261
  and adaptive control, 236
Golden section algorithm, 65
Gradient computation:
  for neurocontroller training, 200
  for plant identification, 137

Hidden states, 60
  identification of, 346
Higher-order units, 47

Identification algorithms:
  batch, 135
  convergence speed of, 143
  incremental, 134
Identification horizon, 121
Incomplete feedback:
  linear, 164
  nonlinear, 175
Incremental learning schemes, 85, 134, 265
  for adaptive control, 227
  for neurocontroller training, 197
  for plant identification, 135
Indirect representation of mappings, 61, 118
Input–output model, nonlinear, 101, 107, 171
Input–output plant model, linear, 95
Integration:
  of control action, 166, 174, 287
  of error, 165, 174, 286
Internal feedback, 55

Jacobian matrix of the plant model, 140

Kalman training algorithm, 89

Learning and adaptiveness, 223
Linear control, 4
Line search, 65, 88
Lyapunov function, 29

analytic, 246
numeric, 247
Locality of representation, 46
Local optimization, 64, 263
  and adaptive control, 236
Lorenz attractor, 128

Model quality criterion, 120
  multistep, 121
  single-step, 121
Model reference adaptive control (MRAC), 10
Model structure:
  classical, 93
  nonlinear, 100
Modulated sigmoid unit, 53
Momentum term, 76
Multilayer perceptron, 39
Multilevel methods, 82
Multistart methods, 82
Multistep criterion, 121

Neural networks:
  feedback, 54
  feedforward, 54
Neural network units:
  higher-order, 47
  modulated sigmoid, 53
  sigmoid, 39
Neurocontroller structure, 171, 184
  conceptual, 173
  operational, 173
Neurocontroller training methods:
  batch, 199
  closed-loop, 197
  incremental, 197
  open-loop, 195
Neurocontroller with separate feedforward mapping, 182, 286
Nonlinear control, 6
Nonlinearity, types of, 38
Nontransitive dependence, 117
Numeric gradient, 71, 137, 200

Observability:
  linear, 94
  nonlinear, 101
  of disturbance, 210
Observability index, 94, 171, 187
Optimal control, 8
Output error model, 108

Parallel model, 108
Plant identification, *see* Identification
Plant inversion, 13

Plant parameter variations, 218
Plant stabilization task:
 linear, 160
 nonlinear, 173, 181, 182
Powell's algorithm, 69
Prefilter matrix, 162

Quasi-Newton methods, 70

Radial basis functions, 43
Random Markov fields, 78
Reference model, 19
Reinforcement learning, 23
Relative degree, 95, 109, 168
Representation of mappings:
 direct, 61, 118
 indirect, 61, 118
Robust control, 9
 and adaptive control, 208
 and equilibrium points, 211
 and stability, 209
 neurocontrol, 207
Robustness:
 against plant variations, 300
 against random disturbances, 216, 220
 against systematic disturbances, 215, 220
 frequency-domain specification, 221
Roessler attractor, 132

Sampled data, 112, 153
 accumulation of, 230
 forgetting of, 233
Second-order local optimization methods, 67
 conjugate gradient, 70
 Powell's algorithm, 69
 quasi-Newton, 70
 variable-metric, 70
Self-tuning regulators (STR), 10

Sigmoid unit, 39
 modulated, 53
Single-step criterion, 121
Software tools for neurocontrol, 266, 274
Stability:
 and robust control, 209
 of identification algorithm, 141
 of nonlinear systems, 7, 243
 of nonlinear systems, linearization, 244
 of nonlinear systems, Ljapunov stability, 245
 under disturbances, 249
State model:
 linear, 93
 nonlinear, 171
State observer, 164
Statistical modeling of cost function, 82
Stochastic approximation, 87
Strange attractor, 127
Strategic utility function, 21, 30
Symmetric networks, 47, 51
Systematic disturbances, 165, 174

Taylor expansion, 61
Template learning, 12
Training examples for neurocontroller training, 194, 216
Trajectory following task:
 case study, 285
 linear, 168
 nonlinear, 173, 182
Transitive dependence, 117
Tunneling methods, 79

Variable-metric methods, 70
Velocity form of I-controller, 165

Wastewater treatment control, biological, 331